Emergence, Complexity and Computation

Volume 47

Series Editors

Ivan Zelinka, Technical University of Ostrava, Ostrava, Czech Republic

Andrew Adamatzky, University of the West of England, Bristol, UK

Guanrong Chen, City University of Hong Kong, Hong Kong, China

Editorial Board

Ajith Abraham, MirLabs, USA

Ana Lucia, Universidade Federal do Rio Grande do Sul, Porto Alegre, Rio Grande do Sul, Brazil

Juan C. Burguillo, University of Vigo, Spain

Sergej Čelikovský, Academy of Sciences of the Czech Republic, Czech Republic

Mohammed Chadli, University of Jules Verne, France

Emilio Corchado, University of Salamanca, Spain

Donald Davendra, Technical University of Ostrava, Czech Republic

Andrew Ilachinski, Center for Naval Analyses, USA

Jouni Lampinen, University of Vaasa, Finland

Martin Middendorf, University of Leipzig, Germany

Edward Ott, University of Maryland, USA

Linqiang Pan, Huazhong University of Science and Technology, Wuhan, China

Gheorghe Păun, Romanian Academy, Bucharest, Romania

Hendrik Richter, HTWK Leipzig University of Applied Sciences, Germany

Juan A. Rodriguez-Aguilar, IIIA-CSIC, Spain

Otto Rössler, Institute of Physical and Theoretical Chemistry, Tübingen, Germany

Yaroslav D. Sergeyev, Dipartimento di Ingegneria Informatica, University of Calabria, Rende, Italy

Vaclav Snasel, Technical University of Ostrava, Ostrava, Czech Republic

Ivo Vondrák, Technical University of Ostrava, Ostrava, Czech Republic

Hector Zenil, Karolinska Institute, Solna, Sweden

The Emergence, Complexity and Computation (ECC) series publishes new developments, advancements and selected topics in the fields of complexity, computation and emergence. The series focuses on all aspects of reality-based computation approaches from an interdisciplinary point of view especially from applied sciences, biology, physics, or chemistry. It presents new ideas and interdisciplinary insight on the mutual intersection of subareas of computation, complexity and emergence and its impact and limits to any computing based on physical limits (thermodynamic and quantum limits, Bremermann's limit, Seth Lloyd limits…) as well as algorithmic limits (Gödel's proof and its impact on calculation, algorithmic complexity, the Chaitin's Omega number and Kolmogorov complexity, non-traditional calculations like Turing machine process and its consequences,…) and limitations arising in artificial intelligence. The topics are (but not limited to) membrane computing, DNA computing, immune computing, quantum computing, swarm computing, analogic computing, chaos computing and computing on the edge of chaos, computational aspects of dynamics of complex systems (systems with self-organization, multiagent systems, cellular automata, artificial life,…), emergence of complex systems and its computational aspects, and agent based computation. The main aim of this series is to discuss the above mentioned topics from an interdisciplinary point of view and present new ideas coming from mutual intersection of classical as well as modern methods of computation. Within the scope of the series are monographs, lecture notes, selected contributions from specialized conferences and workshops, special contribution from international experts.

Indexed by zbMATH.

Andrew Adamatzky
Editor

Fungal Machines

Sensing and Computing with Fungi

Editor
Andrew Adamatzky
Unconventional Computing Laboratory
University of the West of England
Bristol, UK

ISSN 2194-7287   ISSN 2194-7295 (electronic)
Emergence, Complexity and Computation
ISBN 978-3-031-38335-9   ISBN 978-3-031-38336-6 (eBook)
https://doi.org/10.1007/978-3-031-38336-6

© The Editor(s) (if applicable) and The Author(s), under exclusive license to Springer Nature Switzerland AG 2023

This work is subject to copyright. All rights are solely and exclusively licensed by the Publisher, whether the whole or part of the material is concerned, specifically the rights of translation, reprinting, reuse of illustrations, recitation, broadcasting, reproduction on microfilms or in any other physical way, and transmission or information storage and retrieval, electronic adaptation, computer software, or by similar or dissimilar methodology now known or hereafter developed.
The use of general descriptive names, registered names, trademarks, service marks, etc. in this publication does not imply, even in the absence of a specific statement, that such names are exempt from the relevant protective laws and regulations and therefore free for general use.
The publisher, the authors, and the editors are safe to assume that the advice and information in this book are believed to be true and accurate at the date of publication. Neither the publisher nor the authors or the editors give a warranty, expressed or implied, with respect to the material contained herein or for any errors or omissions that may have been made. The publisher remains neutral with regard to jurisdictional claims in published maps and institutional affiliations.

This Springer imprint is published by the registered company Springer Nature Switzerland AG
The registered company address is: Gewerbestrasse 11, 6330 Cham, Switzerland

Preface

The fungi is the largest, widely distributed and the oldest group of living organisms. The smallest fungi are microscopic single cells. The largest mycelium occupies over dozens of hectares and weighs several tons. Fungi possess almost all the senses as humans do. They sense light, chemicals, gases, gravity and electric fields. Fungi exhibit an electrical response to stimulation in a matter of seconds or minutes. The book looks at fungi from an unusual perspective—as sensors, electronic devices and future computers. The book offers fungal electronics and computing as an alternative to currently adopted methodologies used to manufacture electronic devices with a high degree of negative environmental impacts. There are five parts. The first part analyses endogenous fungal electrical activity, its complexity and how it is affected by anaesthesia. The second part is about fungal sensors and reactive wearables. There we discuss reactions of fungi to mechanical, chemical and optical stimuli. We stimulate fungi with heavy weights, white light, ethanol, sugars, stretching and human stress hormones. Living fungal insoles, which could, potentially, detect a posture of a wearer, are presented there as well. The third part is about fungal electronics. There we present experimental prototype of fungal oscillators, capacitors, memristors and low pass filters. Several prototypes of fungal computers are presented in the fourth part of the book. There we present spiking-based computing on fungal colonies, discrimination of electrical frequencies by composites colonised by fungi, electrical fungal gates and mining logical circuits in living fungal substrates. The fifth, and the last, part of the book is devoted to fungal language and cognition. It introduces us to fungal language and fungal cognition. The chapters are written by world-leading experts in computer science, engineering, electronics, biophysics and architecture. The book is the encyclopaedia, the first-ever complete authoritative account, of the theoretical and experimental findings in the sensing and computing of fungi written by the people who pioneered the field of fungal machines. All the chapters are self-contained, and no specialist background is required to appreciate ideas, findings, constructs and designs presented. This treatise in fungal machines appeals to readers from all walks of life, from high-school pupils to university professors, from mathematicians, computers scientists and engineers to chemists and biologists.

Bristol, UK   Andrew Adamatzky

Contents

Fungal Electrical Activity

Action Potential Like Spikes in Oyster Fungi *Pleurotus Djamor* 3
Andrew Adamatzky

On Electrical Spiking of *Ganoderma Resinaceum* 15
Andrew Adamatzky and Antoni Gandia

Electrical Spiking of Psilocybin Fungi 23
Antoni Gandia and Andrew Adamatzky

Complexity of Electrical Spiking of Fungi 33
Mohammad Mahdi Dehshibi and Andrew Adamatzky

Fungi Anaesthesia 61
Andrew Adamatzky and Antoni Gandia

Fungal Sensors and Wearables

Living Mycelium Composites Discern Weights via Patterns
of the Electrical Activity 73
Andrew Adamatzky and Antoni Gandia

Fungal Sensing Skin 83
Andrew Adamatzky, Antoni Gandia, and Alessandro Chiolerio

Reactive Fungal Wearable 93
Andrew Adamatzky, Anna Nikolaidou, Antoni Gandia,
Alessandro Chiolerio, and Mohammad Mahdi Dehshibi

On Stimulating Fungi *Pleurotus Ostreatus* with Hydrocortisone 105
Mohammad Mahdi Dehshibi, Alessandro Chiolerio, Anna Nikolaidou,
Richard Mayne, Antoni Gandia, Mona Ashtari-Majlan,
and Andrew Adamatzky

Fungal Photosensors 123
Alexander E. Beasley, Michail-Antisthenis Tsompanas,
and Andrew Adamatzky

Reactive Fungal Insoles 131
Anna Nikolaidou, Neil Phillips, Michail-Antisthenis Tsompanas,
and Andrew Adamatzky

Electrical Response of Fungi to Changing Moisture Content 149
Neil Phillips, Antoni Gandia, and Andrew Adamatzky

Fungal Electronics

Electrical Resistive Spiking of Fungi 169
Andrew Adamatzky, Alessandro Chiolerio, and Georgios Sirakoulis

Fungal Capacitors 177
Konrad Szaciłowski, Alexander E. Beasley, Krzysztof Mech,
and Andrew Adamatzky

Mem-Fractive Properties of Fungi 193
Alexander E. Beasley, Mohammed-Salah Abdelouahab, René Lozi,
Michail-Antisthenis Tsompanas, and Andrew Adamatzky

Electrical Signal Transfer by Fungi 227
Neil Phillips, Roshan Weerasekera, Nic Roberts, and Andrew Adamatzky

Fungal Computing

Towards Fungal Computer 245
Andrew Adamatzky

On Boolean Gates in Fungal Colony 275
Andrew Adamatzky, Martin Tegelaar, Han A. B. Wosten,
Alexander E. Beasley, and Richard Mayne

Electrical Frequency Discrimination by Fungi *Pleurotus Ostreatus* 293
Dawid Przyczyna, Konrad Szacilowski, Alessandro Chiolerio,
and Andrew Adamatzky

On Electrical Gates on Fungal Colony 301
Alexander E. Beasley, Phil Ayres, Martin Tegelaar,
Michail-Antisthenis Tsompanas, and Andrew Adamatzky

Mining Logical Circuits in Fungi 311
Nic Roberts and Andrew Adamatzky

Fungal Automata 323
Andrew Adamatzky, Eric Goles, Genaro J. Martínez,
Michail-Antisthenis Tsompanas, Martin Tegelaar, and Han A. B. Wosten

Exploring Dynamics of Fungal Cellular Automata 341
Carlos S. Sepúlveda, Eric Goles, Martín Ríos-Wilson,
and Andrew Adamatzky

Computational Universality of Fungal Sandpile Automata 371
Eric Goles, Michail-Antisthenis Tsompanas, Andrew Adamatzky,
Martin Tegelaar, Han A. B. Wosten, and Genaro J. Martínez

Fungal Language and Cognition

Language of Fungi Derived from their Electrical Spiking Activity 389
Andrew Adamatzky

Fungal Minds 409
Andrew Adamatzky, Jordi Vallverdu, Antoni Gandia,
Alessandro Chiolerio, Oscar Castro, and Gordana Dodig-Crnkovic

Index 423

Fungal Electrical Activity

Action Potential Like Spikes in Oyster Fungi *Pleurotus Djamor*

Andrew Adamatzky

Abstract We record extra-cellular electrical potential of fruit bodies of oyster fungi *Pleurotus djamor*. We demonstrated that the fungi generate action potential like impulses of electrical potential. Trains of the spikes are observed. Two types of spiking activity are selected: high-frequency (period 2.6 min) and low-freq (period 14 min); transitions between modes of spiking are illustrated. An electrical response of fruit bodies to short (5 s) and long (60 s) thermal stimulation with open flame is analyses in details. We show that non-stimulated fruit bodies of a cluster react to thermal stimulation, with a single action-potential like spike, faster than the stimulate fruit body does.

1 Introduction

Electricity is one of key factors shaping growth and development of fungi. Polarity and branching of mycelium are induced by electric fields [1]. Hyphae are polarised in electric fields [2]: sites of germ tube formation and branching, the direction of hyphal extension and the frequency of branching and germination could be affected by electric field. Fungi also produce internal electrical currents and fields. Electrical current is generated by a hypha: positive current, more likely carried by protons [3], enters tip of a growing hypha [4, 5]. Current density reported is up to 0.6 μA/cm^2 [3]. Electrostatic repulsion of charged basidiospores propulses the spores from alike charged basidium [6, 7]. The electrical current can be involved or associated with translocation of material in pair with hydraulic pressure [8]. There are evidences of electrical current participation in the interactions between mycelium and plant rots during formation of mycorrhiza [9].

In 1976 Slayman, Long and Gradmann discovered action potential like spikes using intra-cellular recording of mycelium of *Neurospora crassa* [10]. Four types of action potential have been identified: (1) spontaneous quasi-sinusoidal fluctuations of 10–20 mV amplitude, period 3–4 min, (2) as previous but shorter period

---

A. Adamatzky (✉)
Unconventional Computing Lab, UWE, Bristol, UK
e-mail: andrew.adamatzky@uwe.ac.uk

© The Author(s), under exclusive license to Springer Nature Switzerland AG 2023
A. Adamatzky (ed.), *Fungal Machines*, Emergence, Complexity and Computation 47, https://doi.org/10.1007/978-3-031-38336-6_1

of 20–30 s, (3) cyanide induced oscillation of progressively lengthening period, starting with initial depolarisation of 20–60 mV, and (4) damped sinusoidal oscillations with amplitude 50–100 mV, period 0.2–2 mins. In 19995 Olsson and Hansson demonstrated spontaneous action potential like activity in a hypha of *Pleurotus ostreatus* and *Armillaria bulbosa*; they conducted intra-cellular recording with reference electrode in agar substrate [11]. They shown that resting potential is −70 to −100 mV, amplitude of spikes varies from 5 to 50 mV, duration from 20 to 500 ms, frequency 0.5–5 Hz.

Olsson and Hansson shown that frequency of spiking increases in response to stimulating a hypha with a sulphuric acid, malt extract, water, fresh piece of wood. When stimulus is removed the frequency decreases and then increases again if object is re-introduced. Olsson and Hansson [11] speculated that electrical activity could be used for communication with message propagation speed 0.5 mm/s. Changes in frequency of oscillation of a hypha in response to a wide range of stimuli reported in [11] matches results of our personal studies with slime mould *Physarum polycephalum*, see overview in [12]. We established a mapping between volatile chemicals, wavelength of light and tactile stimulation, one side, and changes in frequency of oscillations of electrical potential of slime mould's protoplasmic tubes, on other side [13–16]; and, designed a prototype of a slime mould based sensor devices [17]. To advance our bio-sensing concepts to fungi and evaluate a possibility of using wild fungi *in situ* as sensors we conducted experiments on electrical activity of fungi in conditions more close to natural conditions than experiments [10, 11] on intra-cellular recording of a hypha conducted in laboratory conditions with mycelium growing on a nutrient agar substrate. For our experiments, we chosen oyster mushroom, species *Pleurotus*, family *Tricholomataceae*, they are most widely cultivated family of fungi [18] with proven medicinal properties [19], and they are amongst few species of carnivorous mushrooms [20] which might add some not common sensing properties.

2 Methods

We used commercial mushroom growing kits (© Espresso Mushroom Company, Brighton, UK) of pink oyster mushrooms *Pleurotus djamor*. Each substrate's bag was 22 by 10 by 10 cm, 800–900 g in weight. The bag was placed was cross-sliced 10 cm vertical and 8 cm horizontal and placed in a cardboard box with 8 by 10 cm opening. The fungi kits were kept at room temperature in constant (24 h) ambient lighting of 10 lux.

Electrical potential of fruit bodies was recorded from the second-third day of their emergence. Resistance between cap and stalk of a fruit body was 1.5 MΩ in average, between any two heads in the cluster 2 MΩ (measured by Fluke 8846A). We recorded electrical potential difference between cap and stalk of the fruit body. We used subdermal needle electrodes with twisted cable (© SPES MEDICA SRL Via Buccari 21 16153 Genova, Italy). Recording electrode was inserted into stalk and

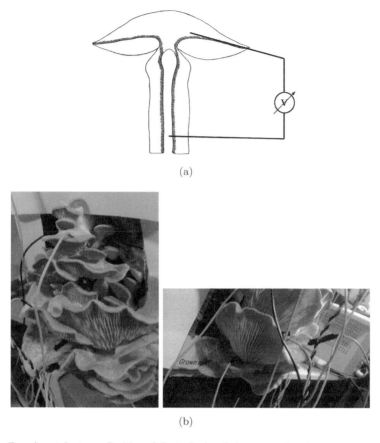

Fig. 1 Experimental setup. **a** Position of electrodes in relation to translocation zone, cross-section of a fruit body showing translocation zone, drawing by Schütte [21]. **b** Photographs of fruit bodies with electrodes inserted photos of experimental setup

reference electrode in the translocation zone of the cap (Fig. 1a); distance between electrodes was 3–5 cm. In each cluster we recorded 4–6 fruit bodies simultaneously (Fig. 1b, c) for 2–3 days.

Electrical activity of fruit bodies was recorded with ADC-24 High Resolution Data Logger (Pico Technology, St Neots, Cambridge shire, UK). The data logger ADC-24 employs differential inputs, galvanic isolation and software-selectable sample rates all contribute to a superior noise-free resolution; its 24-bit A/D converted maintains a gain error of 0.1%. Its input impedance is 2 MΩ for differential inputs, and offset error is 36 μV in $\pm$ 1250 mV range use. We recorded electrical activity one sample per second; during the recording the logger makes as many measurements as possible (typically up 600) per second then saves average value.

3 Results

Here we provide evidence that fruit bodies exhibit spontaneous spiking behaviour, we also characterise types of trains of spikes observed. When calling the spikes spontaneous we mean they are not invoked by an intentional external stimulation, i.e. not expected by an external observer. Otherwise, the spikes indeed reflect physiological and morphological processes ongoing in mycelial networks and growing fruit bodies. We also provide evidence that fruit bodies respond to external stimulation by changing its electrical potential and that neighbours of stimulated fruit bodies might show action-potential like response.

3.1 Spontaneous Spiking

The electrical activity of fruit bodies shows a rich combination of slow (hours) drift of base electrical potential combined with relatively fasts (minutes) oscillations of the potential, see example at Fig. 2a. We observed trains of spikes of electrical potential. Each spike resembles an action potential where all 'classical' part can be found (Fig. 2b): depolarisation, repolarisation, refractory period. The exemplar spike shown in (Fig. 2b) has a period of 130 s, from base level potential to refractory-like period, depolarisation rate 0.05 mV/s, repolarisation rate is 0.02 mV/s, refractory period is c. 360 s.

We observed two types of spike trains: high-frequency (H-spikes), a spike per c. 2.6 min, and low-frequency (L-spikes), a spike per c. 14 min, see examples in Fig. 3. Characteristics of the spikes are shown in Table 1. Period of L-spoke is five time longer than period of H-spikes. Amplitudes of H-spikes are just below 1 mV and of L-spikes is nearly 1.5 mV.

Durations and depolarisation rates of L- and H-spikes are nearly the same. A repolarisation rate of L-spikes is a double of the repolarisation rate of H-spikes. Refractory period of L-spikes is ten times longer than the period of H-spikes. Trains of H-spikes last for up to two-and-half hours while trains of L-spikes for up to six hours. Transitions between trains of H-spikes and L-spikes have been also observed, see example in Fig. 4.

3.2 Response to Stimulation

To check if there will be any changes in electrical potential in response to stimulation we applied 50 μL of 40% ethanol, tap water, polydimethylsiloxane on top of fruit bodies' caps and thermally stimulated edges of the caps with open flame

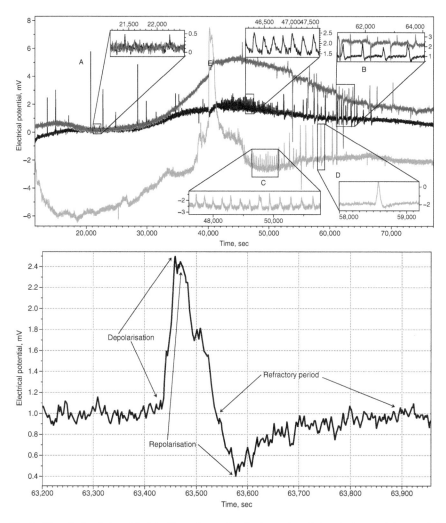

Fig. 2 Electrical activity of fruit bodies. **a** Example of a dynamics of electrical potentials recorded from three fruit bodies of the same cluster during 70 K s, c. 19 h. Some modes of spiking activity are zoomed in the inserts. **b** Analysis of a spike in terms of action potential

for 5 s. Exemplar responses are shown in Fig. 5. All stimuli but polydimethylsiloxane cause positive spike-like responses, parameters are shown in Table 2. There was no response to a drop of polydimethylsiloxane. A fruit body responded to application of water with nearly immediate negative spike with amplitude 0.43 mV and duration 8.2 s, followed (after c. 98 s) by a large positive spike.

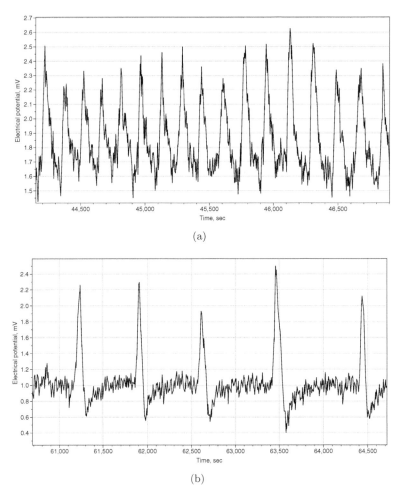

Fig. 3 Examples of spike trains. **a** High frequency. **b** Low frequency

Table 1 Characteristics of spike trains

Parameters	High frequency spikes	Low frequency spikes
Period, s	160.5 (15.1)	838.8 (147.2)
Amplitude, mV	0.88 (0.14)	1.3 (0.21)
Duration, s	115.5 (28.1)	142.6 (33.1)
Depolarisaton rate, mV/s	0.022 (0.006)	0.025 (0.01)
Repolarisation rate, mV/s	0.012 (0.002)	0.024 (0.007)
Refractory like period, s	25.5 (4.2)	256.2 (80.6)
Duration of spike trains, min	80–150	130–360

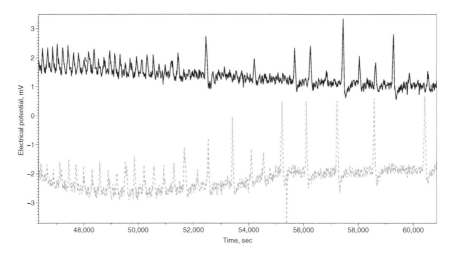

Fig. 4 Example of transition from a train of H-spikes to a train of L-spikes. Electrical potential of two fruit bodies of the same cluster are shown by black solid and green dashed lines

Response of non-stimulated fruit bodies to the short-term thermal stimulation of a member of their cluster is shown in Fig. 5b. While the stimulated fruit body responded to a thermal stimulation after c. 103 s delay, two other fruit bodies in the same cluster shown shorter latency times of their responses. One body responded with a positive spike 26 s after stimulation (green dotted line in Fig. 5b), amplitude c. 1.2 mV, duration 21 s. Another body responded with a negative spike 51 s after stimulation (blue dash-dot line in Fig. 5b), amplitude c. 1 mV, duration 26.3 s.

The fruit bodies' response to a long term thermal stimulation—subjecting an edge of a cap to an open flame for 60 s, was demonstrated to be highly pronounced. A typical response is shown in Fig. 6a. At first we observe an action potential like response of the stimulated fruit body, with depolarisation up c. 1.4 mV, followed by repolarisation by 2 mV. This response lasts 7.6 s. It is immediately followed by a high-amplitude depolarisation. There electrical potential grows by 38.2 mV in 18.4 s followed by slow repolarisation and returning to the base potential in 83 s. Other fruit bodies react with short-living spikes to the long-term thermal stimulation of a member of their cluster. Example is shown in Fig. 6b. Four seconds after start of the stimulation, there is a sharp depolarisation by 5.2 mV reached in 1.38 s. It follows by repolarisation by 6.2 mV reached in 20 s. There is an indication of a refractory period c. 59 s. The potential returns closely to its base (for this fruit body) level after stimulation ends.

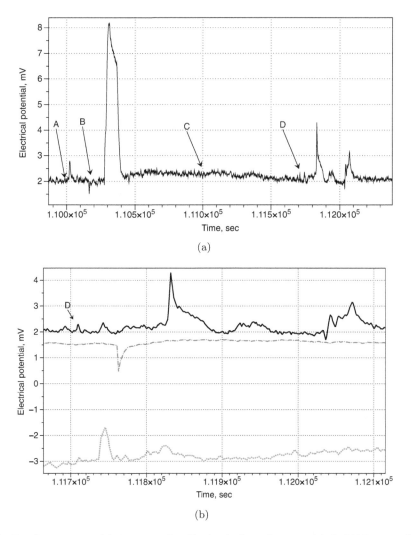

Fig. 5 a Response to spirit (moment of application is shown by arrow labelled (A)), water (B), polydimethylsiloxane (C), thermal stimulation with open flame for 5 s (D). Time between two vertical lines is 500 s. **b** Response to thermal stimulation edge was burned for c. 5 s. Potential of the stimulated fungi is shown by solid black line, two other members of the cluster by green dotted and blue dashed lines. Time between two vertical lines is 100 s

Table 2 Parameters of a fruit body response to stimulation

Stimulus	Amplitude, mV	Depolarisaton rate, mV/s	Repolarisation rate, mV/s	Duration, s
Water	6.1	0.2	0.05	141.5
Spirit	0.8	0.03	0.02	51.2
Thermal	2.1	0.1	0.03	99

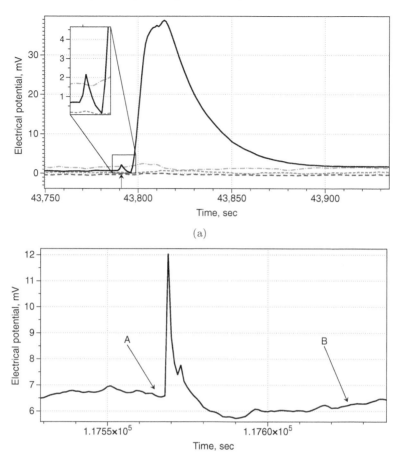

Fig. 6 Electrical response to a long-term thermal stimulation. Electrical potential measured on stimulated fruit body is shown by solid black line. Time between two ticks on horizontal axes on both plots is 50 s. **a** Moment when stimulation was stopped is shown by arrow. Initial action-potential like response is magnified in the insert. **b** Response of an intact fruit body to stimulation of its distant neighbour, c. 6 cm away, by an open flame for c. 60 s. Start of stimulation is show by arrow 'A' and end of stimulation by arrow 'B'

4 Discussion

Fruit bodes, stromata, are made of interwoven hyphae, organic continuation of a mycelium. Thus by inserting electrodes in cap and stalk we measured extracellular electrical potential difference between the cap and the stalk as generated by interwoven hyphae. We observed trains of action potential like spikes. Microtubule bundles observed in basidiomycetae [22] may be responsible for propagation of trains of action potential like spikes. Amplitudes of spikes measured were very low comparing to amplitudes reported in [10, 11] because the works cited used intra-cellular recording while we used extra-cellular.

Periods of spikes evidenced in [10] 0.2–2 min, are comparable with period 2.5–3 min of high frequency spikes in our experiments. High frequency of oscillations of oyster fungi also similar to that recorded in slime mould of *Physarum polycephalum*: electrical potential between two electrodes connected by a protoplasmic tube oscillates with period 1–2 min [23–26]. In slime mould the calcium waves are reflected in oscillations of external membrane potential and periodic reversing of cytoplasmic flow in the tubes. Drawing up analogies between the slime mould and mycelial fungi we speculate that trains of spikes recorded in fruit bodies correlate, or even responsible for, translocation of nutrients and relocation of product of metabolism and communication. There are indication, see e.g. timing of spikes recorded from two fruit bodies of the same cluster in Fig. 4, that trains of spikes are coming from the mycelium, where they most likely originate in way similar to calcium waves in slime mould *P. polycephalum*.

With regards to communicative function of the spikes, fungi responds to stimulation with singular spikes of electrical potential in their fruit bodies. Amplitude of the response is higher in the stimulated body than in its non-stimulated neighbours. However, non-stimulated members of the cluster respond earlier to the stimulation than the stimulated body itself. These response of the non-stimulated bodies to a destructive stimulation of one of the cluster's members might be seen either as a 'byproduct' of an electrical potential deviations propagating from the damaged body through the mycelium network towards intact bodies or a purposeful signal to the intact bodies aimed at speeding up their growth and maturation to shorten a period leading to accelerated production of spores. As shown in [27] (cited by [28]) sporulation could be induced by partial desiccation, and a number of fruit bodies could be larger in proximity of injury. This observation is in line with finding that damaged mycelium responds with branching [29]; and, similar to sprouting response of a slime mould *P. polycephalum* to a dissection of its protoplasmic tubes [30].

References

1. Gow, N.A.R.: Polarity and branching in fungi induced by electrical fields. In: Spatial Organization in Eukaryotic Microbes. IRL Press (1987)
2. Mcgillivray, A.M., Gow, N.A.R.: Applied electrical fields polarize the growth of mycelial fungi. Microbiology **132**(9), 2515–2525 (1986)
3. Mcgillivray, A.M., Gow, N.A.R.: The transhyphal electrical current of *Neuruspua crassa* is carried principally by protons. Microbiology **133**(10), 2875–2881 (1987)
4. Gow, N.A.R.: Transhyphal electrical currents in fungi. Microbiology **130**(12), 3313–3318 (1984)
5. Harold, F.M., Kropf, D.L., Caldwell, J.H.: Why do fungi drive electric currents through themselves? Experimental mycology **9**(3), 3–86 (1985)
6. Savile, D.B.O.: Spore discharge in Basidiomycetes: a unified theory. Science **147**(3654), 165–166 (1965)

7. Leach, C.M.: An electrostatic theory to explain violent spore liberation by *Drechslera turcica* and other fungi. Mycologia 63–86 (1976)
8. Rayner, A.D.M.: The challenge of the individualistic mycelium. Mycologia 48–71 (1991)
9. Berbara, R.L.L., Morris, B.M., Fonseca, H.M.A.C., Reid, B., Gow, N.A.R., Daft, M.J.: Electrical currents associated with arbuscular mycorrhizal interactions. New Phytol. **129**(3), 433–438 (1995)
10. Slayman, C.L., Long, W.S., Gradmann, D.: Action potentials in *Neurospora crassa*, a mycelial fungus. Biochim. Biophys. Acta (BBA)—Biomembr. **426**(4), 732–744 (1976)
11. Olsson, S., Hansson, B.S.: Action potential-like activity found in fungal mycelia is sensitive to stimulation. Naturwissenschaften **82**(1), 30–31 (1995)
12. Adamatzky, A.: Advances in Physarum machines: Sensing and Computing with Slime Mould, vol. 21. Springer, Berlin (2016)
13. Adamatzky, A.: Slime mould tactile sensor. Sens. Actuators, B Chem. **188**, 38–44 (2013)
14. Adamatzky, A.: Towards slime mould colour sensor: recognition of colours by Physarum polycephalum. Org. Electron. **14**(12), 3355–3361 (2013)
15. Whiting, J.G.H., de Lacy Costello, B.P., Adamatzky, A.: Towards slime mould chemical sensor: Mapping chemical inputs onto electrical potential dynamics of *Physarum Polycephalum*. Sens. Actuators B: Chem. **191**, 844–853 (2014)
16. Whiting, J.G.H., de Lacy Costello, B.P., Adamatzky, A.: Sensory fusion in *Physarum polycephalum* and implementing multi-sensory functional computation. Biosystems **119**, 45–52 (2014)
17. Adamatzky, A., Neil, P.: Physarum sensor: biosensor for citizen scientists (2017)
18. Royse, D.J.: Speciality mushrooms and their cultivation. Hortic. Rev. **19**, 59–97 (1997)
19. Md Asaduzzaman Khan and Mousumi Tania: Nutritional and medicinal importance of Pleurotus mushrooms: an overview. Food Rev. Intl. **28**(3), 313–329 (2012)
20. Thorn, R.G., Barron, G.L.: Carnivorous mushrooms. Science **224**(4644), 76–78 (1984)
21. Schütte, K.H.: Translocation in the fungi. New Phytol. **55**(2), 164–182 (1956)
22. Aylmore, R.C., Todd, N.K., Ainsworth, A.M.: Microtubule bundles in Phanerochaete velutina. Trans. Br. Mycol. Soc. **84**(2), 372–374 (1985)
23. Iwamura, T.: Correlations between protoplasmic streaming and bioelectric potential of a slime mold, *Physarum polycephalum*. Shokubutsugaku Zasshi **62**(735–736), 126–131 (1949)
24. Kamiya, N., Abe, S.: Bioelectric phenomena in the myxomycete plasmodium and their relation to protoplasmic flow. J. Colloid Sci. **5**(2), 149–163 (1950)
25. Kishimoto, U.: Rhythmicity in the protoplasmic streaming of a slime mold, *Physarum polycephalum*. I. a statistical analysis of the electric potential rhythm. J. Gen. Physiol. **41**(6), 1205–1222 (1958)
26. Meyer, R., Stockem, W.: Studies on microplasmodia of Physarum polycephalum V: electrical activity of different types of microplasmodia and macroplasmodia. Cell Biol. Int. Rep. **3**(4), 321–330 (1979)
27. Rands, R.D.: The production of spores of Alternaria solani in pure culture. Phytopathology **7**(4), 316–317 (1917)
28. Hawker, L.E.: Physiology of Fungi. University Of London Press Ltd. (1950)
29. Reeves, R.J., Jackson, R.M.: Stimulation of sexual reproduction in *Phytophthora* by damage. Microbiology **84**(2), 303–310 (1974)
30. Adamatzky, A., Jones, J.: Programmable reconfiguration of Physarum machines. Nat. Comput. **9**(1), 219–237 (2010)

On Electrical Spiking of *Ganoderma Resinaceum*

Andrew Adamatzky and Antoni Gandia

Abstract Fungi exhibit action-potential like spiking activity. Up to date most electrical activity of oyster fungi has been characterised in sufficient detail. It remains unclear if there are any patterns of electrical activity specific only for a certain set of species or if all fungi share the same 'language' of electrical signalling. We use pairs of differential electrodes to record extracellular electrical activity of the antler-like sporocarps of the polypore fungus *Ganoderma resinaceum*. The patterns of the electrical activity are analysed in terms of frequency of spiking and parameters of the spikes. The indicators of the propagation of electrical activity are also highlighted.

1 Introduction

Action-potential spikes are an essential component of the information processing system in a nervous system [1–4]. Creatures without a nervous system—plants, slime moulds and fungi—might also use the spikes of electrical potential for coordination and decision-making. There is mounting evidence that plants use the electrical spikes for a long-distance communication aimed to coordinate the activity of their bodies [5–7]. The spikes of electrical potential in plants relate to a motor activity [8–11], responses to changes in temperature [12], osmotic environment [13] and mechanical stimulation [14, 15]. There is evidence of electrical current participation in the interactions between mycelium and plant roots during formation of mycorrhiza [16].

Oscillations of electrical potential in slime mould *Physarum polycephalum* were discovered in the late 1940s [17] and studied extensively [18–20]. It was demonstrated that patterns of electrical activity of the slime mould change in response to stimulation with volatile chemicals, wavelength of light, and tactile stimulation [21–25].

A. Adamatzky (✉)
Unconventional Computing Laboratory, CWE, Bristol, UK
e-mail: andrew.adamatzky@uwe.ac.uk

A. Gandia
Institute for Plant Molecular and Cell Biology, UPV, Valencia, Spain

In 1976 Slayman, Long and Gradmann discovered action potential-like spikes using intra-cellular recording of mycelium of *Neurospora crassa* [26]. Periods of spikes evidenced in [26] were 0.2–2 min.[1] In experiments with recording of electrical potential of oyster fungi *Pleurotus djamor* we discovered two types of spiking activity: high-frequency (period 2.6 min) and low-freq (period 14 min) [28]. In semi-automated analysis of the electrical spiking activity of the hemp substrate colonised by mycelium of *Pleurotus djamor* we found that a predominant spike width is c. 6 min [29].

A question arises—"Are characteristics and patterns of electrical potential spiking the same for all species of fungi or there are some species specific parameters of the fungal electrical activity?". Aiming to find an answer we recorded and analysed electrical activity of *Ganoderma resinaceum* sporocarps. We have been studying the electrical activity of this fungus previously [30] however only electrical responses to stimulation have been analysed and endogenous electrical activity of the mycelium has been ignored.

2 Methods

The *Ganoderma resinaceum* culture used in this experiment was obtained from a wild basidiocarp found on the shores of *Lago di Varese*, Lombardy (Italy) in 2018, and maintained in alternate PDA and MEA slants at MOGU S.r.l. for the last 3 years at 4 °C under the collection code 019–18. For practical purposes, the aforementioned *G. resinaceum* culture was propagated on an sterile mixed substrate of hemp shives and soybean hulls (3:1), with a moisture content (MC) of 65% contained in plastic filter-patch microboxes c. 17×17 cm^2 (SacO$_2$, Belgium). The substrates were incubated in darkness at ambient room temperature c. 22 °C, until antler-shaped sporocarps started to form on the surface c. 14 days post inoculation.

Electrical activity of a forming antler-like sporocarp was recorded using pairs of iridium-coated stainless steel sub-dermal needle electrodes (Spes Medica S.r.l., Italy), with twisted cables and ADC-24 (Pico Technology, UK) high-resolution data logger with a 24-bit A/D converter, galvanic isolation and software-selectable sample rates all contribute to a superior noise-free resolution. Each pair of electrodes $(i, i+1)$, called 'channel', reported a potential difference $p_{i+1} - p_i$ between the electrodes. The pairs of electrodes were pierced into the sporocarp as shown in Fig. 1. The channels were from the top of the antler-like sporocarp to the bottom. Distance between electrodes was 1–2 cm. In each trial, we recorded 8 electrode pairs simultaneously. We recorded electrical activity one sample per second. During the

---

[1] In 1995 Olsson and Hansson demonstrated spontaneous action potential like activity in a hypha of *Pleurotus ostreatus* and *Armillaria bulbosa* (synonymous with *A. gallica* and *A. lutea*) via intra-cellular recording with a reference electrode in an agar substrate [27]. Our present results concern extracellular recordings, therefore we will not compare thee with Olsson and Hansson results.

Fig. 1 Position of electrodes in a sporocarp of *G. resinaceum*

recording, the logger has been doing as many measurements as possible (typically up to 600 s) and saving the average value. The acquisition voltage range was 78 mV.

3 Results

Distribution of spike widths versus amplitudes is shown in Fig. 2a with the majority of spikes having amplitude less than 4 mV and width less than 41 min. Distributions of frequencies of occurrences of the spike amplitudes and widths are shown in Fig. 2b and c. Most common amplitudes lie between 0.1 and 0.4 mV. The frequency distribution of the amplitudes is well described by a power $\frac{0.047}{a}$, where a is an amplitude, shown by a solid line in Fig. 2b. Most common widths of spikes are 300–500 s (5–8 mins) as evidenced in Fig. 2c.

Most common types of nontrivial electrical activity observed are singular spikes, compound spikes, and trains of spikes. Singular spikes are shown by arrows in Fig. 3a. Compound spikes are thought to be composed of several singular spikes, or a spike train which duration is in the range of a single spike width, so several spikes overlap, as labelled by arrows with stars in Fig. 3a. An example of the train of spikes is shown in Fig. 3a. In the first train the average distance between spikes is 731 s and in the second train 2300 s. There are two potential origins of the compound spikes. First, the compound spikes could be seen as trains of spikes with a very short distance between the spikes, i.e. trains of overlapping spikes. Second, the compound spikes

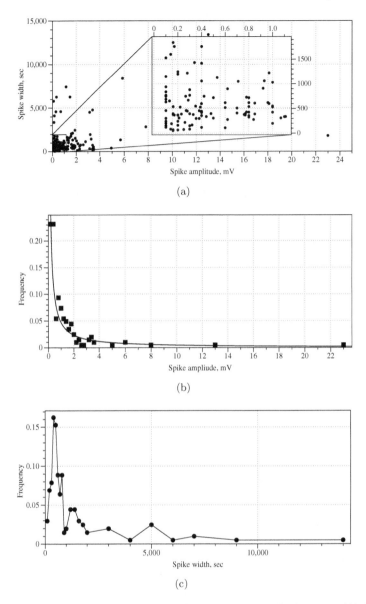

Fig. 2 Quantitative analysis of *G. resinaceum* spiking activity. **a** Amplitude versus width for three sample channels, recorded for 60 h. Distribution of spike **b** amplitudes and **c** widths. Distribution of amplitudes is approximated by a power law $\frac{0.047}{a}$, where a is an amplitude, shown by curve in (**b**)

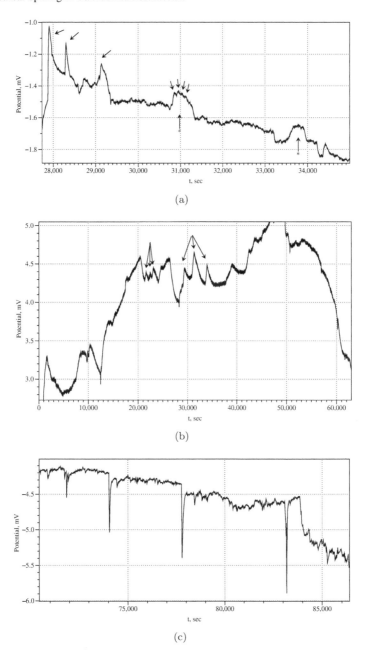

Fig. 3 Examples of an electrical activity of *G. resinaceum* sporocarp. **a** Single and compound spikes. Compound spike is shown by arrow with start. **b** Examples of trains of spikes. Two trains of three spikes each are shown by arrow. **c** Growing type oscillator

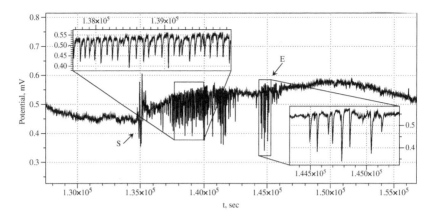

Fig. 4 Example of a long high frequency burst of spikes

could be see as a slow wave potential superimposed with a train of low amplitude spikes.

Single spikes can form analogies of damped or growing type oscillators. The growing type of oscillator with increasing amplitude and period of oscillations is illustrated by four spikes in Fig. 3c. The amplitude increases as 0.35, 0.84, 1.01, 1.27 mV. The width of spikes increases as 88, 210, 252, 273 s. The distance between spikes is changing as 2253, 3739, 5372 s.

A rare but interesting feature observed is a long (up to 2 h) burst of high-frequency (a spike per 7 min) spikes. An example is shown in Fig. 4. The main burst is preceded by a short (13 min) burst of five spikes, shown as 'S' in Fig. 4, and is completed by another short (15 min) burst of seven spikes. There are c. 70 spikes in the main burst: average width of a spike is 29 s and average amplitude 0.1 mV; an average distance spikes is 92 s.

4 Discussion

We found that most common width of an electrical potential spike is 5–8 min. An average distance between electrodes in the pairs of differential electrodes was 1 cm. We speculate it takes a phase-wave, which governs the electrical potential, to propagate between the electrodes 6 min in average, i.e. a speed of the phase-wave is c. 0.028 mm/s. Calcium waves can propagate with the fastest speed up 0.03 mm/s [31]. Thus we can speculate that the spikes of the electrical potential in *Ganoderma resinaceum* antler-like sporocarps might relate to the fast calcium waves. The nature of long burst of high-frequency (a spike per 7 mins) short-width (half-a-minute) spikes is beyond speculations and could not be attributed to calcium waves.

The width of the spike of antler-like sporocarps of *Ganoderma resinaceum* is twice of the width of the spikes detected in sporocarps of *Pleurotus djamor* [28] yet of the same width as spikes detected in a substrate colonised by mycelium of *Pleurotus djamor* [29]. Possible explanations are the following. First, sporocarps show higher growth rate than mycelium. Therefore, spikes detected on sporocarps of *Pleurotus djamor* are twice as narrow as spikes detected on a substrate colonised by mycelium of *Pleurotus djamor*. Second, *Ganoderma resinaceum* might have a slower metabolism than *Pleurotus djamor* and therefore its spiking activity shows wider spikes, and consequently, lower frequency of the spiking.

In the laboratory experiments we have employed a 1–2 cm distance between electrodes in the pair. The distance is optimal for recording electrical impulses in slime moulds and fungi, as have been demonstrated in [29, 32, 33]. With the increasing a distance between active and reference electrode amplitude of spikes decrease due to increased resistance of the shortest path between the electrodes and interference of excitation wave-fronts originated from many disparate sources.

Recalling our initial question—"Do different species of fungi have different parameters of their electrical spiking activity?"—we could answer "Likely, yes". However it would be reckless to base the answer on comparing just two species of fungi. Therefore future research will focus on collecting statistics of spiking of wider range of fungi species.

References

1. Bohte, S.M.: The evidence for neural information processing with precise spike-times: a survey. Nat. Comput. **3**, 195–206 (2004)
2. Cooper, D.C.: The significance of action potential bursting in the brain reward circuit. Neurochem. Int. **41**(5), 333–340 (2002)
3. Maass, W., et al.: Computing with spikes. Special Issue on Found. Inf. Process. TELEMATIK **8**(1), 32–36 (2002)
4. Debanne, D., Bialowas, A., Rama, S.: What are the mechanisms for analogue and digital signalling in the brain? Nat. Rev. Neurosci. **14**(1), 63–69 (2013)
5. Trebacz, K., Dziubinska, H., Krol, E.: Electrical signals in long-distance communication in plants. In: Communication in Plants, pp. 277–290. Springer, Berlin (2006)
6. Fromm, J., Lautner, S.: Electrical signals and their physiological significance in plants. Plant, Cell Environ. **30**(3), 249–257 (2007)
7. Zimmermann, M.R., Mithöfer, A.: Electrical long-distance signaling in plants. In: Long-Distance Systemic Signaling and Communication in Plants, pp. 291–308. Springer, Berlin (2013)
8. Simons, P.J.: The role of electricity in plant movements. New Phytol. **87**(1), 11–37 (1981)
9. Fromm, J.: Control of phloem unloading by action potentials in mimosa. Physiol. Plant. **83**(3), 529–533 (1991)
10. Sibaoka, T.: Rapid plant movements triggered by action potentials. Bot. Mag.= Shokubutsu-gaku-zasshi **104**(1), 73–95 (1991)
11. Volkov, A.G., Foster, J.C., Ashby, T.A., Walker, R.K., Johnson, J.A., Markin, V.S.: Mimosa pudica: electrical and mechanical stimulation of plant movements. Plant, Cell Environ. **33**(2), 163–173 (2010)

12. Minorsky, P.V.: Temperature sensing by plants: a review and hypothesis. Plant, Cell Environ. **12**(2), 119–135 (1989)
13. Volkov, A.G.: Green plants: electrochemical interfaces. J. Electroanal. Chem. **483**(1–2), 150–156 (2000)
14. Roblin, G.: Analysis of the variation potential induced by wounding in plants. Plant Cell Physiol. **26**(3), 455–461 (1985)
15. Pickard, B.G.: Action potentials in higher plants. Bot. Rev. **39**(2), 172–201 (1973)
16. Berbara, R.L.L., Morris, B.M., Fonseca, H.M.A.C., Reid, B., Gow, N.A.R., Daft, M.J.: Electrical currents associated with arbuscular mycorrhizal interactions. New Phytol. **129**(3), 433–438 (1995)
17. Iwamura, T.: Correlations between protoplasmic streaming and bioelectric potential of a slime mold, *Physarum polycephalum*. Shokubutsugaku Zasshi **62**(735–736), 126–131 (1949)
18. Kamiya, N., Abe, S.: Bioelectric phenomena in the myxomycete plasmodium and their relation to protoplasmic flow. J. Colloid Sci. **5**(2), 149–163 (1950)
19. Kishimoto, U.: Rhythmicity in the protoplasmic streaming of a slime mold, *Physarum polycephalum*. I. a statistical analysis of the electric potential rhythm. J. Gen. Physiol. **41**(6), 1205–1222 (1958)
20. Meyer, R., Stockem, W.: Studies on microplasmodia of Physarum polycephalum V: electrical activity of different types of microplasmodia and macroplasmodia. Cell Biol. Int. Rep. **3**(4), 321–330 (1979)
21. Adamatzky, A.: Slime mould tactile sensor. Sens. Actuators, B Chem. **188**, 38–44 (2013)
22. Adamatzky, A.: Towards slime mould colour sensor: recognition of colours by Physarum polycephalum. Org. Electron. **14**(12), 3355–3361 (2013)
23. Whiting, J.G.H., de Lacy Costello, B.P., Adamatzky, A.: Towards slime mould chemical sensor: mapping chemical inputs onto electrical potential dynamics of *Physarum Polycephalum*. Sens. Actuators B: Chem. **191**, 844–853 (2014)
24. Whiting, J.G.H., de Lacy Costello, B.P., Adamatzky, A.: Sensory fusion in *Physarum polycephalum* and implementing multi-sensory functional computation. Biosystems **119**, 45–52 (2014)
25. Adamatzky, A., Neil, P.: Physarum sensor: biosensor for citizen scientists (2017)
26. Slayman, C.L., Long, W.S., Gradmann, D.: "Action potentials" in *Neurospora crassa*, a mycelial fungus. Biochim. Biophys. Acta (BBA)—Biomembr. **426**(4), 732–744 (1976)
27. Olsson, S., Hansson, B.S.: Action potential-like activity found in fungal mycelia is sensitive to stimulation. Naturwissenschaften **82**(1), 30–31 (1995)
28. Adamatzky, A.: On spiking behaviour of oyster fungi pleurotus djamor. Sci. Rep. **8**(1), 1–7 (2018)
29. Dehshibi, M.M., Adamatzky, A.: Electrical activity of fungi: spikes detection and complexity analysis. Biosystems **203**, 104373 (2021)
30. Adamatzky, A., Gandia, A., Chiolerio, A.: Fungal sensing skin. Fungal Biol. Biotechnol. **8**(1), 1–6 (2021)
31. Jaffe, L.F.: Fast calcium waves. Cell Calcium **48**(2–3), 102–113 (2010)
32. Adamatzky, A., Schubert, T.: Slime mold microfluidic logical gates. Mater. Today **17**(2), 86–91 (2014)
33. Whiting, J.G.H., de Lacy Costello, B.P., Adamatzky, A.: Slime mould logic gates based on frequency changes of electrical potential oscillation. Biosystems **124**, 21–25 (2014)

Electrical Spiking of Psilocybin Fungi

Antoni Gandia and Andrew Adamatzky

Abstract Psilocybin fungi, aka "magic" mushrooms, are well known for inducing colourful and visionary states of mind. Such psychoactive properties and the ease of cultivating their basidiocarps within low-tech setups make psilocybin fungi promising pharmacological tools for mental health applications. Understanding of the intrinsic electrical patterns occurring during the mycelial growth can be utilised for better monitoring the physiological states and needs of these species. In this study we aimed to shed light on this matter by characterising the extra-cellular electrical potential of two popular species of psilocybin fungi: *Psilocybe tampanensis* and *P. cubensis*. As in previous experiments with other common edible mushrooms, the undisturbed fungi have shown to generate electric potential spikes and trains of spiking activity. This short analysis provides a proof of intrinsic electrical communication in psilocybin fungi, and further establishes these fungi as a valuable tool for studying fungal electro-physiology.

1 Introduction

Psilocybin fungi, popularly known as "magic" mushrooms, are a group of different species of psychoactive basidiomycetes that have gained an immense popularity since the ethnomycologists Gordon Wasson and his wife Valentina Pavlovna Wasson introduced them to the western audiences in 1957 [1–3]. Psilocybin fungi are remarkably famous for inducing mystical-type experiences thanks to tryptamine alkaloids contained in its hyphae, mainly psilocybin, baeocystin and norbaeocystin [4, 5], to which the community of users and a growing pool of scientific evidence grants different potential benefits, such as treating depression to helping manage alcoholism and drug addiction [6–11].

A. Gandia
Institute for Plant Molecular and Cell Biology, CSIC-UPV, Valencia, Spain

A. Adamatzky (✉)
Unconventional Computing Laboratory, UWE, Bristol, UK
e-mail: andrew.adamatzky@uwe.ac.uk

© The Author(s), under exclusive license to Springer Nature Switzerland AG 2023
A. Adamatzky (ed.), *Fungal Machines*, Emergence, Complexity and Computation 47,
https://doi.org/10.1007/978-3-031-38336-6_3

These organisms have been ever since surrounded of an aura of mysticism and criticism in equal shares, with opinions mostly tied to religious or political beliefs rather than being based in scientific research. Nevertheless, magic mushroom have been used by different human cultures across the globe for millennia, probably since the dawn of mankind, as a tool for exploring and healing psychological and physical disorders, or simply to inspire awe, creativity, introspection, and a better appreciation for nature [12–18]. Considering their cultural and psycho-pharmaceutical importance, the scientific community is trying to make sense of different aspects of their ecology, physiology, pharmacology, and overall, potential biotechnological applications favouring human society and Earth's biosphere.

Recent research suggests that spontaneous electrical low-frequency oscillations (SELFOs) are found across most organisms on Earth, from bacteria to humans, including fungi, playing an important role as electrical organisation signals that guide the development of an organism [19]. Considering its potential function as communication and integration waves, detecting and translating SELFOs in psilocybin fungi species may result of great utility in understanding the growth and behaviour of these organisms, a knowledge that could be added to the toolbox of cultivation and pharmacological optimisation techniques used by the fungal biotech industry.

Thereby, we recorded the extracellular electrical potential in mushrooms and mycelium-colonised substrates as indicators of the fungi intrinsic activity. Action potential-like spikes of electrical potential have been observed using intra-cellular recording of mycelium of *Neurospora crassa* [20] and further confirmed in intra-cellular recordings of action potential in hyphae of *Pleurotus ostreatus* and *Armillaria gallica* [21] and in extra-cellular recordings of basidiocarps of and substrates colonised by mycelium of *P. ostreatus* [22], *Ganoderma resinaceum* [23], and *Omphalotus nidiformis*, *Flammulina velutipes*, *Schizophyllum commune* and *Cordyceps militaris* [24]. While the exact nature of the travelling spikes remains uncertain we can speculate, by drawing analogies with oscillations of electrical potential of slime mould *Physarum polycephalum* [25–28], that the spikes in fungi are triggered by calcium waves, reversing of cytoplasmic flow, and translocation of nutrients and metabolites.

2 Methods

Two widely distributed species of psilocybin fungi, namely *Psilocybe cubensis* strain "B+" (Mondo Mycologicals BV, NL), and *Psilocybe tampanensis* strain "ATL#7" (Mimosa Therapeutics BV, NL), were cultured separately on a mixture of hemp shavings amended with 5% wheat flour, at 60% moisture content, in polypropylene (PP5) filter-patch bags.[1] Electrical activity of the basidiocarps and the colonised substrate was recorded using pairs of iridium-coated stainless steel sub-dermal needle electrodes (Spes Medica S.r.l., Italy), with twisted cables and ADC-24

---

[1] Experiments were conducted at Mimosa Therapeutics BV, The Netherlands.

Fig. 1 Experimental setup. **a** Example of recording from *Psilocybe cubensis* basidiocarps. **b** Example of recording from *Psilocybe tampanensis* mycelium-colonised substrate and view of the experimental setup, in which the electrodes with cables and Pico ADC-24 are seen. (cd) Examples of electrical activity of **c** *Psilocybe cubensis*, two channels, and **d** *Psilocybe tampanensis*, one channel

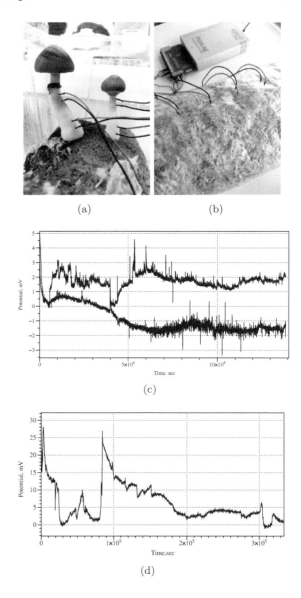

(Pico Technology, UK) high-resolution data logger with a 24-bit analog-to-digital converter (Fig. 1). Resistance of electrodes with cables was 1 Ω. Input impedance of Pico ADC-24 logger was 2 MOhm, ADC input bias current less than 50 nA.

We recorded electrical activity at one sample per second. During the recording, the logger has been doing as many measurements as possible (typically up to 600 s) and saving the average value. We set the acquisition voltage range to 156 mV with an offset accuracy of 9 μV at 1 Hz to maintain a gain error of 0.1%. Each electrode pair was considered independent with the noise-free resolution of 17 bits and conversion time of 60 ms. Each pair of electrodes, called channels, reported a difference of the electrical potential between the electrodes. Distance between electrodes was 1–2 cm. We have conducted eight experiments, in each experiments we recorded electrical activity of the fungi via four channels, i.e. 32 recordings in total.

3 Results

Electrical activity recorded from both species of psilocybin fungi shows a rich dynamics of electrical potential. Examples of the recording conducted for nearly four days are show in Fig. 1c–d. Drift of the base potential can be up to 10–15 mV however rate of the base potential change is measured in days therefore it does affect our ability to recognise spikes. Plots also show that in some cases the signal-to-noise ratio might be substantially low. We omitted such cases from the spike detection pool.

We observed action-potential like spikes of electrical potential. Most expressive spikes, see e.g. Figure 2a, shown very characteristics of action potential recorded in nervous system with distinctive depolarisation and repolarisation phases and a refractory period. In the exemplar action-potential like spike shown in Fig. 2a depolarisation phase is c. 18 s up to 4.5 mV; re-polarisation phase is 97 s; refractory period is rather long c. 450 s.

In some cases, as illustrated in Fig. 2b, two action-potential like spikes can occur at so short interval that they almost merge. In this particular example, an average spike duration is 13 min, and average amplitude is 1.4 mV.

More commonly the spikes emerge in the trains of spikes. A train is a sequence of spikes where distance between two consecutive spikes does not exceed an average duration of a spike. Two trains of spikes are shown in Fig. 2e. Also, spike can stand alone, as shown in Fig. 2d.

Amongst many types of spike classed by their duration we can select very fast spikes, with duration of 1–2 min, and slow spikes, which width can be 15–60 min. An example of very fast spikes directly co-existing with slow spikes is shown in Fig. 2c. In some cases, only very fast spikes can be observed during the whole duration of the recording, see an example in Fig. 2f.

A co-existence of spikes with high, 0.5–1 mV, and low, 0.1–0.3 mV, is evidenced in the recording plotted in Fig. 2g. An amplitude, however, might be not a good characteristic of spikes because it only indicated how far away a wave-front of propagating electrical activity was from a pair of differential electrodes.

Distribution of spike amplitudes version spike width is shown in Fig. 3a. Pearson correlation $R = 0.0753$ calculated on the distribution is technically a positive correlation, however it is low value shows that the relationship between spike width and amplitude is weak.

Distributions of spike width (Fig. 3b), intervals between spikes (Fig. 3c) and spike amplitudes (Fig. 3d) are not normal. This is demonstrated by Kolmogorov-Smirnov test of normality. Values of Kolmogorov-Smirnov statistic is 0.19467 for width dis-

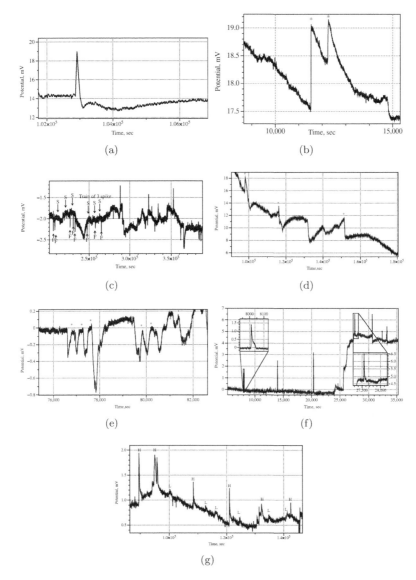

Fig. 2 a Example of an action-potential like voltage spike, *Psilocybe cubensis*. **b** Train of two spikes, *Psilocybe cubensis* **c** Example of fast and slow spike activity, *Psilocybe cubensis*. An average duration of a fast spike is 3 min. An average duration of a slow spike is 16 min. Examples of fast spikes are labelled 'F' and slow 'S'. Train of three slow spikes is also marked. **d** Example of three spikes in electrical potential of *Psilocybe tampanensis*, peaks of the spike are labelled ⋆. **e** Two trains of spikes recorded in *Psilocybe cubensis*: one train comprises of three spikes, another of four spikes; spike are marked by ⋆. **f** Very fast, average 1.5 min, spikes of electrical potential recorded in *Psilocybe cubensis*. **g** Co-existence of high amplitude, labelled 'H', and low amplitude, labelled 'L', spikes in electrical activity of *Psilocybe cubensis*

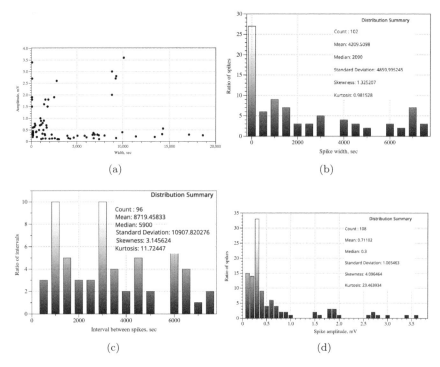

Fig. 3 **a** Spike width versus spike amplitude distribution constructed on recording from bother species of fungi studied. **b** Distribution of spike widths, **c** Distribution of interval between spikes. Bin size is 500 in both distributions. **d** Distribution of spike amplitudes, bin size is 0.1

tribution, 0.21914 for interval distribution, and 0.28243 for amplitude distribution. Corresponding p-values are 0.00072, 0.00016 and 0.00001.

Integrative parameters of spiking behaviour are the following (Table 1). Average duration of a spike is 70 min ($\sigma = 81$ min), median duration is 35 min. Average amplitude of a spike is 0.71 mV ($\sigma = 1.06$ mV), median is 0.3 mV. Average distance between spikes is 145 min ($\sigma = 181$ min), median is 98 min. That is average/median distance between two spikes is a double of the average/median duration of a spike. By the definition of the train, this means that most spikes observed are solitary spikes. Standard deviations of spike duration and amplitude and of interval between spikes are higher than respective average values. This indicates data are more spread out and we should look out for distinct families of spikes.

Let us separate species and—if any—families of spikes in each species. In *Psilocybe tampanensis* spikes are relatively uniform (Table 1): average duration 104 min with $\sigma = 69$ min, average distance between spikes is over three hours and median distance equal to average. In *Psilocybe cubensis* we propose three families of spikes: fast spike, up to 3 min duration, slow spikes, up to 6 h duration, and very slow spikes, up to 2 days (Table 1). Fast spikes rarely form train, an average distance between fast spikes is 17 h. An average duration of a slow spike is 3.9 h with an average

Table 1 Statistical parameters of spiking. In each cell we show average, standard deviation and median

Species	Duration, s	Amplitude, mV	Distance, s
Over all species	4209, 4859, 2090	0.71, 1.06, 0.3	8719, 10907, 5900
Psilocybe tampanensis	6246, 4196, 8800	2.33, 2.22, 1.71	17566, 11183, 17300
Psilocybe cubensis	4082, 4892, 1640	0.48, 0.56, 0.30	8000, 10458, 4930
P. cubensis, fast spikes (1–3 min)	148, 38, 165	0.72, 0.83, 0.35	6128, 4516, 3955
P. cubensis, slow spikes (up to 6 hr duration)	1408, 274, 1394	0.68, 0.60, 0.50	3243, 2125, 3105
P. cubensis, very slow spikes (up to 2 days)	7943, 4658, 6810	0.24, 0.11, 0.25	11948, 13639, 7880

distance between slow spikes is 9 h. Average amplitudes of fast and slow spikes are comparable, 0.72 and 0.68 mV, respectively. An average duration of a very slow spike of *Psilocybe cubensis* is 22 h with an average distance between spikes of 33 h. The very slow spikes have, comparatively to fast and slow spikes, low amplitude of 0.24 mV in average.

4 Discussion

We found that psilocybin fungi exhibit a rich spectrum of oscillations of extracellular electrical potential. We illustrated several types of oscillations and characterised families of fast, slow and very slow oscillations. We demonstrated that several scales—minutes, hours and day—of oscillators states co-exist in basidiocarps and mycelium network of psilocybin fungi. This co-existence is similar to electrical oscillation of a human brain, where fast oscillations might be related to responses to stimulation, including endogenous stimulation by release of nutrients, and slow oscillations might be responsible for memory consolidation [29–32]. Future research could be concerned with decoding and understanding the spiking events to monitor growth, development and physiological states and overall condition of the fungi both in cultivation setups and natural environments. If we were able to decode spiking patterns of fungi we would be able of 'speaking back' to the mycelial network to manipulate the network's morphology, behaviour and, potentially, enhance production of basidiocarps and sclerotia.

References

1. Wasson, R.G.: Seeking the magic mushroom **5** (1957)
2. Wasson, V.P.: I ate the sacred mushroom **5** (1957)
3. Wasson, V.P., Wasson, R.G.: Mushrooms, Russia and History, vol. 2. Pantheon Books (1957)
4. Leung, A.Y., Paul, A.G.: Baeocystin and norbaeocystin: new analogs of psilocybin from psilocybe baeocystis. J. Pharm. Sci. **57** (1968)
5. Blei, F., Dörner, S., Fricke, J., Baldeweg, F., Trottmann, F., Komor, A., Meyer, F., Hertweck, C., Hoffmeister, D.: Simultaneous production of psilocybin and a cocktail of β-carboline monoamine oxidase inhibitors in "magic" mushrooms. Chem.-Eur. J. **26** (2020)
6. Strickland, J.S., Johnson, M.W.: Human behavioral pharmacology of psychedelics. Adv. Pharmacol. **93**(1), 105–132 (2022)
7. Knudsen, G.M.: Sustained effects of single doses of classical psychedelics in humans. Neuropsychopharmacol.: Off. Publ. Am. Coll. Neuropsychopharmacol. **6** (2022)
8. Daws, R.E., Timmermann, C., Giribaldi, B., Sexton, J.D., Wall, M.B., Erritzoe, D., Roseman, L., Nutt, D., Carhart-Harris, R.: Increased global integration in the brain after psilocybin therapy for depression. Nat. Med. **28**(4), 844–851 (2022)
9. Nutt, D., Carhart-Harris, R.: The current status of psychedelics in psychiatry. JAMA Psychiatry **78**(2), 121–122 (2021)
10. Carhart-Harris, R.L.: Trial of psilocybin versus escitalopram for depression. N. Engl. J. Med. **384**, 1402–1411 (2021)
11. Pollan, M.: How to Change Your Mind: The New Science of Psychedelics. Penguin Group, vol. 5, 1st edn. (2019)
12. de Borhegyi, S.F.: Miniature mushroom stones from guatemala. Am. Antiq. **26**(4), 498–504 (1961)
13. Guzmón, G.: Hallucinogenic mushrooms in Mexico: an overview. Econ. Bot. **62**(11), 404–412 (2008)
14. Akers, B.P., Ruiz, J.F., Piper, A., Ruck, C.A.: A prehistoric mural in Spain depicting neurotropic Psilocybe mushrooms?. Econ. Bot. **65**(6), 121–128 (2011)
15. Guerra-Doce, E.: Psychoactive substances in prehistoric times: examining the archaeological evidence. Time Mind **8**(1), 91–112 (2015)
16. de Borhegyi, C., de Borhegyi-Forrest, S.: Mushroom intoxication in mesoamerica. In: History of Toxicology and Environmental Health: Toxicology in Antiquity, vol. 2, pp. 104–115 (2015)
17. Ruck, C.A.P.: Mushroom sacraments in the cults of early Europe. NeuroQuantology **14** (2016)
18. Winkelman, M.: Introduction: evidence for entheogen use in prehistory and world religions. J. Psychedelic Stud. **3**(6), 43–62 (2019)
19. Hanson, A.: Spontaneous electrical low-frequency oscillations: a possible role in hydra and all living systems. Philos. Trans. R. Soc. B: Biol. Sci. **376**, 3 (2021)
20. Slayman, C.L., Long, W.S., Gradmann, D.: Action potentials in *Neurospora crassa*, a mycelial fungus. Biochim. Biophys. Acta (BBA)—Biomembr. **426**(4), 732–744 (1976)
21. Olsson, S., Hansson, B.S.: Action potential-like activity found in fungal mycelia is sensitive to stimulation. Naturwissenschaften **82**(1), 30–31 (1995)
22. Adamatzky, A.: On spiking behaviour of oyster fungi pleurotus djamor. Sci. Rep. **8**(1), 1–7 (2018)
23. Adamatzky, A., Gandia, A.: On electrical spiking of ganoderma resinaceum. Biophys. Rev. Lett. 1–9 (2021)
24. Adamatzky, A.: Language of fungi derived from their electrical spiking activity. R. Soc. Open Sci. **9**(4), 211926 (2022)
25. Iwamura, T.: Correlations between protoplasmic streaming and bioelectric potential of a slime mold, *Physarum polycephalum*. Shokubutsugaku Zasshi **62**(735–736), 126–131 (1949)
26. Kamiya, N., Abe, S.: Bioelectric phenomena in the myxomycete plasmodium and their relation to protoplasmic flow. J. Colloid Sci. **5**(2), 149–163 (1950)

27. Kishimoto, U.: Rhythmicity in the protoplasmic streaming of a slime mold, *Physarum polycephalum*. I. a statistical analysis of the electric potential rhythm. J. Gen. Physiol. **41**(6), 1205–1222 (1958)
28. Meyer, R., Stockem, W.: Studies on microplasmodia of Physarum polycephalum V: electrical activity of different types of microplasmodia and macroplasmodia. Cell Biol. Int. Rep. **3**(4), 321–330 (1979)
29. Mölle, M., Born, J.: Slow oscillations orchestrating fast oscillations and memory consolidation. Prog. Brain Res. **193**, 93–110 (2011)
30. Novikov, N.A., Nurislamova, Y.M., Zhozhikashvili, N.A., Kalenkovich, E.E., Lapina, A.A., Chernyshev, B.V.: Slow and fast responses: two mechanisms of trial outcome processing revealed by EEG oscillations. Front. Hum. Neurosci. **11**, 218 (2017)
31. Mölle, M., Bergmann, T.O., Marshall, L., Born, J.: Fast and slow spindles during the sleep slow oscillation: disparate coalescence and engagement in memory processing. Sleep **34**(10), 1411–1421 (2011)
32. Demanuele, C., Broyd, S.J., Sonuga-Barke, E.J., James, C.: Neuronal oscillations in the EEG under varying cognitive load: a comparative study between slow waves and faster oscillations. Clin. Neurophysiol. **124**(2), 247–262 (2013)

Complexity of Electrical Spiking of Fungi

Mohammad Mahdi Dehshibi and Andrew Adamatzky

Abstract Oyster fungi *Pleurotus djamor* generate action potential like spikes of electrical potential. The trains of spikes might manifest propagation of growing mycelium in a substrate, transportation of nutrients and metabolites and communication processes in the mycelium network. The spiking activity of the mycelium networks is highly variable compared to neural activity and therefore can not be analysed by standard tools from neuroscience. We propose original techniques for detecting and classifying the spiking activity of fungi. Using these techniques, we analyse the information-theoretic complexity of the fungal electrical activity. The results can pave ways for future research on sensorial fusion and decision making by fungi.

1 Introduction

Excitation is an essential property of all living organisms, bacteria [1], Protists [2–4], fungi [5] and plants [6–8] to vertebrates [9–12]. Waves of excitation could be also found in various physical [13–16], chemical [17–19] and social systems [20, 21]. Extracellular (EC) action potential recordings have been widely used to record and measure neural activity in organisms with excitation. When recorded with differential electrodes, the spike manifests a propagating wave of excitation.

In our recent studies [22–24], we have shown that the *Pleurotus djamor* oyster fungi generate action potentials like electrical potential impulses. We observed

---

M. M. Dehshibi
Department of Computer Science and Engineering, Universidad Carlos III de Madrid, Leganés, Spain

A. Adamatzky (✉)
Unconventional Computing Laboratory, University of the West England, Bristol, UK
e-mail: andrew.adamatzky@uwe.ac.uk

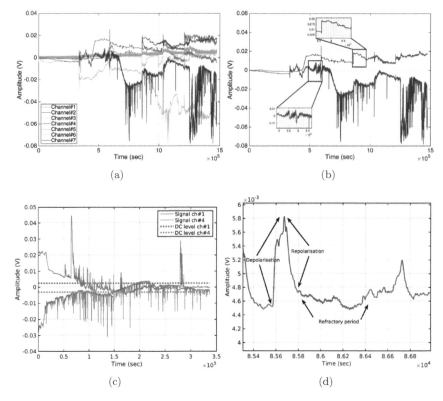

Fig. 1 The electrical behaviour of the mycelium of the grey oyster fungi. **a** Example of electrical potential dynamics recorded in seven channels of the same cluster during 409 h. **b** Two channels are zoomed in the inserts to show the rich combination of slow (hours) drift of base electrical potential combined with relatively fast (minutes) oscillations of the potential. **c** DC levelling for two channels is plotted. The mismatch of DC levels indicates the resistance and different levels of intra-communication in the substrate. **d** All 'classical' parts of the spike, i.e. depolarisation, depolarisation and refractory period, can be found in this sample spike. This spike has a length of 220 s, from the base-level potential to the refractory-like phase, and a refractory period of 840 s. The depolarisation and depolarisation rates are 0.03 and 0.009 mV/s, respectively

spontaneous spike[1] trains with two types of activity, i.e. high-frequency (2.6 min period) and low-frequency (14 min period). However, the proper use of this information is subject to the accurate extraction of the EC spike waveform, separating it from the background activity of the neighbouring cells and sorting the characteristics.

The lack of an algorithmic framework for the exhaustive characterisation of the electrical activity of the substrate colonised by mycelium of oyster fungi *Pleurotus djamor* has inspired us to develop a framework to extract spike patterns, quantify the

---

[1] Calling the spikes spontaneous means that the intentional external stimulus does not invoke them. Otherwise, the spikes actually reflect the ongoing physiological and morphological processes in the mycelial networks.

diversity of spike events and measure the complexity of fungal electrical communication. We evidenced the spiking activity of the mycelium (see an example in Fig. 1), which will enable us to build an experimental prototype of fungi-based information processing devices.

We evaluated the proposed framework in comparison with existing spike detection techniques in neuroscience [25, 26] and observed a significant improvement in the spike activity extraction. The evaluation of the proposed method for detecting spike events compared to the specified spike arrival time by the expert shows true-positive and false-positive rates of 76 and 16%, respectively. We found that the average dominant duration of an action-potential-like spike is 402 s. The amplitude of the spikes ranges from 0.5 to 6 mV and depends on the location of the source of electrical activity (the position of electrodes). We have found that the complexity of the Kolmogorov fungal spike ranges from 11×10^{-4} to 57×10^{-4}. In [27], the human brain's Kolmogorov complexity is measured in normal, pre-ictal and ictal states resulting in 6.01, 5.59 and 7.12 values, respectively. Although the fungi' complexity is considerably smaller than that of the human brain, its changes suggest a degree of intra-communication in the mycelium sub-network. In fact, different parts of the substrate transmit different information to other parts of the mycelium network, where the more prolonged propagation of excitation waves leads to higher levels of complexity.

2 Experimental Set-up

A wood shavings substrate was colonised by the mycelium of the grey oyster fungi, *Pleurotus ostreatus* (Ann Miller's Speciality Mushrooms Ltd, UK). The substrate was placed in a hydroponic growing tent with a silver Mylar lightproof inner lining (Green Box Tents, UK). Recordings were carried out in a stable indoor environment with the temperature remaining stable at $22 \pm 0.5°$ and relative humidity of air $40 \pm 5\%$. The humidity of the substrate colonised by fungi was kept at c. 70–80%. Figure 2 shows examples of the experimental setups.

We inserted pairs of iridium-coated stainless steel sub-dermal needle electrodes (Spes Medica SRL, Italy) with twisted cables into the colonised substrate for recording electrical activity. Using a high-resolution ADC-24 (Pico Technology, UK) data logger with a 24-bit A/D converter, galvanic insulation and software-selectable sample rates all lead to superior noise-free resolution. We recorded electrical activity one sample per second, where the minimum and maximum logging times were 60.04 and 93.45 h, respectively. During recording, the logger makes as many measurements as possible (basically up to 600 per s) and saves the average value. We set the acquisition voltage range to 156 mV with an offset accuracy of 9 µV at 1 Hz to preserve a gain error of 0.1%. Each electrode pair was considered independently with a 17-bit noise-free resolution and a 60 ms conversion time. In our experiments, electrode pairs were placed in one of two configurations: random placement or in-line placement.

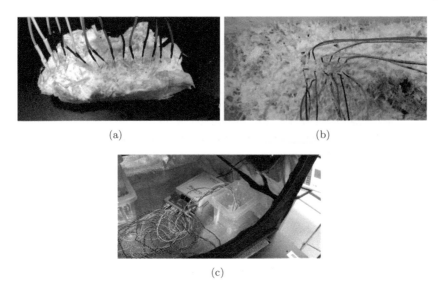

Fig. 2 a In-line placement of electrodes (1 cm distance), **b** random electrode placement, **c** the experimental setup

The distance between the electrodes was between 1–2 cm. In each cluster, we recorded 5–16 pairs of electrodes (channels) simultaneously.

In six trials, we also undertook recordings of the fruit body's resistance, where electrodes were inserted in stalks of the bodies. We measured and logged a range of resistances 1–1.6 kΩ using Fluke 8846A precision multimeter, where the test current being 1 ± 0.0013 μA, once per 10 s, 5×10^4 samples per trial [28]. It should be noted that the placement of the electrodes in two experiments was in-lines with a distance of 1 cm, in two experiments it was in-lines with a distance of 2 cm, and in two experiments it was random with a distance of approximately 2 cm.

3 Proposed Method

A spike event can be formally defined as an extracellular signal that exceeds a simple amplitude threshold and passes through a corresponding pair of user-specific time-voltage boxes. The spike, which involves depolarisation, depolarisation and refractory cycles, represents physiological and morphological processes in mycelial networks. To extract spike events, we proposed an unsupervised approach consisting of three major steps. The pipeline of the proposed approach is shown in Fig. 3

—**Step 1**: We split the entire recording duration ($F(t)$) into k chunks ($f_k(t)$) with respect to the signal transitions. In order to evaluate the transitions, we determined the state level of the signal by its histogram and identified all regions that cross the upper boundary of the low state and the lower boundary of the high state. Then, we

Complexity of Electrical Spiking of Fungi

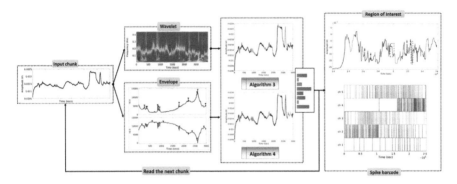

Fig. 3 The pipeline for the identification of spike events

measure scale-to-frequency conversions of the analytic signal in each chunk using Morse wavelet basis [29]. To assess the existence of spike-like events, we scaled the wavelet coefficients at each frequency and obtained the sum of the scales below the threshold specified in Algorithm 6. Finally, we selected regions of interest (ROI) enclosed between a consecutive local minimum and a maximum of more than 30 sec.
—**Step 2**: We used spline interpolation to measure the analytic signal envelopes around local maximum values. To determine the analytical signal, we first applied the discrete approximation of Laplace's differential operator to $f_k(t)$ to obtain a finite sequence of equally-spaced samples. Then, we applied discrete-time Fourier transform to this finite sequence. From the average signal envelope, we extracted regions spanning between a consecutive local minimum and a maximum. These regions created constraints that contributed to the identification of spike events.
—**Step 3**: We retained the ROIs extracted in the first step, which met the constraints of the second step. The signal envelope could direct wavelet decomposition in an unsupervised manner in order to cluster the signal into the spike, pseudo-spike, and background activity of the adjacent cells. In the following sub-sections, we detailed the proposed process.

3.1 Slicing Fungi Electrical Activity

To split the electrical activity of fungi ($F(t)$) with a duration of (t) second into (k) chunks ($fk(t)$, $1 \leq k \leq t - 1$), we used the signal transitions that constitute each pulse. To determine the transitions, we estimated the state level of $F(t)$ using the histogram method [30]. Then, we identified all regions that cross the upper boundary of the low state and the lower boundary of the high state. We followed the following steps to estimate the signal states:

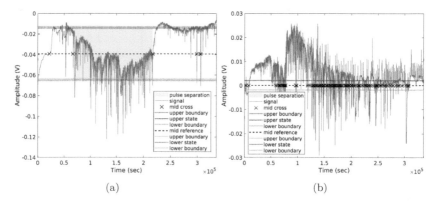

Fig. 4 Slicing electrical potential recordings for two channels

(1) Determine the minimum, maximum and range of amplitudes.
(2) Sort the amplitude values in the histogram bins and determine the width of the bin by dividing the amplitude range by the number of bins.
(3) Identify the lowest- and highest-indexed histogram bins, hb_{low}, hb_{high}, with non-zero counts.
(4) Divide the histogram into two sub-histograms, where the indices of the lower and upper histogram bins are $hb_{low} \leq hb \leq \frac{1}{2}(hb_{high} - hb_{low})$ and $hb_{low} + \frac{1}{2}(hb_{high} - hb_{low}) \leq hb \leq hb_{high}$, respectively.
(5) Calculate the mean of the lower and upper histogram to compute the state levels.

Each chunk is then enclosed between the last negative-going transitions of each positive-polarity pulse and the next positive-going transition. Figure 4 shows the slicing results of two channels.

3.2 Detecting Time-Localised Events by Morse-based Wavelets

The electrical activity of mycelium shows modulated behaviour with changes in amplitude and frequency over time. This feature suggests that the signal can be analysed with analytic wavelets, which are naturally grouped into pairs of even or cosine-like and odd or sine-like pairs, allowing them to capture phase variability. A wavelet ($\psi(t)$) is a finite energy function that projects $f(t)$ to a family of time-scale waveforms through translation and scaling. The Morse wavelet ($\psi_{\beta,\gamma}(t)$) is an analytical wavelet whose Fourier transform is supported only on a positive real axis [29, 31]. This wavelet is defined in the frequency domain using Eq. 1.

$$\psi_{\beta,\gamma}(t) = \frac{1}{2\pi} \int_{-\infty}^{\infty} \Psi_{\beta,\gamma}(\omega) e^{i\omega t} \, d\omega,$$

$$\Psi_{\beta,\gamma}(\omega) \equiv \begin{cases} a_{\beta,\gamma} \omega^{\beta} e^{-\omega^{\gamma}} & \omega > 0 \\ \frac{1}{2}\left(a_{\beta,\gamma} \omega^{\beta} e^{-\omega^{\gamma}}\right) & \omega = 0 \\ 0 & \omega < 0 \end{cases}. \quad (1)$$

where $\beta \geq 0$ and $\gamma > 0$, ω is the angular frequency and $a_{\beta,\gamma} \equiv 2\left(\frac{e\gamma}{\beta}\right)^{\frac{1}{\gamma}}$ is the amplitude coefficient used as a real-valued normalised constant. Here, e is Euler's number, β characterises the low-frequency behaviour, and γ defines the high-frequency decay. We can rewrite Eq. 1 in the Fourier domain, parameterised by β and γ as in Eq. 2.

$$\phi_{\beta,\gamma}(\tau,s) \equiv \int_{-\infty}^{\infty} \frac{1}{s} \psi_{\beta,\gamma}^{*}\left(\frac{t-\tau}{s}\right) f(t) \, dt = \frac{1}{2\pi} \int_{-\infty}^{\infty} e^{i\omega\tau} \Psi_{\beta,\gamma}^{*}(s\omega) F(\omega) \, d\omega. \quad (2)$$

where $F(\omega)$ is the Fourier transform of $f(t)$, and $*$ denotes the complex conjugate. When $\Psi_{\beta,\gamma}^{*}(\omega)$ is real-valued, the conjugation may be omitted. The scale variable s allows the wavelet to stretch or compress in time. In order to reflect the energy of $f(t)$ and to normalise time-domain wavelets to preserve constant energy, $\frac{1}{\sqrt{s}}$ is typically used. However, instead, we used $\frac{1}{s}$ since we define the amplitude of the time-located signals. To recover time-domain representation, we can use the inverse Fourier transform by $f(t) = \frac{1}{2\pi} \int_{-\infty}^{\infty} e^{i\omega t} F(\omega) \, d\omega$ and $\psi_{\beta,\gamma}(t) = \int_{-\infty}^{\infty} e^{i\omega t} \, dt = 2\pi\delta(\omega)$, where $\delta(\omega)$ is the Dirac delta function.

The representation of Morse wavelets would be more oscillatory when both β and γ increase, and more localised with impulses when these parameters decrease. On the other hand, increasing β and holding γ fixed expand the central portion of the wavelet and increase the long-term rate of decay. Whereas, increasing γ by keeping β constant extends the wavelet envelope without affecting the long-term decay rate. Inspired by [32], we set the symmetry parameter γ to 3 and the time-bandwidth product $P^2 = \beta\gamma$ to 60. We have used L_1 normalisation to provide the same magnitude in wavelets when we have the same amplitude oscillatory components at different scales.

Figure 5 displays two randomly chosen 3000-second chunks of fungi electrical activity (namely $Slice_1$ and $Slice_2$) with their Morse wavelet scalograms. We observed that the use of the maximum absolute value at each frequency (level) to normalise coefficients may help to identify events that may involve spikes. We then used Eq. 3 to normalise coefficients and set zero entries to 1.

$$\kappa_{\beta,\gamma}(\tau,s) = |\phi_{\beta,\gamma}(\tau,s)|^{\mathsf{T}},$$

$$g_{\beta,\gamma}(\tau,s) = \left(\eta \times \frac{\kappa_{\beta,\gamma}(\tau,s) - \min_{s}(\kappa_{\beta,\gamma}(\tau,s))}{\max_{s}(\kappa_{\beta,\gamma}(\tau,s))}\right)^{\mathsf{T}}. \quad (3)$$

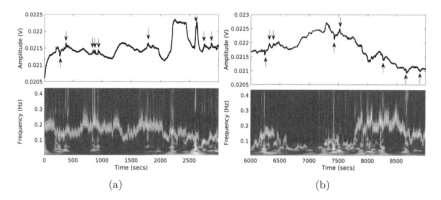

Fig. 5 Annotated spikes by the expert (black arrows) on the Morse wavelet scalogram for **a** $Slice_1$ and **b** $Slice_2$. The scalogram is plotted as a function of time and frequency, where the maximum absolute value at each frequency is used for the normalisation of the coefficient. The frequency axis is shown on a linear scale

---

Algorithm 1: Detecting candidate regions for time-localised events.

Input : $g_{\beta,\gamma}(\tau, s)$ – Scaled wavelets coefficients.
Output: $\mathcal{B}$ – set of candidate regions.

1 **begin**
2     $\epsilon = 0.05 \times (\max(g_{\beta,\gamma}(\tau, s)) - \min(g_{\beta,\gamma}(\tau, s)))$;
3     $max_g \leftarrow$ set of all LocalMaximum($g_{\beta,\gamma}(\tau, s), \epsilon$);
    // LocalMaximum() returns τ^* if $\forall \tau \in (\tau^* \pm \epsilon)$, $g_{\beta,\gamma}(\tau^*, s) \geq g_{\beta,\gamma}(\tau, s)$.
4     $min_g \leftarrow$ set of all LocalMinimum($g_{\beta,\gamma}(\tau, s), \epsilon$);
5     $\mathcal{U} \leftarrow$ **sort**($min_g \cup max_g$);
6     $n = \mathbf{card}(\mathcal{U})$;
    // card(A) returns number of entries in A.
7   **if** $n \equiv 1 \pmod{2}$ **then**
8   slack $\leftarrow$ **mean**(difference of two consecutive entries);
9   Add $\mathbf{min}(\mathcal{U}_n + \text{slack}, \tau)$ to $\mathcal{U}$;
10        $n = n + 1$;
11   **end**
12    $\mathcal{B} \leftarrow (\mathcal{U}_i, \mathcal{U}_{i+1})$, $\forall i \in \{1, 3, \cdots, n-1\}$
13 **end**
14 **return** $\mathcal{B}$

Fig. 6 Detecting candidate regions for time-localised events

---

where $|\bullet|$ and $(\bullet)^\mathsf{T}$ return the absolute value and the matrix transpose, respectively. Here, η is a scaling factor that we empirically set to 240.

We used $g_{\beta,\gamma}(\tau, s)$ in Algorithm (Fig. 6) to extract the candidate ROIs shown in Fig. 7. As shown in Fig. 7c, d, some of the detected regions are either too short[2] or lack repolarisation and depolarisation periods that should be removed from $\mathcal{B}$.

---

[2] We observed in our previous studies [22, 23] that minimum spike length was 5 mins.

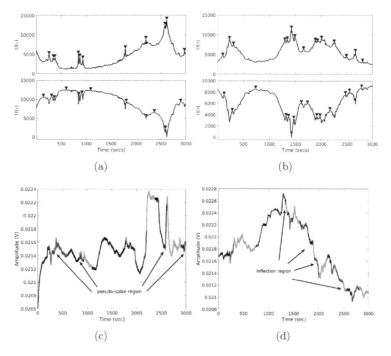

Fig. 7 a, b Identified local maxima and minima over $g_{\beta,\gamma}(\tau,s)$ in **a** $Slice_1$ and **b** $Slice_2$. The second-row of the plot is the inverse of the first row; therefore, the marked maximums are identical to the local minima. **c, d** Candidate regions of interest which are alternately coloured purple and green to ease visual tracking

Algorithm Fig. 8 is proposed to eliminate so-called *pseudo-spike* and *inflection* regions which do not reach the spike characteristics as shown in Fig. 9. Applying Algorithm Fig. 8 resulted in the loss of two spikes in $Slice_1$ (see Fig. 9a) and the failure to eliminate two *pseudo-spike* and two *inflection* regions in $Slice_2$ (see Fig. 9b). We found that the analysis of the analytic signal by its envelope could improve the accuracy of the spike detection.

3.3 Analytical Signal Envelope for Locating Spike Pattern

We calculated the magnitude of the analytic signal to obtain the signal envelope (ξ). The analytic signal is detected using the discrete Fourier Transform as implemented in the Hilbert Transform. In order to highlight effective signal peaks and neutralise *inflection* regions, the second numerical signal derivation ($L = \partial^2 f / 4 \partial t^2$) was calculated. A frequency-domain approach is proposed in [33] to approximately generate a discrete-time analytic signal. In this approach, the negative frequency in half of each spectral period is set to 0, resulting in a periodic one-sided spectrum. The pro-

Algorithm 2: Excluding pseudo-spike and inflation regions form candidate ROI.

Input : B — set of ROI, i.e., Algorithm 1 output,
f — Electrical potential.
Output: C — set of wavelet-based ROIs,
$\mathcal{D}$ — set of *pseudo-spike* and *inflection* regions.

1 **begin**
2   **for** $i = 1$ **to card**(B) **do**
3       $lb \leftarrow B(i, 1)$;
4       $ub \leftarrow B(i, 2)$;
5   **if** $(ub - lb) > 30$ **then**
6          $chunk = f[lb \cdots ub]$;
7          $minima = \mathbf{min}(\text{isLocalMinimum}(chunk))$;
            // isLocalMinimum() and isLocalMaximum() use spline interpolation in locating local extreme Hall and Meyer (1976).
8          $maxima = \mathbf{max}(\text{isLocalMaximum}(chunk))$;
9   **if** $f(minima) < \mathbf{min}(f(lb), f(ub))$ **or** $f(maxima) > \mathbf{max}(f(lb), f(ub))$ **then**
10             $C \leftarrow [lb, ub]$;
11   **else**
12             $\mathcal{D} \leftarrow [lb, ub]$;
13   **end**
14   **end**
15   **end**
16 **end**
17 **return** $C, \mathcal{D}$

Fig. 8 Excluding pseudo-spike and inflation regions form candidate ROI

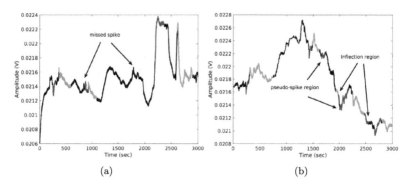

Fig. 9 Results of applying Algorithm 8 to **a** $Slice_1$ and **b** $Slice_2$. Two spike events are missed in $Slice_1$. Two pseudo-spike and two inflection regions still remain in $Slice_2$

cedure for generating a complex-valued N-point (N is even) discrete-time analytic signal ($F(\omega)$) from a real-valued N-point discrete time signal ($L[n]$) is as follows:

(1) Calculate the N-point discrete-time Fourier transform using $F(\omega) = T \sum_{n=0}^{N-1} L[n]e^{-i2\pi\omega Tn}$, where $|\omega| \leq 1/2T$ Hz. $L[n], 0 \leq n \leq N-1$ is obtained by sampling the band-limited real-valued continuous-time signal $L(nT) = L[n]$ at periodic time intervals of T seconds to prevent aliasing.
(2) Calculate the N-point one-sided discrete-time analytic signal transform:

$$Z[m] = \begin{cases} F[0], & \text{for } m = 0 \\ 2F[m], & \text{for } 1 \leq m \leq \frac{N}{2} - 1 \\ F[\frac{N}{2}], & \text{for } m = \frac{N}{2} \\ 0, & \text{for } \frac{N}{2} + 1 \leq m \leq N - 1. \end{cases} \quad (4)$$

(3) Calculate the N-point inverse discrete-time Fourier transform to obtain the complex discrete-time analytic signal of same sample rate as the original $L[n]$

$$z[n] = \frac{1}{NT} \sum_{m=0}^{N-1} Z[m] e^{\frac{i2\pi mn}{N}} \quad (5)$$

Obtaining an analytic signal in this way satisfies two properties: (1) The real part is equivalent to the original discrete-time sequence; (2) the real and imaginary components are orthogonal. Calculating the magnitude of the analytic signal using Eq. 6 results in the signal envelope ($\xi[n]$) containing the upper ($\xi_H[n]$) and lower ($\xi_L[n]$) envelopes of $L[n]$.

$$\xi[n] = |z[n]| \quad (6)$$

Envelopes are calculated using spline interpolation over a local maximum separated by at least $n_p = 60$ samples. We considered $n_p = 60$ because, in our previous studies [22, 23], we did not observe any electrical potential of spikes shorter than 60 s. Algorithm Fig. 10 was proposed to locate candidate regions using a signal envelope.

Figure 11a, d shows the candidate regions in $\mathcal{R}$ before applying Step 13. At this step, while $\mathcal{R}$ includes regions that do not align with the spike definition (pointed by arrow in the plot), the correctly identified spikes are consistent with our findings in [22, 34]. Steps 13 and 14 were used to remove non-spike regions marked in red in Fig. 11b, e. However, the output of Algorithm Fig. 10 (see Fig. 11c, f) still includes regions belonging to either *pseudo-spike/inflection* regions or refractory periods attached to *pseudo-spike* regions.

To fix mis-identified ROIs in Algorithms Figs. 8 and 10, we proposed Algorithm Fig. 12 in which regions in $(\mathcal{C} \cup \mathcal{D})$ are used to update $\mathcal{R}$. Indeed, if the ROI in $\mathcal{R}$ is a subset of $(\mathcal{C} \cup \mathcal{D})$, we add it to the spike event set $(\mathcal{F}_s)$ and update the spike length. If the ROI in $(\mathcal{C} \cup \mathcal{D})$ is a subset of $\mathcal{R}$, we add it to the *pseudo-spike* set $(\mathcal{F}_p)$. In the case of an intersection that does not meet the subset requirement, we concatenate ROIs and divide the new region from the intersection point into two segments. Then, we add the segment with the minimum length to $\mathcal{F}_p$. Finally, regions with a length of less than 60 s are excluded from both $\mathcal{F}_s$ and $\mathcal{F}_p$. The results are shown in Fig. 13.

4 Experimental Results

This section consists of objective and complexity analyses. In the objective analysis, we demonstrated the effectiveness of the spike event detection method compared

Algorithm 3: Detecting candidate spike region from signal envelope.

Input: $\xi[n]$ — Envelope of signal $L[t]$,
$n_p = 60$ — Minimum distance between two consecutive local extreme.
Output: $\mathcal{R}$ — set of envelope-based ROIs.

1 **begin**
2      $\xi_M[n] = (\xi_H[n] + \xi_L[n])/2$;
3      $[val_{min}, ind_{min}]$ = isLocalMinimum$(\xi_M[n], n_p)$;
4      $[val_{max}, ind_{max}]$ = isLocalMaximum$(\xi_M[n], n_p)$;
    // isLocalMinimum() and isLocalMaximum() locate local minimum and maximum, respectively.
5      $j \leftarrow$ index of the first local maximum whose value is greater than the value of the first local minimum;
6   **for** $i = 1$ **to card**(ind_{min}) **do**
7   **if** $j \leq$ **card**(ind_{max}) **then**
8              $\Delta \leftarrow val_{max}(j) - val_{min}(i)$;
9   Add $(ind_{min}(i), ind_{max}(j), \Delta)$ to $\mathcal{R}$;
10             $j \leftarrow j + 1$;
11   **end**
12   **end**
    // $\mathcal{R}$ has j rows and 3 columns, as $\mathcal{R}_1$, $\mathcal{R}_2$, and $\mathcal{R}_3$.
13     $\rho =$ **mean**$(\mathcal{R}_3) -$ **std**$(\mathcal{R}_3)$;
    // **mean**() and **std**() calculate the mean and standard deviation, respectively.
14   Remove the kth entry from $\mathcal{R}$ where $\mathcal{R}_3(k) < \rho$ – see Figure 8(b);
15 **end**
16 **return** $\mathcal{R}$

Fig. 10 Detecting candidate spike region from signal envelope

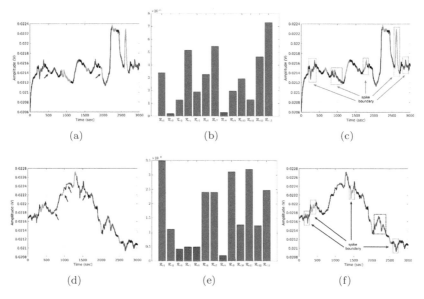

Fig. 11 Results of Algorithm 10 applied to $Slice_1$ (first row) and $Slice_2$ (second row). **a, d** Candidate regions by finding local minima and maxima in the analytic signal envelope. The pointed regions are also highlighted in the bar chart in red. **b, e** The absolute difference in prominence between the successive local minima and maxima. Regions that do not satisfy $\mathcal{R}_3(k) < \rho$ are coloured in red. **c, f** Regions of Interest in $\mathcal{R}$. The grey dashed rectangle shows the correct spike, including repolarisation, depolarisation, and refractory periods. The purple dashed rectangle shows the region whose refractory period attached to the *pseudo-spike* region

to conventional spike detection techniques in neuroscience [25, 26]. We also compared the proposed method with the expert opinion on the location of spikes. In the complexity analysis, we selected the complexity measures used in previous studies [35–41] to quantify spatio-temporal activity patterns.

4.1 Objective Analysis

Various methods have been proposed for detecting and sorting spike events in EC recordings [25, 42–51]. However, only a few of these methods do not require additional details, such as template construction and the supervised setting of thresholds for detecting and sorting spike events [25, 26]. Nenadic and Burdick [25] have developed an unsupervised method for detecting and locating spikes in noisy neural recordings. This approach benefits from the continuous transformation of the wavelet. They applied multi-scale signal decomposition using the 'bior1.3,' 'bior1.5,' 'Haar,' or 'db2' wavelet basis. To determine the presence of spikes, they separated the signal and noise at each scale and performed Bayesian hypothesis testing. Finally, they

Algorithm 4: Extracting fungi *spike* and *pseudo-spike* events.

Input : C, D, R — Regions of interest.
Output: $\mathcal{F}_s, \mathcal{F}_p$ — Fungi *spike* and *pseudo-spike* events, respectively.

1 **begin**
2   **foreach** $r_e \in \mathcal{R}$ **do**
3       $chunk_e \leftarrow [r_e^1 \cdots r_e^2]$;
4   **foreach** $r_w \in (C \cup D)$ **do**
5          $chunk_w \leftarrow [r_w^1 \cdots r_w^2]$;
6   **switch** $chunk_w, chunk_e$ **do**
7   **case** $chunk_e \subset chunk_w$ **do**
8                $chunk_w(end) = chunk_e(end)$;
9                $\mathcal{F}_s \leftarrow chunk_w$;
10   **end**
11   **case** $chunk_w \subset chunk_e$ **do**
12                $\mathcal{F}_p \leftarrow chunk_w$;
13   **end**
14   **case** `intersect`$(chunk_w, chunk_e)$ **do**
                // `intersect()` checks if two chunks have an intersection point.
15   **Split** the concatenation of $chunk_w$ and $chunk_e$ from intersection point into two sub-Chunks;
16                $\mathcal{F}_p \leftarrow$ sub-Chunks;
17   **end**
18   **end**
19   **end**
20   **end**
21   **foreach** $r \in (\mathcal{F}_s \cup \mathcal{F}_p)$ **do**
22   **Remove** r if $|r| < 60$;
23   **end**
24 **end**
25 **return** $\mathcal{F}_s, \mathcal{F}_p$

Fig. 12 Extracting fungi *spike* and *pseudo-spike* events

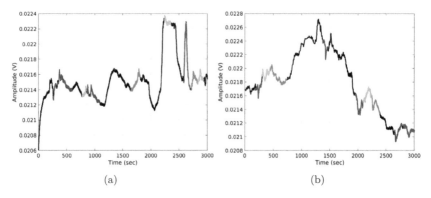

Fig. 13 Results of applying Algorithm 12 to **a** $Slice_1$ and **b** $Slice_2$. We alternatively used red/purple for colouring spike events and blue/cyan for colouring *pseudo-spike* events

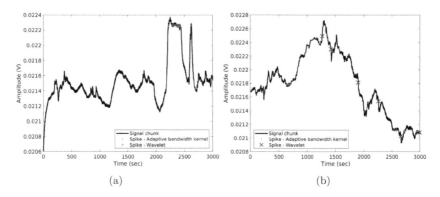

Fig. 14 Results of applying proposed algorithms in [25, 26] to **a** $Slice_1$ and **b** $Slice_2$. Note that the wavelet-based method can only locate spike arrival time. The kernel bandwidth optimisation can, however, extract the spike region

combined decisions on different scales to estimate the arrival times of individual spikes.

Shimazaki and Shinomoto [26] proposed an optimisation technique for the timing-histogram bin width selection. This optimisation minimised the mean integrated square in the kernel density estimation. This method benefits from variable kernel width, which allowed grasping non-stationary phenomena. Also, this method used stiffness constant to avoid possible overfitting due to excessive freedom in the bandwidth variability. The calculated bandwidth was then used as a proxy for filtering spike event regions. Figure 14 shows the results of applying these methods to two chunks with a length of 3000 s.

Both compete methods could not correctly detect all spike events that were located by the expert. The wavelet-based method could locate three spikes in Fig. 14b without detecting any spike in Fig. 14a. The adaptive bandwidth kernel-based method could detect one spike in Fig. 14b and one pseudo-spike in Figs. 14a, b. While our proposed method misidentified one spike event in Fig. 13a and three spikes in Fig. 13b.

We also compared the proposed method with the expert opinion on a randomly selected 36,000 s chunk, i.e., 10 h of electrical activity recordings. In this quantitative comparison, the proposed approach could correctly locate 21 spikes and four pseudo-spike events. Our method also overestimated two refractory periods, resulting in the true-positive and false-positive rates of 76 and 16%, respectively. Figure 15a shows located spikes by the expert, and Fig. 15b demonstrates the results of the proposed spike detection method.

We applied the proposed method to six experiments where the statistical results are shown in Figs. 16 and 17 and summarised in Table 1. It should be noted that the placement of the electrodes in two experiments was in lines with a distance of 1 cm, in two experiments it was in lines with a distance of 2 cm, and in two experiments it was random with a distance of approximately 2 cm. The proposed method is implemented

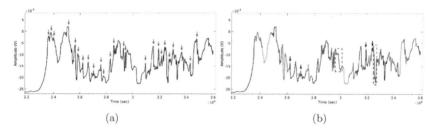

Fig. 15 a Spike arrival time located by the expert. Here we used augmented pink arrow to point to these spikes. **b** Spike regions extracted by the proposed method. Spike regions are alternatively coloured in orange and violet. The green areas point to *pseudo-spike* regions that are mistaken for spikes. Blue rectangles with dash edge show overestimated refractory periods. We used black arrows to point to the missed spikes

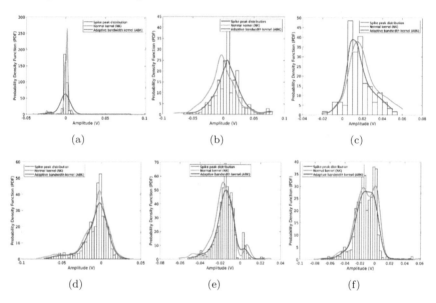

Fig. 16 Distribution of spike event lengths with superimposed Gaussian and Adaptive bandwidth kernels [26]. **a, b** In-line electrode arrangements with a distance of 1 cm. **c, d** In-line electrode arrangements with a distance of 2 cm. **e, f** Random electrode arrangements with an approximate distance of 2 cm

in MATLAB R2020a, where the code and details of the experiments can be found in [52].

These results are consistent with the experiments carried out on the electrical activity of *Physarum polycephalum* [34, 53] where it has been reported that the length of the Physarum spike is between 60 and 120 s. Physarum is faster than fungi in terms of growth. Now, with further observations, we can hypothesise that the length of the fungal spikes cannot be less than 60 s.

Table 1 The dominant value and bandwidth for the length and amplitude of the spike in each experiment across all recording channels. Duration and amplitude of spikes are estimated by the probability density function (PDF) and the adaptive bandwidth kernel (ABK) [26]. The bold-faced **blue** and *red* entries show the absolute minimum and maximum values, respectively. As we have bi-directional potential changes, we have considered absolute value

	#Channels	#Spikes	Length (s)				Amplitude (V)			
			PDF		ABK		PDF		ABK	
			Dominant	Bandwidth	Dominant	Bandwidth	Dominant	Bandwidth	Dominant	Bandwidth
#1	8	565	**84.00**	75.61	**84.00**	60.22	**0.00003**	0.00048	**-0.00117**	0.00576
#2	5	447	366.80	154.31	625.60	126.47	0.00642	0.00544	0.00642	0.00667
#3	4	124	84.00	75.61	84.00	60.22	0.00003	0.00048	-0.00117	0.00576
#4	5	951	534.12	80.09	534.12	84.80	-0.00239	0.00301	-0.00239	0.00508
#5	5	573	334.25	74.52	334.25	80.9	*-0.01536*	0.00218	*-0.01462*	0.00357
#6	15	862	*1014.72*	99.53	*1014.72*	92.67	-0.00172	0.00381	-0.01277	0.00591

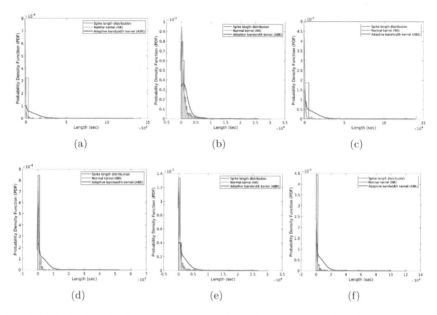

Fig. 17 Distribution of spike maximum amplitudes with superimposed Gaussian and Adaptive bandwidth kernels [26] for **a, b** in lines electrode placement with a distance of 1 cm. **c, d** in lines electrode placement with a distance of 2 cm. **e, f** random electrode placement with an approximate distance of 2 cm

4.2 Complexity Analysis

To quantify the complexity of the electrical signalling recorded, we used the following measurements:

(1) The Shannon entropy (H) is calculated as $H = -\sum_{w \in W}(\nu(w)/\eta \cdot ln(\nu(w)/\eta))$, where $\nu(w)$ is a number of times the neighbourhood configuration w is found in configuration W, and η is the total number of spike events found in all channels of the experiment.
(2) Simpson's diversity (S) is calculated as $S = \sum_{w \in W}(\nu(w)/\eta)^2$. It linearly correlates with Shannon entropy for $H < 3$ and the relationship becomes logarithmic for higher values of H. The value of S ranges between 0 and 1, where 1 represents infinite diversity and 0, no diversity.
(3) Space filling (D) is the ratio of non-zero entries in W to the total length of string.
(4) Expressiveness (E) is calculated as the Shannon entropy H divided by space-filling ratio D, where it reflects the 'economy of diversity'.
(5) Lempel–Ziv complexity (LZ) is used to assess temporal signal diversity, i.e., compressibility. Here, we represented the spiking activity of mycelium with a binary string where '1s' indicates the presence of a spike and '0s' otherwise. Formally, as both the barcode and the channels' electrical activity are stored as

PNG images, LZ is the ratio of the barcode image size to the size of the electrical activity image (see Figs. 18 and 19).
(6) Perturbation complexity index $PCI = LZ/H$.

To calculate Lempel–Ziv complexity, we saved each signal as a PNG image (see two examples in Fig. 19), where the 'deflation' algorithm used in PNG lossless compression [54–56] is a variation of the classical LZ77 algorithm [57]. We employed this approach as the recorded signal is a non-binary string. We take the largest PNG file size to normalise this measurement.

In order to assess signal diversity across all channels and observations, each experiment was represented by a binary matrix with a row for each channel and a column for each observation. This binary matrix is then concatenated by observation to form a single binary string. We used Kolmogorov complexity algorithm [58] to measure the Lempel–Ziv complexity (LZc) across channels. LZc captures both temporal diversity on a single channel and spatial diversity across channels. We also normalised LZc by dividing the raw value by the randomly shuffled value obtained for the same binary input sequence. Since the value of LZ for a fixed-length binary sequence is maximum if the sequence is absolutely random, the normalised values represent the degree of signal diversity on a scale from 0 to 1. The results of the calculation of these complexity measurements for all six configurations are shown in Fig. 20 and summarised in Table 2.

We calculated the aforementioned complexity criteria for three forms of writing to illustrate the communication complexity of the mycelium substrate, including (1) news items,[3] (2) random alphanumeric sequences[4] and (3) periodic alphanumeric sequences encoded with Huffman code [59], see barcode in Fig. 21. Table 3 presents the results of the comparison. We also considered two podcasts in English (387 s) and Chinese (385 s) to compare the complexity of fungal spiking with human speech. Both podcasts were in MP3 format at a sampling rate of 44 100 Hz. We randomly selected two chunks from two electrical activity channels with a duration of 388 and 342 s to compare with the English and Chinese podcasts, respectively. We observed that, in both cases, the Kolmogorov complexity of the fungal is lower than the human speech, implying the fact that the amount of information transmitted by the fungi is less than the human voice. From a technical point of view, we computed the DC level of each signal and binarised the signal with respect to that level [60]. To binaries the signal, we set the inputs with values less than or equal to the DC level to 0 and the rest of the inputs to 1. The findings are shown in Fig. 22.

---

[3] https://www.sciencemag.org/news/2020/07/meet-lizard-man-reptile-loving-biologist-tackling-some-biggest-questions-evolution.

[4] We used available service at https://www.random.org/.

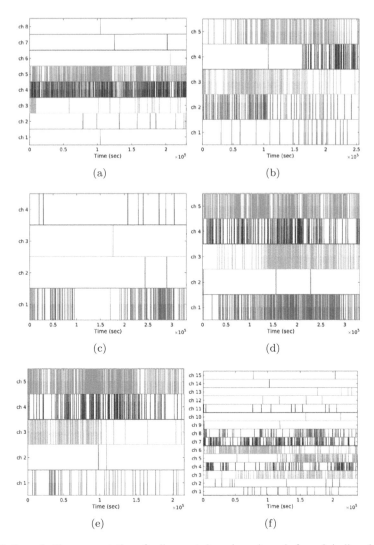

Fig. 18 Barcode-like representation of spike events in various channels for **a**, **b** in-line electrode arrangements at a distance of 1 cm, **c**, **d** in-line electrode arrangements at a distance of 2 cm, and **e**, **f** random electrode arrangements at an approximate distance of 2 cm

5 Discussion

We developed algorithmic framework for exhaustive characterisation of electrical activity of a substrate colonised by mycelium of oyster fungi *Pleurotus djamor*. We evidenced spiking activity of the mycelium. We found that average dominant duration of an action-potential like spike is 402 sec. The spikes amplitudes' depends on the location of the source of electrical activity related to the position of electrodes, thus

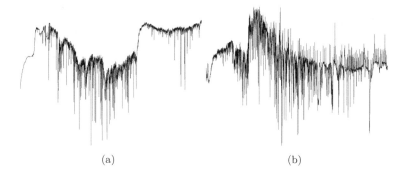

Fig. 19 Two samples from input channels, which are saved in black and white PNG format without axes and annotations

the amplitudes provide less useful information. The amplitudes vary from 0.5 to 6 mV. This is indeed low compared to 50–60 mV of intracellular recording, nevertheless understandable due to the fact the electrodes are inserted not even in mycelium strands but in the substrate colonised by mycelium. The shift of the distribution to higher values of spike amplitude in experiments with a distance of 2 cm between the electrodes might indicate that the width of the propagation of the excitation wave front exceeds 1 cm and might even be close to 2 cm.

The spiking events have been characterised with several complexity measures. Most measures, apart of Kolmogorov complexity shown a low degree of variability between channels (different sites of the recordings). The Kolmogorov complexity of fungal spiking varies from 11×10^{-4} to 57×10^{-4}. This might indicated mycelium sub-networks in different parts of the substrate have been transmitting different information to other parts of the mycelium network. This is somehow echoes experimental results on communication between ants analysed with Kolmogorov complexity: longer paths communicated ants corresponds to higher values of complexity [61].

LZ complexity of fungal language (Table 2) is much higher than of news, random or periodic sequences (Table 3). The same can be observed for Shannon entropy. Kolmogorov complexity of the fungal language is much lower than that of news sampler or random or periodic sequences. Complexity of European languages based on their compressibility [62] is shown in Fig. 23, French having lowest LZ complexity 0.66 and Finnish highest LZ complexity 0.79. Fungal language of electrical activity has minimum LZ complexity 0.61 and maximum 0.91 (media 0.85, average 0.83). Thus, we can speculate that a complexity of fungal language is higher than that of human languages (at least for European languages).

Table 2 The mean of complexity measurements for six experiments

	#Channel	#Spike	Lempel-Ziv complexity	Shannon entropy	Simpson's diversity	Space filling	Kolmogorov	PCI	Expressiveness
#1	8	565	0.79	45.81	0.76	30.68×10^{-5}	30.36×10^{-4}	0.365	20.8×10^{4}
#2	5	447	0.91	63.27	0.98	35.20×10^{-5}	35.78×10^{-4}	0.021	18.6×10^{4}
#3	4	124	0.75	22.57	0.61	48.10×10^{-5}	10.94×10^{-4}	0.333	29.71
#4	5	951	0.93	123.11	0.89	57.30×10^{-5}	56.05×10^{-4}	0.072	23.8×10^{4}
#5	5	573	0.88	75.75	0.79	53.02×10^{-5}	52.80×10^{-4}	0.077	16.4×10^{4}
#6	15	862	0.69	39.96	0.71	24.20×10^{-5}	25.06×10^{-4}	0.207	20.4×10^{4}

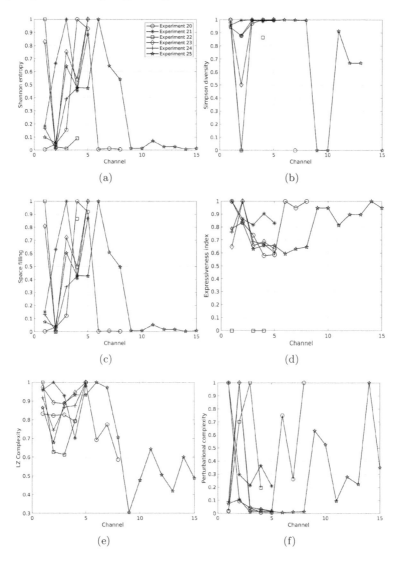

Fig. 20 **a** Shannon entropy, **b** Simpson's diversity, **c** Space filling, **d** Expressiveness, **e** Lempel-Ziv complexity, and **f** Perturbation complexity index. All measurements are scaled to the range of [0, 1]

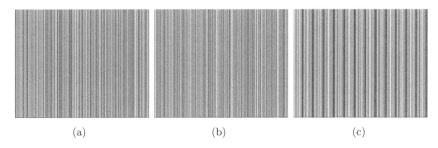

(a) (b) (c)

Fig. 21 Binary representation of **a** pieces of news, **b** random sequence of alphanumeric, and **c** periodic sequence of alphanumeric after applying Huffman coding

Table 3 The complexity measurements for pieces of news, a random sequence of alphanumeric, a periodic sequence of alphanumeric along with three chunks randomly selected from our experiments

	Length	Lempel-Ziv complexity	Shannon entropy	Simpson's diversity	Space filling	Kolmogorov	PCI	Expressiveness
News	36187	0.127919	4.421728	0.999941	0.465996	0.765382	0.173096	9.49
Random sequence	36002	0.125465	5.770331	0.999941	0.469835	1.001850	0.173621	12.28
Periodic sequence	36006	0.127090	3.882058	0.999937	0.442426	0.076508	0.019708	8.77
Chunk 1	36000	0.067611	16.194914	0.947368	0.000556	0.006307	0.000389	29150.84
Chunk 2	36000	0.007250	15.478087	0.944444	0.000528	0.006727	0.000435	29326.90
Chunk 3	36000	0.068417	31.680374	0.976190	0.001194	0.012613	0.000398	26523.10

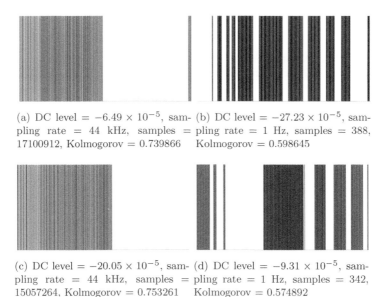

(a) DC level = -6.49×10^{-5}, sampling rate = 44 kHz, samples = 17100912, Kolmogorov = 0.739866

(b) DC level = -27.23×10^{-5}, sampling rate = 1 Hz, samples = 388, Kolmogorov = 0.598645

(c) DC level = -20.05×10^{-5}, sampling rate = 44 kHz, samples = 15057264, Kolmogorov = 0.753261

(d) DC level = -9.31×10^{-5}, sampling rate = 1 Hz, samples = 342, Kolmogorov = 0.574892

Fig. 22 Comparison of the human voice in **a–c** English/Chinese with the electrical activity of fungi with the duration of **b–d** 388/342 s

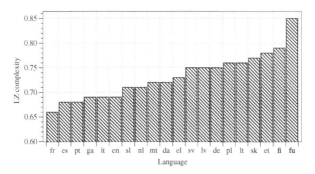

Fig. 23 Lempel-Ziv complexity of European languages (data from [62]) with average complexity of fungal ('fu') electrical activity language added

References

1. Masi, E., Ciszak, M., Santopolo, L., Frascella, A., Giovannetti, L., Marchi, E., Viti, C., Mancuso, S.: Electrical spiking in bacterial biofilms. J. R. Soc. Interface **12**(102), 20141036 (2015)
2. Eckert, R., Brehm, P.: Ionic mechanisms of excitation in paramecium. Annu. Rev. Biophys. Bioeng. **8**(1), 353–383 (1979)
3. Hansma, H.G.: Sodium uptake and membrane excitation in paramecium. J. Cell Biol. **81**(2), 374–381 (1979)
4. Bingley, M.S.: Membrane potentials in amoeba proteus. J. Exp. Biol. **45**(2), 251–267 (1966)
5. McGillviray, A.N., Gow, N.A.R.: The transhyphal electrical current of *N eurusрua crassa* is carried principally by protons. Microbiology **133**(10), 2875–2881 (1987)
6. Trebacz, K., Dziubinska, H., Krol, E.: Electrical signals in long-distance communication in plants. In: Communication in Plants, pp. 277–290. Springer, Berlin (2006)
7. Fromm, J., Lautner, S.: Electrical signals and their physiological significance in plants. Plant, Cell Environ. **30**(3), 249–257 (2007)
8. Zimmermann, M.R., Mithöfer, A.: Electrical long-distance signaling in plants. In: Long-Distance Systemic Signaling and Communication in Plants, pp. 291–308. Springer, Berlin (2013)
9. Hodgkin, A.L., Huxley, A.F.: A quantitative description of membrane current and its application to conduction and excitation in nerve. J. Physiol. **117**(4), 500–544 (1952)
10. Aidley, D.J., Ashley, D.J.: The Physiology of Excitable Cells, vol. 4. Cambridge University Press Cambridge, Cambridge (1998)
11. Nelson, P.G., Lieberman, M.: Excitable Cells in Tissue Culture. Springer Science & Business Media (2012)
12. Davidenko, J.M., Pertsov, A.V., Salomonsz, R., Baxter, W., Jalife, J.: Stationary and drifting spiral waves of excitation in isolated cardiac muscle. Nature **355**(6358), 349 (1992)
13. Kittel, Ch.: Excitation of spin waves in a ferromagnet by a uniform RF field. Phys. Rev. **110**(6), 1295 (1958)
14. Tsoi, M., Jansen, A.G.M., Bass, J., Chiang, W.-C., Seck, M., Tsoi, V., Wyder, P.: Excitation of a magnetic multilayer by an electric current. Phys. Rev. Lett. **80**(19), 4281 (1998)
15. Slonczewski, J.C.: Excitation of spin waves by an electric current. J. Magn. Magn. Mater. **195**(2), L261–L268 (1999)
16. Gorbunov, L.M., Kirsanov, V.I.: Excitation of plasma waves by an electromagnetic wave packet. Sov. Phys. JETP **66**(290–294), 40 (1987)
17. Belousov, B.P.: A periodic reaction and its mechanism. Compil. Abstr. Radiat. Med. **147**(145), 1 (1959)
18. Zhabotinsky, A.M.: Periodic processes of malonic acid oxidation in a liquid phase. Biofizika **9**(306–311), 11 (1964)
19. Zhabotinsky, A.M.: Belousov-zhabotinsky reaction. Scholarpedia **2**(9), 1435 (2007)
20. Farkas, I., Helbing, D., Vicsek, T.: Social behaviour: Mexican waves in an excitable medium. Nature **419**(6903), 131 (2002)
21. Farkas, I., Helbing, D., Vicsek, T.: Human waves in stadiums. Phys. A: Stat. Mech. Its Appl. **330**(1–2), 18–24 (2003)
22. Adamatzky, A., Tuszynski, J., Pieper, J., Nicolau, D.V., Rinaldi, R., Sirakoulis, G.C., Erokhin, V., Schnauss, J., Smith, D.M.: Towards cytoskeleton computers: a proposal. In: Adamatzky, A., Akl, S. Sirakoulis, G.C. (eds.) From Parallel to Emergent Computing. CRC Group/Taylor & Francis (2019)
23. Adamatzky, A.: Plant leaf computing. Biosystems (2019)
24. Adamatzky, A., Nikolaidou, A., Gandia, A., Chiolerio, A., Dehshibi, M.M.: Reactive fungal wearable. Biosystems **199**, 104304 (2020)
25. Nenadic, Z., Burdick, J.W.: Spike detection using the continuous wavelet transform. IEEE Trans. Biomed. Eng. **52**(1), 74–87 (2004)
26. Shimazaki, H., Shinomoto, S.: Kernel bandwidth optimization in spike rate estimation. J. Comput. Neurosci. **29**(1–2), 171–182 (2010)

27. Vicnesh, J., Hagiwara, Y.: Accurate detection of seizure using nonlinear parameters extracted from EEG signals. J. Mech. Med. Biol. **19**(01), 1940004 (2019)
28. Adamatzky, A., Gandia, A.: On electrical spiking of ganoderma resinaceum. Biophys. Rev. Lett. 1–9 (2021)
29. Lilly, J.M., Olhede, S.C.: Generalized morse wavelets as a superfamily of analytic wavelets. IEEE Trans. Signal Process. **60**(11), 6036–6041 (2012)
30. IEEE standard for transitions, pulses, and related waveforms. IEEE Std 181-2011 (Revision of IEEE Std 181-2003), pp, 1–71 (2011)
31. Lilly, J.M.: Element analysis: a wavelet-based method for analysing time-localized events in noisy time series. Proc. R. Soc. A: Math., Phys. Eng. Sci. **473**(2200), 20160776 (2017)
32. Lilly, J.M., Olhede, S.C.: Higher-order properties of analytic wavelets. IEEE Trans. Signal Process. **57**(1), 146–160 (2008)
33. Marple, L.: Computing the discrete-time analytic signal via FFT. IEEE Trans. Signal Process. **47**(9), 2600–2603 (1999)
34. Adamatzky, A.: On spiking behaviour of oyster fungi pleurotus djamor. Sci. Rep. **8**(1), 1–7 (2018)
35. Minoofam, S.A.H., Dehshibi, M.M., Bastanfard, A., Eftekhari, P.: Ad-hoc ma'qeli script generation using block cellular automata. J. Cell. Autom. **7**(4), 321–334 (2012)
36. Minoofam, S.A.H., Dehshibi, M.M., Bastanfard, A., Shanbehzadeh, J.: Pattern formation using cellular automata and l-systems: a case study in producing islamic patterns. In: Cellular Automata in Image Processing and Geometry, pp. 233–252. Springer, Berlin (2014)
37. Parsa, S.S., Sourizaei, M., Dehshibi, M.M., Esmaeilzadeh Shateri, R., Parsaei, M.R.: Coarse-grained correspondence-based ancient Sasanian coin classification by fusion of local features and sparse representation-based classifier. Multimed. Tools Appl. **76**(14), 15535–15560 (2017)
38. Taghipour, N., Javadi, H.H.S., Dehshibi, M.M., Adamatzky, A.: On complexity of persian orthography: L-systems approach. Complex Syst. **25**(2), 127–156 (2016)
39. Dehshibi, M.M., Shirmohammadi, A., Adamatzky, A.: On growing persian words with l-systems: visual modeling of neyname. Int. J. Image Graph. **15**(03), 1550011 (2015)
40. Dehshibi, M.M., Shanbehzadeh, J., Pedram, M.M.: A robust image-based cryptology scheme based on cellular nonlinear network and local image descriptors. Int. J. Parallel, Emergent Distrib. Syst. **35**(5), 514–534 (2020)
41. Gholami, N., Dehshibi, M.M., Adamatzky, A., Rueda-Toicen, A., Zenil, H., Fazlali, M., Masip, D.: A novel method for reconstructing CT images in gate/geant4 with application in medical imaging: a complexity analysis approach. J. Inf. Process. **28**, 161–168 (2020)
42. Quiroga, R.Q., Nadasdy, Z., Ben-Shaul, Y.: Unsupervised spike detection and sorting with wavelets and superparamagnetic clustering. Neural Comput. **16**(8), 1661–1687 (2004)
43. Obeid, I., Wolf, P.D.: Evaluation of spike-detection algorithms fora brain-machine interface application. IEEE Trans. Biomed. Eng. **51**(6), 905–911 (2004)
44. Wilson, S.B., Emerson, R.: Spike detection: a review and comparison of algorithms. Clin. Neurophysiol. **113**(12), 1873–1881 (2002)
45. Gotman, J., Wang, L.Y.: State-dependent spike detection: concepts and preliminary results. Electroencephalogr. Clin. Neurophysiol. **79**(1), 11–19 (1991)
46. Wilson, S.B., Turner, C.A., Emerson, R.G., Scheuer, M.L.: Spike detection ii: automatic, perception-based detection and clustering. Clin. Neurophysiol. **110**(3), 404–411 (1999)
47. Franke, F., Natora, M., Boucsein, C., Munk, M.H., Obermayer, K.: An online spike detection and spike classification algorithm capable of instantaneous resolution of overlapping spikes. J. Comput. Neurosci. **29**(1–2), 127–148 (2010)
48. Rácz, M., Liber, C., Németh, E., Fiáth, R., Rokai, J., Harmati, I., Ulbert, I., Márton, G.: Spike detection and sorting with deep learning. J. Neural Eng. **17**(1), 016038 (2020)
49. Wang, Z., Duanpo, W., Dong, F., Cao, J., Jiang, T., Liu, J.: A novel spike detection algorithm based on multi-channel of BECT EEG signals. In: Express Briefs, IEEE Transactions on Circuits and Systems II (2020)
50. Sablok, S., Gururaj, G., Shaikh, N., Shiksha, I., Choudhary, A.R.: Interictal spike detection in EEG using time series classification. In: 2020 4th International Conference on Intelligent Computing and Control Systems (ICICCS), pp. 644–647. IEEE (2020)

51. Liu, Z., Wang, X., Yuan, Q.: Robust detection of neural spikes using sparse coding based features. Math. Biosci. Eng. **17**(4), 4257 (2020)
52. Dehshibi, M.M., Adamatzky, A.: Supplementary material for "Electrical activity of fungi: spikes detection and complexity analysis" 08 (2020). (Accessed on 24 Aug 2020). https://doi.org/10.5281/zenodo.3997031
53. Adamatzky, A.: Tactile bristle sensors made with slime mold. IEEE Sens. J. **14**(2), 324–332 (2013)
54. Deutsch, P., Gailly, J.: Zlib compressed data format specification version 3.3. Technical report, RFC 1950 (1996)
55. Howard, P.G.: The Design and Analysis of Efficient Lossless Data Compression Systems. Ph.D. thesis, Citeseer (1993)
56. Roelofs, G., Koman, R.: PNG: The Definitive Guide. O'Reilly & Associates, Inc. (1999)
57. Ziv, J., Lempel, A.: A universal algorithm for sequential data compression. IEEE Trans. Inf. Theory **23**(3), 337–343 (1977)
58. Kaspar, F., Schuster, H.G.: Easily calculable measure for the complexity of spatiotemporal patterns. Phys. Rev. A **36**(2), 842 (1987)
59. Huffman, D.A.: A method for the construction of minimum-redundancy codes. Proc. IRE **40**(9), 1098–1101 (1952)
60. Huang, H., Lin, F.: A speech feature extraction method using complexity measure for voice activity detection in WGN. Speech Commun. **51**(9), 714–723 (2009)
61. Ryabko, B., Reznikova, Z.: Using Shannon entropy and Kolmogorov complexity to study the communicative system and cognitive capacities in ants. Complexity **2**(2), 37–42 (1996)
62. Sadeniemi, M., Kettunen, K., Lindh-Knuutila, T., Honkela, T.: Complexity of European union languages: a comparative approach. J. Quant. Linguist. **15**(2), 185–211 (2008)

Fungi Anaesthesia

Andrew Adamatzky and Antoni Gandia

Abstract Electrical activity of fungus *Pleurotus ostreatus* is characterised by slow (hours) irregular waves of baseline potential drift and fast (minutes) action potential likes spikes of the electrical potential. An exposure of the mycelium colonised substrate to a chloroform vapour lead to several fold decrease of the baseline potential waves and increase of their duration. The chloroform vapour also causes either complete cessation of spiking activity or substantial reduction of the spiking frequency. Removal of the chloroform vapour from the growth containers leads to a gradual restoration of the mycelium electrical activity.

1 Introduction

Most living cells are sensitive to anaesthetics [1–3]. First experiments on anaesthesia of plants have be done by Claude Bernard in late 1800s [4]. Later experiments on amoeba [5] shown that weak concentration of narcotics causes the amoebae to spread out and propagate in a spread condition while narcotic concentrations led to cessation of movements. During last century the experimental evidences mounted up including anaesthesia of yeasts [1], various aquatic invertebrates [6], plants [3, 7], protists [8], bronchial ciliated cells [9]. A general consensus now is that any living substrate can be anaesthetised [3]. The question remains, however, how exactly species without a nervous system would respond to exposure to anaesthetics.

In present chapter we focus on fungi anaesthesia. Why fungi? Fungi are the largest, most widely distributed, and oldest group of living organisms [10]. Smallest fungi are microscopic single cells. The largest (15 hectares) mycelium belongs to *Armillaria gallica* (synonymous with *A. bulbosa*, *A. inflata*, and *A. lutea*) [11] and the largest fruit body belongs to *Phellinus ellipsoideus* (formerly *Fomitiporia ellipsoidea*) which weighs half-a-ton [12].

A. Adamatzky (✉)
Unconventional Computing Laboratory, UWE, Bristol, UK
e-mail: andrew.adamatzky@uwe.ac.uk

A. Gandia
Institute for Plant Molecular and Cell Biology, CSIC-UPV, Valencia, Spain

© The Author(s), under exclusive license to Springer Nature Switzerland AG 2023
A. Adamatzky (ed.), *Fungal Machines*, Emergence, Complexity and Computation 47,
https://doi.org/10.1007/978-3-031-38336-6_5

Fungi exhibit a high degree of protocognitive abilities. For example, they are capable for efficient exploration of confined spaces [13–17]. Moreover, optimisation of the mycelial network [18] is similar to that of the slime mould *Physarum polycephalum* [19] and transport networks [20]. Therefore, we can speculate that the fungi can solve the same range of computational problems as *P. polycephalum* [21], including shortest path [22–26], Voronoi diagram [27], Delaunay triangulation, proximity graphs and spanning tree, concave hull and, possibly, convex hull, and, with some experimental efforts, travelling salesman problem [28]. The fungi's protocognitive abilities and computational potential make them fruitful substrates for anaesthesia because they might show us how non-neuron awareness is changing under effects of narcotics.

We use extracellular electrical potential of mycelium as indicator of the fungi activity. Action potential-like spikes of electrical potential have been discovered using intra-cellular recording of mycelium of *Neurospora crassa* [29] and further confirmed in intra-cellular recordings of action potential in hyphae of *Pleurotus ostreatus* and *A. gallica* [30] and in extra-cellular recordings of fruit bodies of and substrates colonized by mycelium of *P. ostreatus* [31]. While the exact nature of the travelling spikes remains uncertain we can speculate, by drawing analogies with oscillations of electrical potential of slime mould *Physarum polycephalum* [32–35], that the spikes in fungi are triggered by calcium waves, reversing of cytoplasmic flow, translocation of nutrients and metabolites. Studies of electrical activity of higher plants can brings us even more clues. Thus, the plants use the electrical spikes for a long-distance communication aimed to coordinate the activity of their bodies [36–38]. The spikes of electrical potential in plants relate to a motor activity [39–42], responses to changes in temperature [43], osmotic environment [44], and mechanical stimulation [45, 46].

2 Methods

A commercial strain of the fungus *Pleurotus ostreatus* (collection code 21–18, Mogu S.r.l., Italy), previously selected for its superior fitness growing on the targeted substrate, was cultured on sterilised hemp shives contained in plastic (PP5) filter patch microboxes ($SacO_2$, Belgium) that were kept in darkness at ambient room temperature c. 22 °C. After one week of incubation, a hemp brick well colonised by the fungus was manually crumbled and spread on rectangular fragments, c. 12×12 cm^2, of moisturised non-woven hemp pads. When these fragments were colonised, as visualised by white and healthy mycelial growth on surface, they were used for experiments.

Electrical activity of the colonised hemp pads was recorded using pairs of iridium-coated stainless steel sub-dermal needle electrodes (Spes Medica S.r.l., Italy), with twisted cables and ADC-24 (Pico Technology, UK) high-resolution data logger with a 24-bit A/D converter. To keep electrodes stable we have been placing a polyurethane

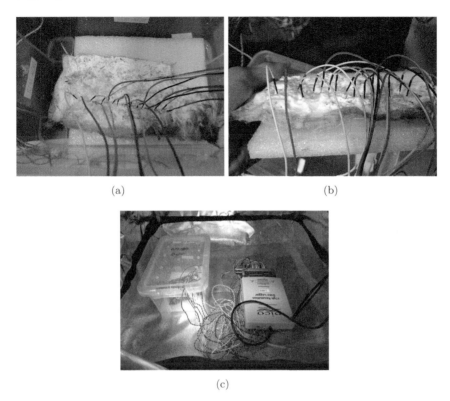

Fig. 1 Experimental setup. **a, b** Exemplar locations of electrodes. **a** Top view. **b** Side view. **c** Setup in the grow tent

pad under the fabric. The electrodes were arranged in a line (Fig. 1a, b). The pairs of electrodes were pierced through the fabric and into the polyurethane pad.

The fungal substrates pierced with electrodes was placed into 20 cm by 10 cm by 10 cm plastic boxes with tight lids.

We recorded electrical activity at one sample per second. During the recording, the logger has been doing as many measurements as possible (typically up to 600 per second) and saving the average value. We set the acquisition voltage range to 156 mV with an offset accuracy of 9 μV at 1 Hz to maintain a gain error of 0.1%. Each electrode pair was considered independent with the noise-free resolution of 17 bits and conversion time of 60 ms. Each pair of electrodes, called channels, reported a difference of the electrical potential between the electrodes. Distance between electrodes was 1–2 cm. In each trial, we recorded eight electrode pairs, channels, simultaneously.

To study the effect of chloroform we soaked a piece of filter paper c. 4 cm by 4 cm in chloroform (Sigma Aldrich, analytical standard) and placed the piece of paper inside the plastic container with the recorded fungal substrate.

The humidity of the fungal colonies was 70–80% (MerlinLaser Protimeter, UK). The experiments were conducted in a room with ambient temperature 21 °C and in the darkness of protective growing tents (Fig. 1c).

We have conducted ten experiments, in each experiments we recorded electrical activity of the fungi via eight channels, i.e. 80 recordings in total.

3 Results

Mycelium colonised hemp pad exhibit patterns of electrical activity similar to that of spiking neural tissue. Examples of action potential like spikes, solitary and in trains, are shown in Fig. 2.

Application of the chloroform to the container with fungi substantially affected the electrical activity of the fungi. An example of an extreme, i.e. where almost all electrical activity of mycelium seized, response is shown in Fig. 3. In this example, the introduction of the chloroform leads to the suppression of the spiking activity and reduction of deviation in values of the electrical potential differences recorded on the channels.

The intact mycelium composite shows median amplitude of the irregular movements of the baseline potential is 0.45 mV (average 0.64 mV, $\sigma = 0.64$), median duration 29850 s (average 67507 s, $\sigma = 29850$). After exposure to chloroform the baseline potential movements show median amplitude reduced to 0.16 mV (average 0.18 mV, $\sigma = 0.12$) and median duration increased to 38507 s (average 38114 s, $\sigma = 38507$). For the eight channels (pairs of differential electrodes) recorded exposure to chloroform led to nearly three times decrease in amplitude of the drifts of baseline potential and nearly 1.3 increase in duration of the drifts. Before exposure to chloroform the mycelium composite produced fast (i.e. less than 10–20 mins) spikes. Median amplitude of the spikes was 0.48 mV (average 0.52 mV, $\sigma = 0.2$). Median duration of spikes was 62 s (average 63 s, $\sigma = 18$), median distance between the

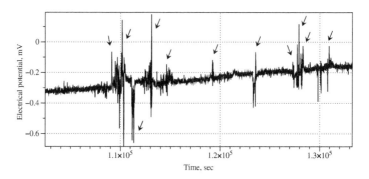

Fig. 2 Example of spiking activity recorded from intact mycelium colonised hemp pad. Spikes are shown by arrows

Fungi Anaesthesia

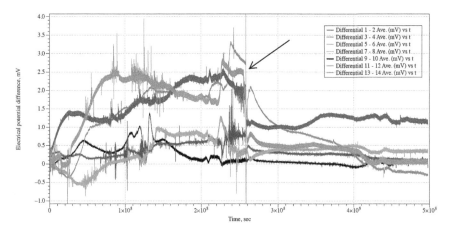

Fig. 3 An example showing how the electrical activity of fungi changes when chloroform is introduced. The moment of the chloroform introduction is shown by arrow

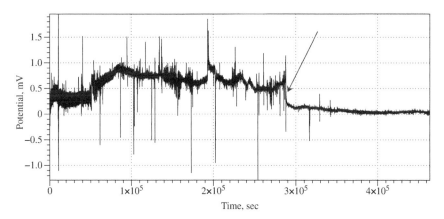

Fig. 4 Example of reduced frequency of spiking under effect of chloroform vapour. Moment when a source of chloroform vapour was added into the container is shown by arrow

spike 214 s (average 189 s, $\sigma = 90$). After exposure to chloroform the mycelium composite did not show any spiking activity above level of background noise, which was for this particular recording c. 0.05 mV.

In some cases the spiking activity is diminished gradually with decreased frequency and lowered amplitude, as exemplified in Fig. 4. Typically, the intact spiking frequency is a spike per 70 min while after inhalation of chloroform a spike per 254 min in the first 40–50 h and decreased to nearly zero after. The median amplitude of intact mycelium spikes is 0.51 mV, average 0.74 mV ($\sigma = 0.59$). Anaesthetised mycelium shows, spikes with median amplitude 0.11 mV, average 0.2 mV ($\sigma = 0.2$). Spikes are not distributed uniformly but gathered in trains of spikes. In the intact mycelium there is a media of 3 spikes in the train, average number of

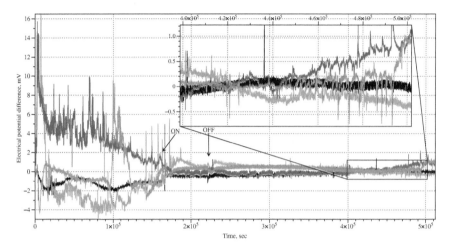

Fig. 5 Example of electrical activity of mycelium colonised hemp pad before, during and after stimulation with chloroform vapour. Arrow labelled 'ON' shows moment when a source of chloroform vapour was added was added into the enclosure, 'OFF' when the source of chloroform was removed. The spiking activity of the mycelium recovering from anaesthesia is zoomed in

spikes is 4.2 ($\sigma = 4.4$). Median duration of a spike train is 84 min, average 112 min ($\sigma = 32$). Media interval between trains is 53 min, average 55 s ($\sigma = 29$). Anaesthetised mycelium emits trains with median number of 2 spikes, average 2.5 spikes, average 2.5 spikes ($\sigma = 0.84$). A median duration of such trains is 29 min, average 51 min ($\sigma = 22$). The trains appear much more rarely than the trains in the intact mycelium: median interval between trains is 227 min.

In all ten but one experiment the container remained closed for over two-three days. By that time all kinds of electrical activity in mycelium bound substrate extinguished and the mycelium never recovered to a functional state. In experiment illustrated in Fig. 5 we removed a source of chloroform after 16 h and kept the container open and well ventilated for an hour to remove any traces of chloroform from the air. The intact mycelium shows median frequency of spiking as one spike per 27 min, average 24 min. Median amplitude of the spikes is 3.4 mV, average 3.25 mV ($\sigma = 1.45$). The anaesthetised mycelium demonstrates electrical spiking activity reduced in amplitude: median amplitude of spikes is 0.24 mV, average 0.32 mV ($\sigma = 0.2$), and low frequency of spiking: median distance between spikes is 38 min, average 40 min. Electrical activity of the mycelium restores to above noise level c. 60 h after the source of the chloroform is removed from the enclosure (insert in Fig. 5). Frequency of spikes is one spike per 82 min (median), average 88 min. The amplitudes of recovering spikes are 0.96 mV in median (average 0.93 mV, $\sigma = 0.08$) which are three times less than of the spikes in the mycelium before the narcosis but nearly five times higher than of the spike of the anaesthetised mycelium.

4 Discussion

We demonstrated that the electrical activity of the fungus *Pleurotus ostreatus* is a reliable indicator of the fungi anaesthesia. When exposed to a chloroform vapour the mycelium reduces frequency and amplitude of its spiking and, in most cases, cease to produce any electrical activity exceeding the noise level (Table 1a). When the chloroform vapour is eliminated from the mycelium enclosure the mycelium electrical activity restores to a level similar to that before anaesthesia (Table 1b).

The fungal responses to chloroform are similar to that recorded by us with slime mould *Physarum polycephalum* (unpublished results). A small concentration of anaesthetic leads to reduced frequency and amplitude of electrical potential oscillation spikes of the slime mould, and some irregularity of the electrical potential spikes (Fig. 6a). Large amounts of anaesthetic causes the electrical activity to cease completely and never recover (Fig. 6b).

With regards to directions of future research, as far as we are aware, the present paper is the first in the field, and therefore it rather initiates the research than brings any closure or conclusions. We know that anaesthetics block electrical activity of fungi (as well as slime moulds) however we do not know exact biophysical mechanisms of these actions. The study of biophysics and molecular biology of fungi anaesthesia could be a scope for future research. Another direction of studies could be the analysis of the decision making abilities of fungi under the influence of anaesthetics. An experiment could be constructed when fungal hyphae are searching for an optimal path in a labyrinth when subjected to increasing doses of chloroform vapour. There may be an opportunity to make a mapping from concentrations of anaesthetic to geometry of the mycelium search path.

Table 1 Anaesthesia induced changes in electrical activity of fungi. (a) Containers with chloroform remain sealed. (b) Lid from the container is removed after 16 h

(a)

	Intact	Anaesthetised	
Baseline potential (mV)	0.64	0.18	
Duration of oscillations	$29 \cdot 10^3$	$28 \cdot 10^3$	
Spike amplitude	0.4 mV	0	
Spike duration	62 s	n/a	
Spike frequency	214 s	n/a	

(b)

	Intact	Anaesthetised	Recovered
Spike amplitude (mV)	3.25	0.32	0.93
Spike frequency (min)	24	40	88

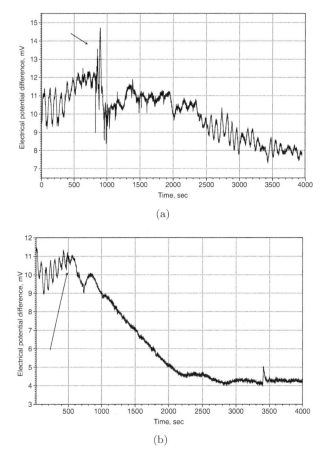

Fig. 6 Response of slime mould *Physarum polycephalum* to trifluoroethane (Sigma Aldrich, UK). Electrical potential difference between two sites of 10 mm long protoplasmic tube was measured using aluminium electrodes, amplified and digitised with ADC-20 (Pico Technology, UK). **a** 5 μL of trifluoroethane applied to a 5 × 5 mm piece of filter paper placed in the Petri dish with slime mould. **b** 25 μL applied. Arrows indicate moments when the piece of paper soaked in trifluoroethane was placed in a Petri dish with the slime mould

References

1. Sonner, J.M.: A hypothesis on the origin and evolution of the response to inhaled anesthetics. Anesth. Analg. **107**(3), 849 (2008)
2. Eckenhoff, R.G.: Why can all of biology be anesthetized? Anesth. Analg. **107**(3), 859–861 (2008)
3. Grémiaux, A., Yokawa, K., Mancuso, S., Baluška, F.: Plant anesthesia supports similarities between animals and plants: claude bernard's forgotten studies. Plant Signal. Behav. **9**(1), e27886 (2014)
4. Bernard, C., et al.: Lectures on the Phenomena of Life Common to Animals and Plants. Translation by Hoff, H.E., Guillemin, R., Guillemin, L., (1974)

5. Hiller, Stanislaw: Action of narcotics on the ameba by means of microinjection and immersion. Proc. Soc. Exp. Biol. Med. **24**(5), 427–428 (1927)
6. Oliver, A.E., Deamer, D.W., Akeson, M.: Sensitivity to anesthesia by pregnanolone appears late in evolution. Ann. N. Y. Acad. Sci. **625**, 561–565 (1991)
7. Milne, A., Beamish, T.: Inhalational and local anesthetics reduce tactile and thermal responses in Mimosa pudica. Can. J. Anesth. **46**(3), 287–289 (1999)
8. Nunn, J.F., Sturrock, J.E., Wills, E.J., Richmond, J.E., McPherson, C.K.: The effect of inhalational anaesthetics on the swimming velocity of tetrahymena pyriformis. J. Cell Sci. **15**(3), 537–554 (1974)
9. Verra, F., Escudier, E., Pinchon, M.-C., Fleury, J., Bignon, J., Bernaudin, J.-F.: Effects of local anaesthetics (lidocaine) on the structure and function of ciliated respiratory epithelial cells. Biol. Cell **69**, 99–105 (1990)
10. Carlile, M.J., Watkinson, S.C., Gooday, G.W.: The Fungi. Gulf Professional Publishing (2001)
11. Smith, M.L., Bruhn, J.N., Anderson, J.B.: The fungus Armillaria bulbosa is among the largest and oldest living organisms. Nature **356**(6368), 428 (1992)
12. Dai, Y.-C., Cui, B.-K.: Fomitiporia ellipsoidea has the largest fruiting body among the fungi. Fungal Biol. **115**(9), 813–814 (2011)
13. Hanson, K.L., Nicolau Jr, D.V., Filipponi, L., Wang, L., Lee, A.P., Nicolau, D.V.: Fungi use efficient algorithms for the exploration of microfluidic networks. Small **2**(10), 1212–1220 (2006)
14. Held, M., Edwards, C., Nicolau, D.V.: Examining the behaviour of fungal cells in microconfined mazelike structures. In: Imaging, Manipulation, and Analysis of Biomolecules, Cells, and Tissues VI, vol. 6859, p. 68590U. International Society for Optics and Photonics (2008)
15. Marie, M., Edwards, C., Nicolau, D.V.: Fungal intelligence; or on the behaviour of microorganisms in confined micro-environments. In: Journal of Physics: Conference Series, vol. 178, p. 012005. IOP Publishing (2009)
16. Held, M., Lee, A.P., Edwards, C., Nicolau, D.V.: Microfluidics structures for probing the dynamic behaviour of filamentous fungi. Microelectron. Eng. **87**(5–8), 786–789 (2010)
17. Held, M., Edwards, C. and Nicolau, D.V.: Probing the growth dynamics of neurospora crassa with microfluidic structures. Fungal Biol. **115**(6), 493–505 (2011)
18. Boddy, L., Hynes, J., Bebber, D.P., Fricker, M.D.: Saprotrophic cord systems: dispersal mechanisms in space and time. Mycoscience **50**(1), 9–19 (2009)
19. Adamatzky, A.: Developing proximity graphs by Physarum polycephalum: does the plasmodium follow the Toussaint hierarchy? Parallel Process. Lett. **19**(01), 105–127 (2009)
20. Adamatzky, A. (ed.): Bioevaluation of World Transport Networks. World Scientific (2012)
21. Adamatzky, A. (ed.): Advances in Physarum Machines: Sensing and Computing with Slime Mould. Springer, Berlin (2016)
22. Nakagaki, T., Yamada, H., Tóth, Á.: Intelligence: maze-solving by an amoeboid organism. Nature **407**(6803), 470 (2000)
23. Nakagaki, T.: Smart behavior of true slime mold in a labyrinth. Res. Microbiol. **152**(9), 767–770 (2001)
24. Nakagaki, T., Yamada, H., Toth, A.: Path finding by tube morphogenesis in an amoeboid organism. Biophys. Chem. **92**(1–2), 47–52 (2001)
25. Nakagaki, T., Iima, M., Ueda, T., Nishiura, Y., Saigusa, T., Tero, A., Kobayashi, R., Showalter, K.: Minimum-risk path finding by an adaptive amoebal network. Phys. Rev. Lett. **99**(6), 068104 (2007)
26. Tero, A., Takagi, S., Saigusa, T., Ito, K., Bebber, D.P., Fricker, M.D., Yumiki, K., Kobayashi, R., Nakagaki, T.: Rules for biologically inspired adaptive network design. Science **327**(5964), 439–442 (2010)
27. Shirakawa, T., Adamatzky, A., Gunji, Y.-P., Miyake, Y.: On simultaneous construction of Voronoi diagram and Delaunay triangulation by Physarum polycephalum. Int. J. Bifurc. Chaos **19**(09), 3109–3117 (2009)
28. Jones, J., Adamatzky, A.: Computation of the travelling salesman problem by a shrinking blob. Nat. Comput. **13**(1), 1–16 (2014)

29. Slayman, C.L., Long, W.S., Gradmann, D.: "Action potentials" in Neurospora crassa, a mycelial fungus. Biochim. et Biophys. Acta (BBA)—Biomembr. **426**(4), 732–744 (1976)
30. Olsson, S., Hansson, B.S.: Action potential-like activity found in fungal mycelia is sensitive to stimulation. Naturwissenschaften **82**(1), 30–31 (1995)
31. Adamatzky, A.: On spiking behaviour of oyster fungi Pleurotus djamor. Sci. Rep. **8**(1), 1–7 (2018)
32. Iwamura, T.: Correlations between protoplasmic streaming and bioelectric potential of a slime mold. Physarum polycephalum. Shokubutsugaku Zasshi **62**(735–736), 126–131 (1949)
33. Kamiya, N., Abe, S.: Bioelectric phenomena in the myxomycete plasmodium and their relation to protoplasmic flow. J. Colloid Sci. **5**(2), 149–163 (1950)
34. Kishimoto, U.: Rhythmicity in the protoplasmic streaming of a slime mold, Physarum polycephalum. I. a statistical analysis of the electric potential rhythm. J. Gen. Physiol. **41**(6), 1205–1222 (1958)
35. Meyer, R., Stockem, W.: Studies on microplasmodia of Physarum polycephalum V: electrical activity of different types of microplasmodia and macroplasmodia. Cell Biol. Int. Rep. **3**(4), 321–330 (1979)
36. Trebacz, K., Dziubinska, H., Krol, E.: Electrical signals in long-distance communication in plants. In: Communication in Plants, pp. 277–290. Springer, Berlin (2006)
37. Fromm, J., Lautner, S.: Electrical signals and their physiological significance in plants. Plant, Cell Environ. **30**(3), 249–257 (2007)
38. Zimmermann, M.R., Mithöfer, A.: Electrical long-distance signaling in plants. In: Long-Distance Systemic Signaling and Communication in Plants, pp. 291–308. Springer, Berlin (2013)
39. Simons, P.J.: The role of electricity in plant movements. New Phytol. **87**(1), 11–37 (1981)
40. Fromm, J.: Control of phloem unloading by action potentials in mimosa. Physiol. Plant. **83**(3), 529–533 (1991)
41. Sibaoka, T.: Rapid plant movements triggered by action potentials. Bot. Mag. Shokubutsugaku-zasshi **104**(1), 73–95 (1991)
42. Volkov, A.G., Foster, J.C., Ashby, T.A., Walker, R.K., Johnson, J.A., Markin, V.S.: Mimosa pudica: electrical and mechanical stimulation of plant movements. Plant, Cell Environ. **33**(2), 163–173 (2010)
43. Minorsky, P.V.: Temperature sensing by plants: a review and hypothesis. Plant, Cell Environ. **12**(2), 119–135 (1989)
44. Volkov, A.G.: Green plants: electrochemical interfaces. J. Electroanal. Chem. **483**(1–2), 150–156 (2000)
45. Roblin, G.: Analysis of the variation potential induced by wounding in plants. Plant Cell Physiol. **26**(3), 455–461 (1985)
46. Pickard, B.G.: Action potentials in higher plants. Bot. Rev. **39**(2), 172–201 (1973)

Fungal Sensors and Wearables

Living Mycelium Composites Discern Weights via Patterns of the Electrical Activity

Andrew Adamatzky and Antoni Gandia

Abstract Fungal construction materials—substrates colonised by mycelium—are getting increased recognition as viable ecologically friendly alternatives to conventional building materials. A functionality of the constructions made from fungal materials would be enriched if blocks with living mycelium, known for their ability to respond to chemical, optical and tactile stimuli, were inserted. We investigated how large blocks of substrates colonised with mycelium of *Ganoderma resinaceum* responded to stimulation with heavy weights. We analysed details of the electrical responses to the stimulation with weights and show that ON and OFF stimuli can be discriminated by the living mycelium composites and that a habituation to the stimulation occurs. Novelty of the results in the reporting on changes in electrical spiking activity of mycelium bound composites in response to a heavy loads.

1 Introduction

Current practices in the construction industry make a substantial contribution to global climate change with potential damages to natural environment, natural resources supplies, agriculture, and human health [1–3]. Therefore recyclable biomaterials are emerging as potential key players in the alternative construction industry [4–6]. Materials produced from and with fungi are amongst most promising candidates.

Mycelium bound composites—masses of organic substrates colonised by fungi—are considered to be future environmentally sustainable growing biomaterials [7–9]. The fungal materials are used in acoustic insulation panels [10–12], thermal insulation wall cladding [13–18], packaging materials [19–21] and wearables [7, 22–25].

A. Adamatzky (✉)
Unconventional Computing Laboratory, University of the West of England, Bristol, UK
e-mail: andrew.adamatzky@uwe.ac.uk

A. Gandia
Institute for Plant Molecular and Cell Biology, CSIC-UPV, Valencia, Spain

© The Author(s), under exclusive license to Springer Nature Switzerland AG 2023
A. Adamatzky (ed.), *Fungal Machines*, Emergence, Complexity and Computation 47,
https://doi.org/10.1007/978-3-031-38336-6_6

In [26] we proposed to develop a structural substrate by using live fungal mycelium, functionalise the substrate with nanoparticles and polymers to make mycelium-based electronics [27–29], implement sensorial fusion and decision making in the mycelium networks [30]. The structural substrate—the mycelium bound composites—will be used to grow monolithic buildings from the functionalised fungal substrate [31]. Fungal buildings would self-grow, build, and repair themselves subject to substrate supplied, use natural adaptation to the environment, sense all that humans can sense. Whilst major parts of a building will be made from dried and cured mycelium composites there is an opportunity to use blocks with living mycelium as embedded sensorial elements. On our venture to investigate sensing properties of the mycelium composite blocks, called 'fungal blocks' further, we decided to study how large fungal blocks respond to pressure via changes in their electrical activity. The electrical activity have been chosen as indicator because fungi are known to respond to chemical and physical stimuli by changing patterns of their electrical activity [32–34] and electrical properties [29].

2 Methods and Materials

Mycelium bound composites have been prepared as follows. A pre-selected strain of the filamentous polypore fungus *Ganoderma resinaceum* (stock culture #19-18, Mogu S.r.l., Italy) was cultured on a block shaped substrate based of hemp shives and soybean hulls (mixture ratio 3:1, moisture content 65%) in plastic filter-patch microboxes in total darkness and at ambient room temperature c. 22 °C. After 7 days of incubation, the colonised substrate produced living blocks c. $20 \times 20 \times 10$ cm that were immediately used for the experiments.

Electrical activity of the colonised fungal blocks was recorded using pairs of iridium-coated stainless steel sub-dermal needle electrodes (Spes Medica S.r.l., Italy), with twisted cables and ADC-24 (Pico Technology, UK) high-resolution data logger with a 24-bit A/D converter, galvanic isolation and software-selectable sample rates all contribute to a superior noise-free resolution. An overall scheme of recording is shown in Fig. 1a. The pairs of electrodes were pierced, 5 mm deep, into sides of the blocks as shown in Fig. 1bc, two pairs per side. Distance between electrodes was 1–2 cm. In each trial, we recorded 8 electrode pairs, channels, simultaneously. We recorded electrical activity one sample per second. During the recording, the logger has been doing as many measurements as possible (typically up to 600 per second) and saving the average value. The humidity of the fungal blocks was 70–80% (MerlinLaser Protimeter, UK). The experiments were conducted in a growing tent with constant ambient temperature at 21 °C in absence of light.

We stimulated the fungal blocks by placing a 8 and 16 kg cast iron weights on the tops of the blocks (Fig. 1d). The surface of the fungal blocks was insulated from the cast iron weight by a polyethylene film.

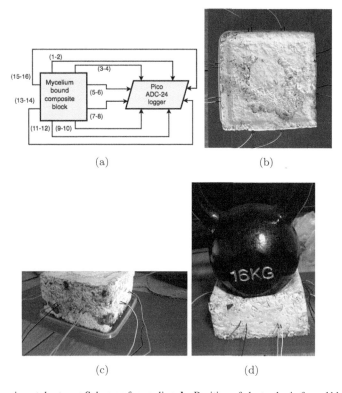

Fig. 1 Experimental setup. **a** Scheme of recording. **bc** Position of electrodes in fungal blocks: **b** top view and **c** side view. **d** Pairs of differential electrodes inserted in a fungal block and 16 kg kettle bell placed on top of the fungal block. Channels are from the top right clockwise (1–2), (3–5), ..., (15–16)

3 Results and Discussion

An example of fungal block's electrical responses to 8 kg load is shown in Fig. 2a. The responses are characterised by an immediate response, i.e. occurring in 10–20 min of the stimulation, and a delayed, in 1–4 h after beginning of the stimulation, response. The immediate responses were manifested in spikes of electrical potential recorded on the electrodes. Average amplitude of an immediate response to the loading with 8 kg was 3.05 mV, $\sigma = 2.5$ and the spikes' average duration 489 s, $\sigma = 273$. Average amplitude of the immediate response to lifting the weight was 4 mV, $\sigma = 4.4$ and average duration of 217 s, $\sigma = 232$. Delayed responses were manifested in trains of spikes with average amplitude 1.7 mV, $\sigma = 1$, average duration of 161 s, $\sigma = 74$ and distance between spikes 125 s, $\sigma = 34$. The responses are clearly attributed to the electrical activity of the living mycelium. The recording of electrical activity of the dried mycelium bound composite block (with dead mycelium) is shown in Fig. 3. The electrical activity of the dry block does not exceed a level of

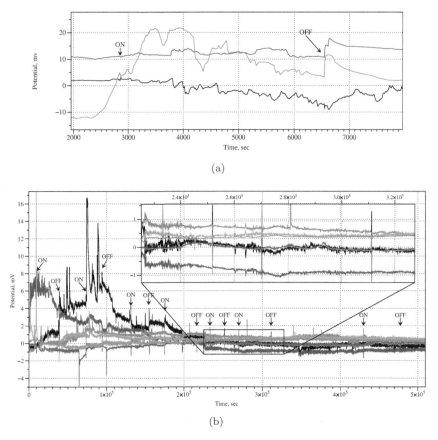

Fig. 2 Electrical activity of the fungal blocks, stimulated with heavy loads. **a** The activity of the block stimulated with 8 kg load. **b** The activity of the block stimulated with 16 kg load. Moments of the loads applications are labelled by 'ON' and lifting the loads by 'OFF'. Channels are colour coded as (1–2)—black, (3–4)—red, (5–6)—blue, (7–8)—green, (9–10)—magenta, (11–12)—orange, (13–14)—yellow

noise [−0.002, 0.002] mV. When a load is applied an amplitude of the noise slightly decreases and stays in the interval [−0.0015, 0.0015] mV.

Fungal blocks shown responses, with patterns of electrical activity different in amplitude and frequency of oscillations from that observed in the non-stimulated blocks, to the 8 kg loads only for 1–2 cycles of loading and unloading, no significant responses to further cycles of the stimulation have been observed. The fungal blocks responded to stimulation with 16 kg weight for at least 8 cycles of loading and unloading. Let us discuss these responses in details.

An example of electrical activity recorded on 8 channels, during the stimulation with 16 kg weight, is shown in Fig. 2b. Distributions of 'spikes amplitudes versus spikes duration for spike-responses to application of the weight (ON spikes) and

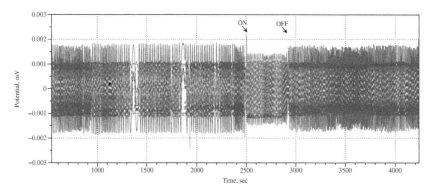

Fig. 3 Recording of electrical activity of the dry block of mycelium bound composite stimulated with 16 kg load. Moments of the loads applications are labelled by 'ON' and lifting the loads by 'OFF'. Channels are colour coded as (1–2)—black, (3–4)—red, (5–6)—blue, (7–8)—green, (9–10)—magenta, (11–12)—orange, (13–14)—yellow

lifting of the weight (OFF spikes) are shown in Fig. 4. In response to application of 16 kg weight the fungal blocks produced spikes with median amplitude 1.4 mV and median duration 456 s; average amplitude of ON spikes was 2.9 mV, $\sigma = 4.9$ and average duration 880 s, $\sigma = 1379$. OFF spikes were characterised by median amplitude 1 mV and median duration 216 s; average amplitude 2.1 mV, $\sigma = 4.6$, and average duration 453 s, $\sigma = 559$. ON spikes are 1.4 higher than and twice as longer as OFF spikes. Based on this comparison of the response spikes we can claim that fungal blocks recognise when a weight was applied or removed.

Would living fungal materials habituate to the stimulation with weights? Yes, as evidenced in Fig. 5. Amplitudes of ON and OFF spikes decline with iterations of stimulation as shown in Fig. 5a. The duration of spikes also decreases, in overall, with iterations of stimulation, Fig. 5b, albeit not monotonously.

As previously demonstrated in [32–34] mycelium networks exhibit action-potential like spikes. An example of spiking activity recorded in present experiments is shown in Fig. 6, where spikes are shown by arrows. The plot demonstrates an extent of the variability of the spike in amplitude and duration. The spikes have an average amplitude of 0.02 mV, $\sigma = 0.01$. The amplitudes of spikes depend on a distance of an excitation wave-front from the electrodes and therefore will be ignored here, and we will focus only on frequencies of spiking. We found that median frequency of spiking of the non-stimulated fungal blocks is $\frac{1}{702}$ Hz while the fungal blocks loaded with the weights spike with a median frequency $\frac{1}{958}$ Hz. Average spiking frequency of the unloaded fungal blocks is $\frac{1}{793}$ Hz and of the loaded blocks $\frac{1}{1031}$ Hz. Thus, we can speculate that fungal blocks loaded with weights spike 1.4 times more frequently than unloaded blocks.

Let us overview the findings presented. We applied heavy weights to large blocks of mycelium bound composites, fungal blocks, and recorded electrical activity of the fungal blocks. We found that the fungal block respond to application and removal of

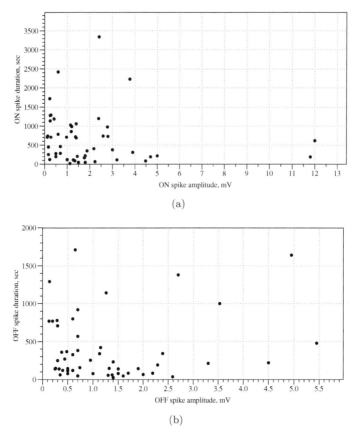

Fig. 4 Distribution of spike-response amplitude versus duration in response to **a** application of the 16 kg weight and **b** removal of the 16 kg weight

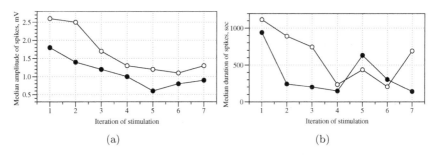

Fig. 5 Responses of living fungal blocks to stimulation as functions of iterations of stimulation. Median amplitude (**a**) and median duration (**b**) of ON (circles) and OFF (discs) spikes

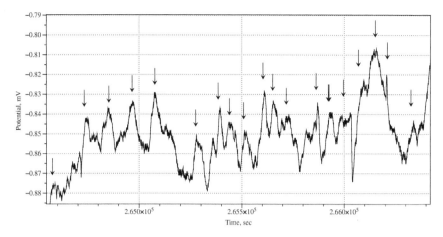

Fig. 6 Example of spiking activity of living mycelium composite. The spikes of electrical potential are shown by arrows

the weights with spikes of electrical potential. The results complement our studies on tactile stimulation of fungal skin (mycelium sheet with no substrate) [35]: the fungal skin responds to application and removal of pressure with spikes of electrical potential. The fungal blocks can discern whether a weight was applied or removed because the blocks react to application of the weights with higher amplitude and longer duration spikes than the spikes responding to the removal of the weights. The fungal responses to stimulation show habituation. This is in accordance with previous studies on stimulation of plants, fungi, bacteria, and protists [36–40]. An additional finding was that loading of the fungal blocks with weights increase frequency of electrical potential spiking. This increase in the spiking frequency might be due to physiological responses to a mild mechanical damage caused by heavy loads; the responses involve calcium waves and lead to regeneration processes and sprouting [41].

Further studies in stimulation of living mycelium bound composites with weights could focus on studying whether shapes of the weights could be recognised by mycelium networks. A possible scenario would be to map a set of basic shapes into sets of electrical responses recorded on pairs of differential electrodes inserted into sides of the fungal blocks. Another promising direction of the research will be to study electrical responses of the fungal blocks to changes in ambient temperature.

4 Conclusion

Live mycelium composites exhibit a range of electrical activity. The patterns of their electrical activity change when a pressure (in the form of a weight) is applied to the composites. Whilst it is still unknown if the composites can accurately reflect an amount of pressure developed via their electrical activity patterns, it is proved that they can act as ON/OFF sensors. The findings open new horizons into reactive biomaterials.

References

1. Onat, N.C., Kucukvar, M.: Carbon footprint of construction industry: a global review and supply chain analysis. Renew. Sustain. Energy Rev. **124**, 109783 (2020)
2. Schwartz, Y., Raslan, R., Mumovic, D.: The life cycle carbon footprint of refurbished and new buildings-a systematic review of case studies. Renew. Sustain. Energy Rev. **81**, 231–241 (2018)
3. Omer, A.M.: Energy, environment and sustainable development. Renew. Sustain. Energy Rev. **12**(9), 2265–2300 (2008)
4. Pellicer, E., Nikolic, D., Sort, J., Baró, M., Zivic, F., Grujovic, N., Grujic, R., Pelemis, S.: Advances in Applications of Industrial Biomaterials. Springer (2017)
5. Zeller, P., Zocher, D.: Ecovative's breakthrough biomaterials. Fungi Mag. **5**(1), 51–56 (2012)
6. Williams, D.: Essential Biomaterials Science. Cambridge University Press (2014)
7. Karana, E., Blauwhoff, D., Hultink, E.-J., Camere, S.: When the material grows: a case study on designing (with) mycelium-based materials. Int. J. Design **12**(2) (2018)
8. Jones, M., Mautner, A., Luenco, S., Bismarck, A., John, S.: Engineered mycelium composite construction materials from fungal biorefineries: a critical review. Mat. & Design **187**, 108397 (2020)
9. Cerimi, K., Akkaya, K.C., Pohl, C., Schmidt, B., Neubauer, P.: Fungi as source for new bio-based materials: a patent review. Fungal Biol. Biotechnol. **6**(1), 1–10 (2019)
10. Pelletier, M.G., Holt, G.A., Wanjura, J.D., Bayer, E., McIntyre, G.: An evaluation study of mycelium based acoustic absorbers grown on agricultural by-product substrates. Ind. Crops Prod. **51**, 480–485 (2013)
11. Elsacker, E., Vandelook, S., Van Wylick, A., Ruytinx, J., De Laet, L., Peeters, E.: A comprehensive framework for the production of mycelium-based lignocellulosic composites. Sci. Total Environ. **725**, 138431 (2020)
12. Robertson, O. et al. Fungal future: a review of mycelium biocomposites as an ecological alternative insulation material. In: DS 101: Proceedings of NordDesign 2020, Lyngby, Denmark, 12–14th August 2020, pp. 1–13 (2020)
13. Yang, Z., Zhang, F., Still, B., White, M., Amstislavski, P.: Physical and mechanical properties of fungal mycelium-based biofoam. J. Mat. Civil Eng. **29**(7), 04017030 (2017)
14. Xing, Y., Brewer, M., El-Gharabawy, H., Griffith, G., Jones, P.: Growing and testing mycelium bricks as building insulation materials. In: IOP Conference Series: Earth and Environmental Science, vol. 121, p. 022032. IOP Publishing (2018)
15. Girometta, C., Picco, A.M., Baiguera, R.M., Dondi, D., Babbini, S., Cartabia, M., Pellegrini, M., Savino, E.: Physico-mechanical and thermodynamic properties of mycelium-based bio-composites: a review. Sustainability **11**(1), 281 (2019)
16. Dias, P.P., Jayasinghe, L.B., Waldmann, D.: Investigation of mycelium-miscanthus composites as building insulation material. Results Mat. **10**, 100189 (2021)
17. Wang, F., Li, H.-Q., Kang, S.-S., Bai, Y.-F., Cheng, G.-Z., Zhang, G.-Q.: The experimental study of mycelium/expanded perlite thermal insulation composite material for buildings. Sci. Technol. Eng. **2016**, 20 (2016)

18. Cárdenas-R, J.P.: Thermal insulation biomaterial based on hydrangea macrophylla. In: Bio-Based Materials and Biotechnologies for Eco-Efficient Construction, pp. 187–201. Elsevier (2020)
19. Holt, G.A., Mcintyre, G., Flagg, D., Bayer, E., Wanjura, J.D., Pelletier, M.G.: Fungal mycelium and cotton plant materials in the manufacture of biodegradable molded packaging material: evaluation study of select blends of cotton byproducts. J. Biobased Mat. Bioenergy **6**(4), 431–439 (2012)
20. Sivaprasad, S., Byju, S.K., Prajith, C., Shaju, J., Rejeesh, C.R.: Development of a novel mycelium bio-composite material to substitute for polystyrene in packaging applications. Proc. Mat. Today (2021)
21. Mojumdar, A., Behera, H.T., Ray, L.: Mushroom mycelia-based material: an environmental friendly alternative to synthetic packaging. Microb. Polym. 131–141 (2021)
22. Adamatzky, A., Nikolaidou, A., Gandia, A., Chiolerio, A., Dehshibi, M.M.: Reactive fungal wearable. Biosystems **199**, 104304 (2021)
23. Silverman, J., Cao, H., Cobb, K.: Development of mushroom mycelium composites for footwear products. Cloth. Text. Res. J. **38**(2), 119–133 (2020)
24. Appels, F.V.W.: The use of fungal mycelium for the production of bio-based materials. Ph.D. Thesis, Universiteit Utrecht (2020)
25. Jones, M., Gandia, A., John, S., Bismarck, A.: Leather-like material biofabrication using fungi. Nat. Sustain. 1–8 (2020)
26. Adamatzky, A., Ayres, P., Belotti, G., Wösten, H.: Fungal architecture position paper. Int. J. Unconv. Comput. **14** (2019)
27. Beasley, A.E., Powell, A.L., Adamatzky, A.: Capacitive storage in mycelium substrate (2020). arXiv:2003.07816
28. Beasley, A.E., Abdelouahab, M.-S., Lozi, R., Powell, A.L., Adamatzky, A.: Mem-fractive properties of mushrooms (2020). arXiv:2002.06413
29. Beasley, A.E., Powell, A.L., Adamatzky, A.: Fungal photosensors (2020). arXiv:2003.07825
30. Adamatzky, A., Tegelaar, M., Wosten, H.A.B., Powell, A.L., Beasley, A.E., Mayne, R.: On boolean gates in fungal colony. Biosystems **193**, 104138 (2020)
31. Adamatzky, A., Gandia, A., Ayres, P., Wösten, H., Tegelaar, M.: Adaptive fungal architectures. LINKs-Series **5**, 66–77
32. Olsson, S., Hansson, B.S.: Action potential-like activity found in fungal mycelia is sensitive to stimulation. Naturwissenschaften **82**(1), 30–31 (1995)
33. Adamatzky, A.: On spiking behaviour of oyster fungi pleurotus djamor. Sci. Rep. **8**(1), 1–7 (2018)
34. Adamatzky, A., Tuszynski, J., Pieper, J., Nicolau, D.V., Rinalndi, R., Sirakoulis, G., Erokhin, V., Schnauss, J., Smith, D.M.: Towards cytoskeleton computers. A proposal. In: Adamatzky, A., Akl, S., Sirakoulis, G. (eds.) From Parallel to Emergent Computing. CRC Group/Taylor & Francis (2019)
35. Adamatzky, A., Gandia, A., Chiolerio, A.: Fungal sensing skin. Fungal Biol. Biotechnol. **8**(1), 1–6 (2021)
36. Applewhite, P.B.: Learning in bacteria, fungi, and plants. Invertebrate Learn. **3**, 179–186 (1975)
37. Fukasawa, Yu., Savoury, M., Boddy, L.: Ecological memory and relocation decisions in fungal mycelial networks: responses to quantity and location of new resources. ISME J. **14**(2), 380–388 (2020)
38. Ginsburg, S., Jablonka, E.: Evolutionary transitions in learning and cognition. Philos. Trans. R. Soc. B **376**(1821), 20190766 (2021)
39. Boussard, A., Delescluse, J., Pérez-Escudero, A., Dussutour, A.: Memory inception and preservation in slime moulds: the quest for a common mechanism. Philos. Trans. R. Soc. B **374**(1774), 20180368 (2019)
40. Yokochi, K. et al.: An investigation on the habituation of amoeba. Aichi Igakkwai Zasshi= Jl. Aichi Med. Soc. **33**(3) (1926)
41. Hernández-Oñate, M.A., Herrera-Estrella, A.: Damage response involves mechanisms conserved across plants, animals and fungi. Curr. Genet. **61**(3), 359–372 (2015)

Fungal Sensing Skin

Andrew Adamatzky, Antoni Gandia, and Alessandro Chiolerio

Abstract A fungal skin is a thin flexible sheet of a living homogeneous mycelium made by a filamentous fungus. The skin could be used in future living architectures of adaptive buildings and as a sensing living skin for soft self-growing/adaptive robots. In experimental laboratory studies we demonstrate that the fungal skin is capable for recognising mechanical and optical stimulation. The skin reacts differently to loading of a weight, removal of the weight, and switching illumination on and off. These are the first experimental evidences that fungal materials can be used not only as mechanical 'skeletons' in architecture and robotics but also as intelligent skins capable for recognition of external stimuli and sensorial fusion.

1 Background

Flexible electronics, especially electronic skins [1–3] is amongst the most rapidly growing and promising fields of novel and emergent hardware. The electronic skins are made of flexible materials where electronics capable of tactile sensing [4–7] are embedded. The electronic skins are capable of low level perception [8, 9] and could be developed as autonomous adaptive devices [10]. Typical designs of electronic skins include thin-film transistor and pressure sensors integrated in a plastic substrate [11], micro-patterned polydimethylsiloxane with carbon nanotube ultra-thin films [12, 13], a large-area film synthesised by sulfurisation of a tungsten film [14], multilayered graphene [15], platinum ribbons [3], Polyethylene terephthalate (PET) based silver electrodes [16], digitally printed hybrid electrodes for electromyographic recording [17] or for piezoresistive pressure sensing [18], or channels filled with conductive polymer [19].

A. Adamatzky (✉) · A. Chiolerio
Unconventional Computing Laboratory, UWE, Bristol, UK
e-mail: andrew.adamatzky@uwe.ac.uk

A. Gandia
Mogu S.r.l., Inarzo, Italy

A. Chiolerio
Center for Sustainable Future Technologies, Istituto Italiano di Tecnologia, Torino, Italy

© The Author(s), under exclusive license to Springer Nature Switzerland AG 2023
A. Adamatzky (ed.), *Fungal Machines*, Emergence, Complexity and Computation 47,
https://doi.org/10.1007/978-3-031-38336-6_7

Whilst the existing designs and implementations are highly impactful, the prototypes of electronic skins lack a capacity to self-repair and grow. Such properties are useful, and could be necessary, when an electronic skin is used in e.g. unconventional living architecture [20], soft and self-growing robots [21–24] and development of intelligent materials from fungi [25–28]. Based on our previous experience with designing tactile, colour sensors from slime mould *Physarum polycephalum* [29–31] and our recent results on fungal electrical activity [32–34], as well as following previously demonstrated thigmotropic and phototropic response (Fig. 1) in higher fungi [35], we decided to propose a thin layer of homogeneous mycelium of the trimitic polypore species *Ganoderma resinaceum* as a live electronic skin and thus investigate its potential to sense and respond to tactile and optical stimuli. We call the fungal substrate, used in present paper, 'fungal skin' due to its overall appearance and physical feeling. In fact, several species of fungi have been proposed as literal skin substitutes and tested in wound healing [36–41].

2 Methods

Potato dextrose agar (PDA), malt extract agar (MEA) and malt extract (ME) were purchased from Sigma-Aldrich (USA). The *Ganoderma resinaceum* culture used in this experiment was obtained from a wild basidiocarp found at the shores of *Lago di Varese*, Lombardy (Italy) in 2018 and maintained in alternate PDA and MEA slants at MOGU S.r.l. for the last 3 years at 4 °C under the collection code 019-18.

The fungal skin was prepared as follows. *G. resinaceum* was grown on MEA plates and a healthy mycelium plug was inoculated into an Erlenmeyer flask containing 200 ml of 2% ME broth (MEB). The liquid culture flask was then incubated in a rotary shaker at 200 rpm and 28 °C for 5 days. Subsequently, this liquid culture was homogenised for 1 minute at max. speed in a sterile 1L Waring laboratory blender (USA) containing 400 mL of fresh MEB, the resulting 600 mL of living slurry were then poured into a 35 by 35 cm static fermentation tray. The slurry was let to incubate undisturbed for 15 days to allow the fungal hyphae to inter-mesh and form a floating mat or skin of fungal mycelium. Finally, a living fungal skin circa 1.5 mm thick was harvested (see texture of the skin in Fig. 1a), washed in sterile demineralised water, cut to the size 23 cm by 11 cm and placed onto a polyurethane base to keep electrodes stable during the electrical characterisation steps (Fig. 1b).

The electrical activity of the skin was measured as follows. We used iridium-coated stainless steel sub-dermal needle electrodes (Spes Medica S.r.l., Italy), with twisted cables. The pairs of electrode were inserted in the fungal skin as shown (Fig. 1b): the first placed in position 2×5 cm from a vertex, the following placed at 1 cm distance each. In each pair we recorded a difference in electrical potential between the electrodes. We used ADC-24 (Pico Technology, UK) high-resolution data logger with a 24-bit Analog to Digital converter, galvanic isolation and software-selectable sample rates. We recorded electrical activity with a frequency of one sample per second. We set the acquisition voltage range to 156 mV with an offset accuracy of

Fig. 1 Phototropism is one of the leading guiding factors in the formation of basidiocarps in *Ganoderma* spp

9 μV to maintain a gain error of 0.1%. For mechanical stimulation with 30 g nylon cylinder placed at 3 cm from the long edge and 3 from the electrodes, and aligned with electrode number 5, contact area with the fungal skin was circa 35 mm disc. For optical stimulation we used an aquarium light, array of LEDs, 36 white LEDs and 12 blue LEDs, 18 W, illumination on the fungal skin was 0.3 Lux.

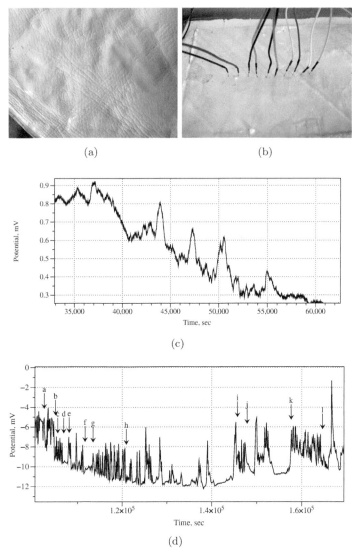

Fig. 2 Recording of electrical activity of fungal skin. **a** Close-up texture detail of a fungal skin. **b** A photograph of electrodes inserted into the fungal skin. **c** Train of three low-frequency spikes, average width of spikes there is 1500 s, a distance between spike peaks is 3000 s and average amplitude is 0.2 mV. **d** Example of several train of high-frequency spikes. Each train $T_{xy} = (A_{xy}, W_{xy}, P_{xy})$ is characterised by average amplitude of spikes A_{xy} mV, width of spikes W_{xy} s and average distance between neighbouring spikes' peaks P_{xy} s: $T_{ab} = (2.6, 245, 300)$, $T_{cd} = (1.7, 160, 220)$, $T_{ef} = (1.6, 340, 340)$, $T_{gh} = (2.5, 240, 350)$, $T_{ij} = (2.5, 220, 590)$, $T_{kl} = (2.6, 290, 440)$

3 Results

Endogenous electrical activity of the fungal material is polymorphic. Low and high frequency oscillations patterns can emerge intermittently. A train of four spikes in Fig. 1c is an example of low frequency oscillations. By measuring the electrical response with multiple electrodes, positioned along coordinated axes like row and columns of a matrix, and connecting them to a differential operational amplifier, it is possible to exclude singularities and enhance coordinated responses, which is indicated as a filtering procedure to exclude endogenous polymorphic activity.

Electrical responses to tactile loading and illumination are distinctive and can be easily recognized from endogenous activity. An example of several rounds of stimulation is shown in Fig. 3a. The fungal skin responds to loading of a weight with a high-amplitude wide spike of electrical potential sometimes followed by a train of high-frequency spikes. The skin also responds to removal of the weight by a high-amplitude spike of electrical potential (Fig. 2).

An exemplar response to loading and removal of weight is shown in Fig. 3b. The parameters of the fungal skin responses to the weight being placed on the skin are the following. An average delay of the response (the time from weight application to a peak of the high-amplitude spike) is 911.4 s ($\sigma = 1280.1$, minimum 25 s and maximum 3200 s). An average amplitude of the response spike (marked 's' in the example Fig. 3b) is 0.4 mV ($\sigma = 0.2$, minimum 0.1 mV and maximum 0.8 mV). An average width of the response spike is 1261.8 s ($\sigma = 1420.3$, minimum 199 s and maximum 4080 s), meaning that the average energy consumed per current unit, associated to the response, is approximately 0.5 J/A. A train of spikes (marked 'r' in the example Fig. 3b), if any, following the response spike usually has 4 or 5 spikes. The fungal skin responds to removal of the weight (the response is marked 'p' in the example Fig. 3b) with a spike which average amplitude is 0.4 mV ($\sigma = 0.2$, minimum 0.2 mV and maximum 0.85 mV). Amplitudes are less indicative than frequencies because an amplitude depends on the position of electrodes with regards to propagating wave of excitation. An average width of the spike is 774 s ($\sigma = 733.1$, minimum 100 s and maximum 2000 s). A response of the fungal skin to removal of the weight was not observed in circa 20% of differential electrode pairs. The average response time is 385.5 s ($\sigma = 693.3$ s, minimum 77 s and maximum 1800 s). By taking into account inter-electrode distance it could be possible to weigh temporal delays and further strengthen the rejection circuits based on operational amplifiers, as per above suggestion to discard endogenous activity.

The response of the fungal skin to illumination is manifested in the raising of the baseline potential, as illustrated in the exemplar recordings in Fig. 3c. In contrast to mechanical stimulation response the response-to-illumination spike does not subside but the electrical potential stays raised until illumination is switched off. An average amplitude of the response is 0.61 mV ($\sigma = 0.27$, minimum 0.2 mV and maximum 1 mV). The raise of the potential starts immediately after the illumination is switched on. The potential saturation time is 2960 s in average ($\sigma = 2201$, minimum 879 s and maximum 9530 s); the potential relaxation time is 8700 s in average

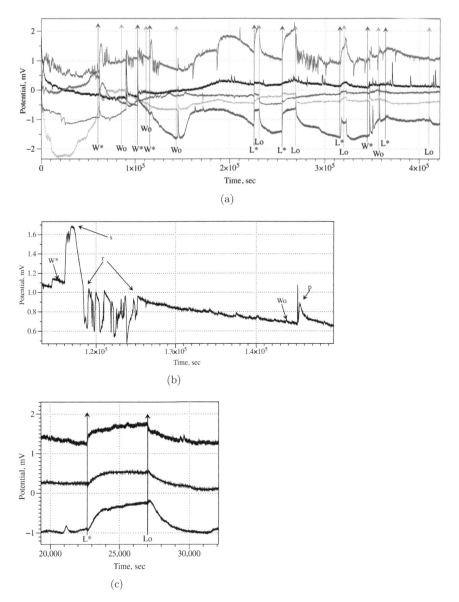

Fig. 3 Fungal skin response to mechanical and optical stimulation. **a** Exemplar recording of fungal skin electrical activity under tactile and optical stimulation. Moments of applying and removing a weight are shown as 'W*' and 'Wo' and switching light ON and OFF as 'L*' and 'Lo'. **b** Exemplar response to mechanical stimulation. Moments of applying and removing a weight are shown as 'W*' and 'Wo'. High-amplitude response is labelled 's'. This response is followed by a train of spikes 'r'. A response to the removal of the weight is labelled 'p'. **c** Exemplar response of fungal skin to illumination, recorded on three pairs of differential electrodes. 'L*' indicates illumination is applied, 'Lo' illumination is switched off. **d** A train of spikes on the raised potential as a response to illumination

($\sigma = 4500$, minimum 962 s and maximum 24790 s). Typically, we did not observe any spike trains after the illumination switched off however in a couple of trials we witnessed spike trains on top of the raised potential, as shown in Fig. 3d.

In the case of illumination it is particularly easy to imagine how effective a rejection stage could be, since all the responses are well synchronized (Fig. 3c).

4 Discussion

We demonstrated that a thin sheet of homogeneous living mycelium of *Ganoderma resinaceum*, which we named 'fungal skin', shows pronounced electrical responses to mechanical and optical stimulation. Can we differentiate between the fungal skin's response to mechanical and optical stimulation? Definitely, see Fig. 4a. The fungal skin responds to mechanical stimulation with a 15 min spike of electrical potential, which diminishes even if the applied pressure on the skin remains. The skin responds to optical stimulation by raising its electrical potential and keeping it raised till the light is switched off.

Can we differentiate the responses to loading and removal of the weight? Yes. Whilst amplitudes of 'loading' and 'removal' spikes are the same (0.4 mV in average) the fungal skin average reaction time to removal of the weight is 2.4 times shorter than the reaction to loading of the weight (385 s vs. 911 s). Also 'loading' spikes are 1.6 times wider than 'removal' spikes (1261 s vs. 774 s).

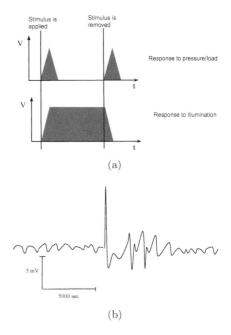

Fig. 4 **a** A scheme of the fungal skin responses to mechanical load and optical stimulations. **b** Slime mould *P. polycephalum* response to application of 0.01 g glass capillary tube. Redrawn from [30]

Fungal skin response to weight application is, in some cases, esp. Figure 3b, similar to response of slime mould to application of the light weight [30]. The following events are observed (Fig. 4b): oscillatory activity before stimulation, immediate response to stimulation, prolonged response to stimulation as a train of high-amplitude spikes, return to normal oscillatory activity. This might indicate some universal principles of sensing and information processing in fungi and slime moulds.

The sensing fungal skin proposed has a range of advantages comparing to other living sensing materials, e.g. slime mould sensors [29–31] electronic sensors with living cell components [42], chemical sensors using living taste, olfactory, and neural cells and tissues [43] and tactile sensor from living cell culture [44]. The advantages are low production costs, simple maintenance and durability. The last but not least advantage is scalability: a fungal skin patch can be as small as few milimeters or it can be grown to several metres in size.

In future studies we will aim to answer the following questions. Would it be possible to infer a weight of the load applied to the fungal skin from patterns of its electrical activity? Would the fungal skin indicate directionality of the load movement by its spiking activity? Would it be possible to locate the position of the weight within the fungal network? Would it be possible to map a spectrum of the light applied to the skin onto patterns of the skin's electrical activity?

References

1. Soni, M., Dahiya, R.: Soft eskin: distributed touch sensing with harmonized energy and computing. Phil. Trans. R. Soc. A **378**(2164), 20190156 (2020)
2. Ma, M., Zhang, Z., Liao, Q., Yi, F., Han, L., Zhang, G., Liu, S., Liao, X., Zhang, Y.: Self-powered artificial electronic skin for high-resolution pressure sensing. Nano Energy **32**, 389–396 (2017)
3. Zhao, S., Zhu, R.: Electronic skin with multifunction sensors based on thermosensation. Adv. Mater. **29**(15), 1606151 (2017)
4. Chou, H.-H., Nguyen, A., Chortos, A., To, J.W.F., Lu, C., Mei, J., Kurosawa, T., Bae, W.-G., Tok, J.B.-H., Bao, Z.: A chameleon-inspired stretchable electronic skin with interactive colour changing controlled by tactile sensing. Nat. Commun. **6**(1), 1–10 (2015)
5. Yang, T., Wang, W., Zhang, H., Li, X., Shi, J., He, Y., Zheng, Q., Li, Z., Zhu, H.: Tactile sensing system based on arrays of graphene woven microfabrics: electromechanical behavior and electronic skin application. ACS Nano **9**(11), 10867–10875 (2015)
6. Wang, X., Dong, L., Zhang, H., Yu, R., Pan, C., Wang, Z.L.: Recent progress in electronic skin. Adv. Sci. **2**(10), 1500169 (2015)
7. Pu, X., Liu, M., Chen, X., Sun, J., Du, C., Zhang, Y., Zhai, J., Hu, W., Wang, Z.L.: Ultra-stretchable, transparent triboelectric nanogenerator as electronic skin for biomechanical energy harvesting and tactile sensing. Sci. Adv. **3**(5), e1700015 (2017)
8. Chortos, A., Liu, J., Bao, Z.: Pursuing prosthetic electronic skin. Nat. Mater. **15**(9), 937–950 (2016)
9. Park, S., Kim, H., Vosgueritchian, M., Cheon, S., Kim, H., Koo, J.H., Kim, T.R., Lee, S., Schwartz, G., Chang, H., et al.: Stretchable energy-harvesting tactile electronic skin capable of differentiating multiple mechanical stimuli modes. Adv. Mat. **26**(43), 7324–7332 (2014)
10. Núñez, C.G., Manjakkal, L., Dahiya, R.: Energy autonomous electronic skin. npj Flex. Electr. **3**(1), 1–24 (2019)

11. Wang, C., Hwang, D., Zhibin, Yu., Takei, K., Park, J., Chen, T., Ma, B., Javey, A.: User-interactive electronic skin for instantaneous pressure visualization. Nat. Mater. **12**(10), 899–904 (2013)
12. Wang, X., Yang, G., Xiong, Z., Cui, Z., Zhang, T.: Silk-molded flexible, ultrasensitive, and highly stable electronic skin for monitoring human physiological signals. Adv. Mater. **26**(9), 1336–1342 (2014)
13. Sekitani, T., Someya, T.: Stretchable organic integrated circuits for large-area electronic skin surfaces. MRS Bull. **37**(3), 236–245 (2012)
14. Guo, H., Lan, C., Zhou, Z., Sun, P., Wei, D., Li, C.: Transparent, flexible, and stretchable ws 2 based humidity sensors for electronic skin. Nanoscale **9**(19), 6246–6253 (2017)
15. Qiao, Y., Wang, Y., Tian, H., Li, M., Jian, J., Wei, Y., Tian, Y., Dan-Yang Wang, Yu., Pang, X.G., et al.: Multilayer graphene epidermal electronic skin. ACS Nano **12**(9), 8839–8846 (2018)
16. Zhao, X., Hua, Q., Ruomeng, Yu., Zhang, Y., Pan, C.: Flexible, stretchable and wearable multifunctional sensor array as artificial electronic skin for static and dynamic strain mapping. Adv. Electr. Mat. **1**(7), 1500142 (2015)
17. Scalisi, R.G., Paleari, M., Favetto, A., Stoppa, M., Ariano, P., Pandolfi, P., Chiolerio, A.: Inkjet printed flexible electrodes for surface electromyography. Org. Electron. **18**, 89–94 (2015)
18. Chiolerio, A., Rivolo, P., Porro, S., Stassi, S., Ricciardi, S., Mandracci, P., Canavese, G., Bejtka, K., Pirri, C.F.: Inkjet-printed pedot: pss electrodes on plasma modified pdms nanocomposites: quantifying plasma treatment hardness. RSC Advances **4**, 51477 (2014)
19. Chiolerio, A., Adamatzky, A.: Tactile sensing and computing on a random network of conducting fluid channels. Flex. Printed Electr. (2020)
20. Adamatzky, A., Ayres, P., Belotti, G., Wösten, H.: Fungal architecture position paper. Int. J. Unconv. Comput. **14** (2019)
21. El-Hussieny, H., Mehmood, U., Mehdi, Z., Jeong, S.-G., Usman, M., Hawkes, E.W., Okarnura, A.M., Ryu, J.-H.: Development and evaluation of an intuitive flexible interface for teleoperating soft growing robots. In: 2018 IEEE/RSJ International Conference on Intelligent Robots and Systems (IROS), pp. 4995–5002. IEEE (2018)
22. Sadeghi, A., Mondini, A., Mazzolai, B.: Toward self-growing soft robots inspired by plant roots and based on additive manufacturing technologies. Soft Rob. **4**(3), 211–223 (2017)
23. Rieffel, J., Knox, D., Smith, S., Trimmer, B.: Growing and evolving soft robots. Artif. Life **20**(1), 143–162 (2014)
24. Greer, J.D., Morimoto, T.K., Okamura, A.M., Hawkes, E.W.: A soft, steerable continuum robot that grows via tip extension. Soft Robot. **6**(1), 95–108 (2019)
25. Meyer, V., Basenko, E.Y., Benz, J.P., Braus, G.H., Caddick, M.X., Csukai, M., de Vries, R.P., Endy, D., Frisvad, J.C., Gunde-Cimerman, N., Haarmann, T., Hadar, Y., Hansen, K., Johnson, R.I., Keller, N.P., Kraševec, N., Mortensen, U.H., Perez, R., Ram, A.F.J., Record, E., Ross, P., Shapaval, V., Steiniger, C., van den Brink, H., van Munster, J., Yarden, O., Wösten, H.A.B.: Growing a circular economy with fungal biotechnology: a white paper. Fungal Biol. Biotechnol. **7**(1), 5 (2020)
26. Haneef, M., Ceseracciu, L., Canale, C., Bayer, I.S., Heredia-Guerrero, J.A., Athanassiou, A.: Advanced materials from fungal mycelium: fabrication and tuning of physical properties. Sci. Rep. **7**(1), 1–11 (2017)
27. Jones, M., Mautner, A., Luenco, S., Bismarck, A., John, S.: A critical review, Engineered mycelium composite construction materials from fungal biorefineries (2020)
28. Wösten, H.A.B.: Filamentous fungi for the production of enzymes, chemicals and materials. Curr. Opin. Biotechnol. **59**, 65–70 (2019)
29. Adamatzky, A.: Towards slime mould colour sensor: recognition of colours by Physarum polycephalum. Org. Electron. **14**(12), 3355–3361 (2013)
30. Adamatzky, A.: Slime mould tactile sensor. Sens. Actuators B Chem. **188**, 38–44 (2013)
31. Whiting, J.G.H., de LacyCostello, B.P.J., Adamatzky, A.: Towards slime mould chemical sensor: Mapping chemical inputs onto electrical potential dynamics of *Physarum Polycephalum*. Sens. Actuators B: Chem. **191**, 844–853 (2014)

32. Adamatzky, A.: On spiking behaviour of oyster fungi pleurotus djamor. Sci. Rep. **8**(1), 1–7 (2018)
33. Beasley, A.E., Powell, A.L., Adamatzky, A.: Capacitive storage in mycelium substrate (2020). arXiv:2003.07816
34. Beasley, A.E., Abdelouahab, M.-S., Lozi, R., Powell, A.L., Adamatzky, A.: Mem-fractive properties of mushrooms (2020). arXiv:2002.06413
35. Moore, D.: Perception and response to gravity in higher fungi - a critical appraisal. New Phytol. (1991)
36. Hamlyn, P.F.: Fabricating fungi. In: Glasman, I., Lennox-Kerr, P. (eds.) Textile Technology International, chapter New applications, pp. 254–257. Sterling Publications Ltd, London (1991)
37. Hamlyn, P.F., Schmidt, R.J.: Potential therapeutic application of fungal filaments in wound management. Mycologist **8**(4), 147–152 (1994)
38. Ching-Hua, S., Sun, C.-S., Juan, S.-W., Chung-Hong, H., Ke, W.-T., Sheu, M.-T.: Fungal mycelia as the source of chitin and polysaccharides and their applications as skin substitutes. Biomaterials **18**(17), 1169–1174 (1997)
39. Ching-Hua, S., Sun, C.-S., Juan, S.-W., Ho, H.-O., Chung-Hong, H., Sheu, M.-T.: Development of fungal mycelia as skin substitutes: effects on wound healing and fibroblast. Biomaterials **20**(1), 61–68 (1999)
40. Hui, X., Liu, L., Cao, C., Weisheng, L., Zhu, Z., Guo, Z., Li, M., Wang, X., Huang, D., Wang, S., et al.: Wound healing activity of a skin substitute from residues of culinary-medicinal winter mushroom flammulina velutipes (agaricomycetes) cultivation. Int. J. Med. Mushrooms **21**(7) (2019)
41. Narayanan, K.B., Zo, S.M., Han, S.S.: Novel biomimetic chitin-glucan polysaccharide nano/microfibrous fungal-scaffolds for tissue engineering applications. Int. J. Biol. Macromol. **149**, 724–731 (2020)
42. Kovacs, G.T.A.: Electronic sensors with living cellular components. Proc. IEEE **91**(6), 915–929 (2003)
43. Wu, C., Lillehoj, P.B., Wang, P.: Bioanalytical and chemical sensors using living taste, olfactory, and neural cells and tissues: a short review. Analyst **140**(21), 7048–7061 (2015)
44. Minzan, K., Shimizu, M., Miyasaka, K., Ogura, T., Nakai, J., Ohkura, M., Hosoda, K.: Toward living tactile sensors. In: Conference on Biomimetic and Biohybrid Systems, pp. 409–411. Springer (2013)

Reactive Fungal Wearable

Andrew Adamatzky, Anna Nikolaidou, Antoni Gandia, Alessandro Chiolerio, and Mohammad Mahdi Dehshibi

Abstract Smart wearables sense and process information from the user's body and environment and report results of their analysis as electrical signals. Conventional electronic sensors and controllers are commonly, sometimes augmented by recent advances in soft electronics. Organic electronics and bioelectronics, especially with living substrates, offer a great opportunity to incorporate parallel sensing and information processing capabilities of natural systems into future and emerging wearables. Nowadays fungi are emerging as a promising candidate to produce sustainable textiles to be used as ecofriendly biowearables. To assess the sensing potential of fungal wearables we undertook laboratory experiments on electrical response of a hemp fabric colonised by oyster fungi *Pleurotus ostreatus* to mechanical stretching and stimulation with attractants and repellents. We have shown that it is possible to discern a nature of stimuli from the fungi electrical responses. The results paved a way towards future design of intelligent sensing patches to be used in reactive fungal wearables.

A. Adamatzky (✉) · A. Nikolaidou · A. Chiolerio · M. M. Dehshibi
Unconventional Computing Laboratory, UWE, Bristol, UK
e-mail: andrew.adamatzky@uwe.ac.uk

A. Nikolaidou
Department of Architecture, UWE, Bristol, UK

A. Gandia
Institute for Plant Molecular and Cell Biology, CSIC-UPV, Valencia, Spain

A. Chiolerio
Center for Bioinspired Soft Robotics, Istituto Italiano di Tecnologia, Genova, Italy

M. M. Dehshibi
Department of Computer Science and Engineering, Universidad Carlos III de Madrid, Leganés, Spain

© The Author(s), under exclusive license to Springer Nature Switzerland AG 2023
A. Adamatzky (ed.), *Fungal Machines*, Emergence, Complexity and Computation 47, https://doi.org/10.1007/978-3-031-38336-6_8

1 Introduction

Smart wearables are devices that extend the functionality of clothes and gadgets, they are responsive to the wearer and can act as an interface between the wearer and the environment producing a user responsive symbiotic system. The smart wearables have been developed as a result of the convergence between textiles and electronics (e-textiles). They integrate a high level of technology to provide complex functions and an easy operation and maintenance [1]. They can be divided into three subgroups: (1) passive smart wearables: able to sense the environment/user, (2) active or reactive smart wearables: able to sense the environment/user, and react performing some actions, therefore integrating an actuator, (3) advanced smart wearables: able to sense, react and adapt their behaviour to given circumstances. Sensors provide means to detect signals, actuators react upon stimuli either autonomously or after commands received from a central processing unit [2]. Textile-embedded sensing systems have been developed and commercially exploited in both the biomedical and safety communities [3]. Smart wearables have been used to record electrocardiography signals [4], electromyography signals [5], electroencephalography signals [6], temperature [7], biophotonic sensing [8], movement [9], oxygen content, salinity, moisture, or contaminants [10, 11]. Active functionalities might include power generation or storage capabilities [12], machine to human interface elements [13], radio frequency communication capabilities [14]. Wearable intelligent systems, intrinsically soft and better compliant with extension, deformation and skin stiffness have been developed since a long time [15].

The electronic wearables cannot self-grow and self-repair. This deficiency limits their application in the field of soft robotics and self-growing robots [16–19]. We can overcome these limits by incorporating living fabric in the smart wearables. One of the solutions, explored by us previously, would be to grow slime mould *P. polycephalum* on a surface of the cloths or a body of a robot [20]. The slime mould is proven to be a biosensor for the chemical, mechanical and optical stimuli [21–23]. Despite the sufficient sensorial abilities, the slime mould is rather fragile, highly dependent on environmental conditions and requires particular sources of nutrients.

Fungi could, however, make a feasible alternative to the slime mould. Fungal materials—grow substrates colonised with mycelium of filamentous fungi—are emerging to be robust, reliable and ecologically friendly replacement for conventional building materials and fabrics [24–33]. Fungi "possess almost all the senses used by humans" [34]. Fungi sense light, chemicals, gases, gravity and electric fields. Fungi show a pronounced response to changes in a substrate pH [35], mechanical stimulation [36], toxic metals [37], CO_2 [38], stress hormones [39]. Thus, wearables made of or incorporating fabric colonised by fungi might act as a large distributed sensorial network. Fungi is known to respond to chemical and physical stimuli by changing patterns of its electrical activity [40–42] and electrical properties [43]. This feature would allow to interface fungal wearables with conventional electronics. In view of their extension and interconnectivity, fungal networks represent certainly a sustainable infrastructure-forming substrate, able to wire loci separated by considerable space. Moreover, there are indications that mycelium networks not just sense the external stimuli but also process information, and that there is feasible opportunity to convert fungal responses into Boolean circuits, thus making fungal wearables parallel biological processing networks [44]. Previously conducted experiments on sensorial properties of fungi have been using experimental laboratory setups where substrates colonised by fungi have been kept in 'comfortable' conditions of closed containers with preserved humidity [45]. To assess the feasibility of a fungal wearable prototype in the real world we conducted experiments with a thin hemp-mycelium composite fabric incorporated on a t-shirt wore by a mannequin.

2 Methods

A commercial strain of the fungus *P. ostreatus* (Mogu's collection code 21–18), previously selected for its superior fitness growing on the targeted substrate, was cultured on a hemp bedding substrate in plastic boxes c. 35×20 cm^2 in darkness at ambient room temperature c. 22 °C.

A hemp substrate well colonised by the fungus was spread on rectangular fragments, c. 12×12 cm^2, of moisturised hemp fabric. When the fragments were colonised, as visualised by white mycelial growth on surface, they were used for experiments. The colonised fabric was attached to a cloth, which in turn was placed on a mannequin (Fig. 1a). The mannequin was covered by a plastic sheet to prevent a quick decrease of moisture in the fungal fabric. The fabric was sprayed with distilled water once per two days. The humidity of the fungal fabric was 70–80% (Merlin-Laser Protimeter, UK). The experiments were conducted in a room with ambient temperature 21 °C and illumination 30-150 LUX (ISO-Tech ILM 1332A).

Electrical activity of the colonised fabric was recorded using pairs of iridium-coated stainless steel sub-dermal needle electrodes (Spes Medica S.r.l., Italy), with twisted cables and ADC-24 (Pico Technology, UK) high-resolution data logger with a 24-bit A/D converter, galvanic isolation and software-selectable sample rates all contribute to a superior noise-free resolution. To keep electrodes stable we have been placing a polyurethane pad under the fabric. The pairs of electrodes were pierced through the fabric and into the polyurethane pad (Fig. 1b and c). We recorded electrical activity one sample per second, where the minimum and maximum logging times were 60.04 and 93.45 h, respectively. During the recording, the logger has been doing as many measurements as possible (typically up to 600 per second) and saving the average value. We set the acquisition voltage range to 156 mV with an offset accuracy of 9 μV at 1 Hz to maintain a gain error of 0.1%. Each electrode pair was considered independently with the noise-free resolution of 17 bits and conversion time of 60 ms. Each pair of electrodes, further called a Channel (Ch), reported a difference of the electrical potential between the electrodes. Distance between electrodes was 1–2 cm. In each trial, we recorded 5–8 electrode pairs (Ch) simultaneously. We stimulated the fungus with 96% ethanol, malt extract powder (Sigma Aldrich, UK) dissolved in

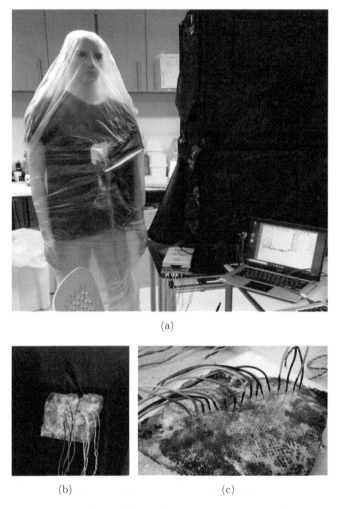

Fig. 1 Experimental setup. **a** Overall view of the experimental setup. **b** Close up of the fungal wearable incorporated into real cloth. **c** Exemplar locations of electrodes

distilled water, dextrose (Ritchie Products Ltd, UK) and by attaching weights (using foldback clips) to the hemp pads.

3 Results

A response of the fungal wearable to a chemo-attractant was studied using malt extract and dextrose. An exemplar response to application of malt extract is shown in Fig. 2a. The immediate, first 3 h, response is manifested in the spikes up to 15 mV and duration up to 140 min, and is due to a sudden increase in humidity of the substrate. Further response is attributed to fungi sensing malt extract as a chemo-attractant and a source of nutrients. The onset of the response is characterised by low frequency trains of spikes (Fig. 2b). There are 3–4 spikes, with amplitude over 2 mV, in each train. Average distance between spikes in each train is 291 s, $\sigma = 90$, average spike width is 273 s, $\sigma = 110$, average spike amplitude 2.6 mV, $\sigma = 0.15$. Average duration of a train is 31 min, $\sigma = 3$, a distance is up to one hour. Typically, a frequency of spike trains increases with time, a distance between trains decreases to 15 min in average, $\sigma = 5$ (Fig. 2c). Average amplitude of spike trains is 4.6 mV, $\sigma = 2.5$.

Results of experiments with malt extract echo in the experiments with application of dextrose (Fig. 2d). The fungi show low frequency oscillatory activity before stimulation: average distance between spikes is 197 min, $\sigma = 13.9$, average amplitude 0.3 mV, $\sigma = 0.2$. In first 5 h after the dextrose application the frequency of spikes substantially increases: average distance between spikes becomes 22 min, $\sigma = 13$ and amplitude increases to average 0.43 mV, $\sigma = 0.27$. In next 5 h average ampli-

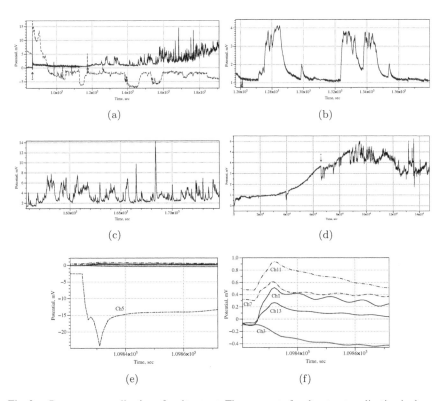

Fig. 2 a Response to application of malt extract. The moment of malt extract application is shown by asterisk. **b** Low frequency trains of spikes. **c** High frequency trains of spikes. **d** Response to application of dextrose. The moment of malt extract application is shown by asterisk. **e, f** Response to stimulation with ethanol. An overall response is shown in (**e**) with some channels zoomed in (**f**)

tude of spikes increases to 1.3 mV, $\sigma = 0.35$, and distance between spikes 20 min, $\sigma = 7$.

To assess a response to chemo-repellents we used ethanol. We applied 1 ml of 96% ethanol on the colonised fabric near loci of Ch5. The response on one of the channel close to the application loci is shown in Fig. 2e. We observed a drop by nearly 8 mV followed by further drop of the electrical potential by nearly 8 mV. The time to the peak of the response is c. 7 s. The drop in potential followed by repolarisation phase, which lasted c. 14 s. The potential remained c. 11 mV lower than that before stimulation. The response on channels further from the application loci is less pronounced, as seen in Fig. 2f. The spikes of electrical potential recorded on the channels have the following amplitudes: 0.65 mV on Ch1, 0.34 mV on Ch3, 0.31 mV on Ch7, 0.5 mV on Ch11, and 0.3 mV on Ch13, where Ch1 is the closest to the application loci and Ch13 is the farthest one.

To uncover the fungal wearable's response to stretching we attached 50 and 200 g weights to the bottom part of the fabric colonised by the fungus. A typical response to the application of 50 g weight is shown in Fig. 3a. The response duration is 97 s in average, $\sigma = 37$ s with average response amplitudes is 1.3 mV, $\sigma = 0.74$ mV. Differential electrode pairs, labelled as channels in Fig. 3a have been arranged in a line from the top to the bottom (Fig. 1), with Ch1 being closest to the top of the fabric and Ch13 to the bottom part. Most of the response spikes show action potential like depolarisation, repolarisation and hyperpolarisation phases. Ch1 and Ch13 show hyperpolarisation phases set up at higher, compared to that before stimulation, base potential.

On application of 200 g weight to lower part of the fabric, variety of response from differential electrodes pairs have been recorded. An exemplar response is shown in Fig. 3b. The response has an average duration 38 min, $\sigma = 2$ min, and average amplitude 1.56 mV, $\sigma = 1.24$ mV. The response in the example consists of two trains of high ('high' in the frameworks of fungal temporal activity) frequency spikes. Average spike width is 80 s ($\sigma = 50$), average amplitude 0.31 mV ($\sigma = 0.32$), average distance between spike in each train is 71 s ($\sigma = 47$ s).

Overall reaction to the removal of the stretching stimuli is in the drift of the base potentials on electrodes towards zero, e.g. in Fig. 3c we see that the average based potential is -1.17, $\sigma = 2$ mV, before stimulation was removed, and -0.8, $\sigma = 1.3$ mV, after the weight was removed. A typical response, to removal of 200 g, recorded on a single Ch is shown in Fig. 3d). The spike there has a duration of 9 s and amplitude 11 mV.

4 Discussion

We demonstrated that a fabric colonised by the fungus *P. ostreatus* shows distinctive sets of responses to chemical and mechanical stimulation. The response to 50 g load, Fig. 3a, is in the range of c. 1.5 min which might indicate that rather purely electro-mechanical events take place than reactions involving propagation of cal-

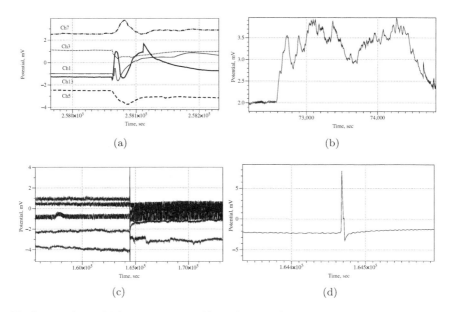

Fig. 3 Fungal wearable's response to stretching. **a** An exemplar response to stretching of the fabric by attaching 50 g weight to it. **b** Response to removal of 200 g weight recorded from five pairs of differential electrodes. **c** An exemplar response to removal of the weight recorded on one pair of differential electrodes. **d** Typical response to removal of 200 g weight, recorded on a one pair of differential electrodes

cium waves [46]. A difference d between timing of the response spikes peaks at the electrodes pairs in the line is as follows: d(Ch1, Ch3)=3.6 s, d(Ch1, Ch5)=20 s, d(Ch1, Ch7)=16 s, d(Ch1, Ch13)=3 s. This might indicate that the mycelium networks closer to the fixed end (Ch1) and the end where the load is attached (Ch13) react to the stretching first, the reaction then propagates further into the interior parts of the fabric, thus delayed reactions are recorded on the channels Ch5 and Ch7.

The response to stimulation with ethanol is in a range of 7 s. This would rather indicate physico-chemical damages to hyphae walls and corresponding electrical responses. Would amplitude of a response spike reflect a distance to the stimulation loci? As seen in Fig. 2f), on most channel the response amplitudes slightly decrease with increasing distance to the stimulation loci however more studies are required to give a certain answer. The response on the channels remote to the stimulation loci happens at the same time, as on the channel in proximity of the stimulation loci. This indicates that the response might be purely electrical (due to damage to cell walls impulse) and not due to diffusion in the fabric or volatile ethanol.

The increase of frequency of electrical potential oscillation in a response to application of chemo-attractants or nutrients is consistent with previous studies, where intracellular electrical potential of stimulated fungi was measured [40]. Even if in the case of malt extract solution increased spiking could be attributed to a water the experiments with dextrose, which was applied dry, show that the spiking shown

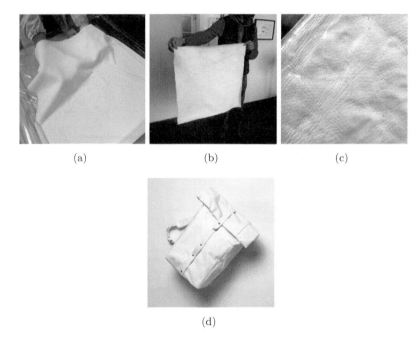

Fig. 4 Example of pure fungal flexible materials grown by Mogu S.rl. and branded as PURA Flex™. **a** Harvesting of a fungal skin, **b** size comparison with a human being, **c** texture detail resembling animal skin and **d** backpack prototype made with PURA Flex™ material

increased frequency, and often amplitude, due to reaction to a chemoattractant or nutrient. The increase in amplitude of spiking five hours after the application of malt or dextrose might be due to the fungus ingesting the nutrients and transposing them across the wide mycelial network.

In laboratory conditions the fungal wearables survive for several months being kept in high humidity conditions. For practical future applications of the fungal wearables, preserving the moisture is fundamental. For example, the fragments of fungal materials could be coated with a breathable bioplastic.

Future developments in the field of fungal wearables may be along the following directions.

First direction is a computational one. We demonstrate in computational models that a fungal colony can implement a range of Boolean function [44]. It might be possible to implement an experimental mapping between a set of stimuli and distribution of Boolean gates implemented by fungal wearables, as we demonstrated on sensing and computing organic liquid skin [47]. In other words, in a response to a particular stimuli the fungal wearable will generate a unique set of Boolean function.

Second direction is in development of a large scale fabric made purely from mycelium—fungal skin (Fig. 4) and tailoring the fabric into wearables. Such mycelial tissue can be prepared using trimitic polypore fungal cultures, which are appar-

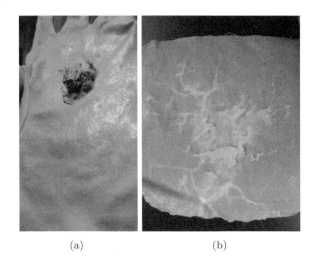

Fig. 5 **a** Part of the hemp glove colonised by fungus is visible in reflected light. **b** Stands of hyphae on the hemp fabric

ently preferred for the production of sturdy fungal skins, such as fungal leather or *mycoleather* [33]. More specifically, a fungal fabric can be prepared by pouring a homogenised slurry of a liquid culture of *Ganoderma resinaceum* into a static fermentation tray and incubated for two weeks to allow the fungal hyphae to intermesh, forming a floating mat or skin [42]. Examples of such type of fungal fabrics are shown in Fig. 4.

Third direction would be to culture fungi directly onto the pieces of clothing (Fig. 5a). This will allow us to achieve full response cloths and garments.

Fourth direction in the development of fungal wearables could be in using fungal hyphae (Fig. 5b) as wires and programmable (with e.g. light) resistor or electrically activated resistive switching devices in hybrid architectures incorporating conventional flexible electronics [48] and live fungi. Routing the direction of the fungal wires can be done by arranging sources of attractants and repellents. Isolation of fungal wires, as well as localized connections when ordered arrays like the cross-bar array arrangement are required, could be done using inorganic materials, such as metal oxides of the proper work function deposited by means of atomic layer deposition [49], or digitally printed over a large scale, also in case of uneven surfaces [50].

References

1. Stoppa, M., Chiolerio, A.: Wearable electronics and smart textiles: a critical review. Sensors **14**, 11957–11992 (2014)
2. Langereis, G.R., Bouwstra, S., Chen, W.: Sensors, actuators and computing architecture systems for smart textiles. In: Smart Textiles for Protection, 7th ed. Woodhead Publishing (2012)
3. Custodio, V., Herrera, F.J., López, G., Moreno, J.I.: A review on architectures and communications technologies for wearable health-monitoring systems. Sensors **12**, 13907–13946 (2012)

4. Coosemans, J., Hermans, B., Puers, R.: Integrating wireless ECG monitoring in textiles. Sens. Actuators A: Phys. **130**, 48–53 (2003)
5. Scalisi, R.G., Paleari, M., Favetto, A., Stoppa, M., Ariano, P., Pandolfi, P., Chiolerio, A.: Inkjet printed flexible electrodes for surface electromyography. Org. Electron. **18**, 89–94 (2015)
6. Löfhede, J., Martinez, S.F., Thordstein, M.: Soft textile electrodes for EEG monitoring. In: Proceedings of 2010 the 10th IEEE International Conference on Information Technology and Applications in Biomedicine (ITAB), pp. 1–4. IEEE (2010)
7. Sibinski, M., Jakubowska, M., Sloma, M.: Flexible temperature sensors on fibers. Sensors **10**, 7934–7946 (2010)
8. Omenetto, F., Kaplan, D., Amsden, J., Dal Negro, L.: Silk Based Biophotonic Sensors (2013)
9. Meyer, J., Lukowicz, P., Tröster, G.: Textile pressure sensor for muscle activity and motion detection. In: Proceeding of the 10th IEEE International Symposium on Wearable Computers, pp. 11–14. IEEE (2006)
10. Coyle, S., Lau, K.-T., Moyna, N., O'Gorman, D., Diamond, D., Di Francesco, F., Costanzo, D., Salvo, P., Trivella, M.G., De Rossi, D.E.: Flexible temperature sensors on fibers. IEEE Trans. Inf. Technol. Biomed. **14**, 364–370 (2010)
11. Zadeh, E.: Flexible biochemical sensor array for laboratory-on-chip applications. In: Proceeding of the International Workshop on Computer Architecture for Machine Perception and Sensing, pp. 65–66, (2006)
12. Vatansever, D., Siores, E., Hadimani, R., Shah, T.: Smart Woven Fabrics in Renewable Energy Generation. InTech (2011)
13. Baurley, S.: Interactive and experiential design in smart textile products and applications. Pers. Ubiquitous Comput. **8**, 274–281 (2004)
14. Black, S.: Trends in smart medical textiles. In: Smart Textiles for Medicine and Healthcare: Materials, Systems and Applications. University of Ghent (2007)
15. Rajan, K., Garofalo, E., Chiolerio, A.: Wearable intrinsically soft, stretchable, flexible devices for memories and computing. Sensors **18**(2), 367 (2018)
16. Mazzolai, B.: Plant-inspired growing robots. In: Soft Robotics: Trends, Applications and Challenges, pp. 57–63. Springer (2017)
17. Sadeghi, A., Mondini, A., Mazzolai, B.: Toward self-growing soft robots inspired by plant roots and based on additive manufacturing technologies. Soft Robot. **4**(3), 211–223 (2017)
18. Del Dottore, E., Sadeghi, A., Mondini, A., Mattoli, V., Mazzolai, B.: Toward growing robots: a historical evolution from cellular to plant-inspired robotics. Front. Robot. AI **5**, 16 (2018)
19. Sadeghi, A., Del Dottore, E., Mondini, A., Mazzolai, B.: Passive morphological adaptation for obstacle avoidance in a self-growing robot produced by additive manufacturing. Soft Robot. **7**(1), 85–94 (2020)
20. Schubert, T., Markert, M., Dreßler, M., Adamatzky, A.: Bodymetries. mapping the human body through amorphous intelligence. In: Experiencing the Unconventional: Science in Art, pp. 315–327. World Scientific (2015)
21. Whiting, J.G.H., de Lacy Costello, B.P.J., Adamatzky, A.: Towards slime mould chemical sensor: mapping chemical inputs onto electrical potential dynamics of *Physarum Polycephalum*. Sens. Actuators B: Chem. **191**:844–853 (2014)
22. Adamatzky, A.: Slime mould tactile sensor. Sens. Actuators B: Chem. **188**, 38–44 (2013)

23. Adamatzky, A.: Towards slime mould colour sensor: recognition of colours by Physarum polycephalum. Org. Electron. **14**(12), 3355–3361 (2013)
24. Travaglini, S., Dharan, C.K.H., Ross, P.: Manufacturing of mycology composites. In: Proceedings of the American Society for Composites: Thirty-First Technical Conference (2016)
25. Haneef, M., Ceseracciu, L., Canale, C., Bayer, I.S., Heredia-Guerrero, J.A., Athanassiou, A.: Advanced materials from fungal mycelium: fabrication and tuning of physical properties. Sci. Rep. **7**(1), 1–11 (2017)
26. Ross, P.: Method for producing fungus structures, Apr. 2018. US Patent 9,951,307
27. Appels, F.V.W., Camere, S., Montalti, M., Karana, E., Jansen, K.M.B., Dijksterhuis, J., Krijgsheld, P., Wösten, H.A.B.: Fabrication factors influencing mechanical, moisture-and water-related properties of mycelium-based composites. Mater. Des. **161**, 64–71 (2019)
28. Islam, M.R., Tudryn, G., Bucinell, R., Schadler, L., Picu, R.C.: Morphology and mechanics of fungal mycelium. Sci. Rep. **7**(1), 1–12 (2017)
29. Dahmen, J.: Soft futures: mushrooms and regenerative design. J. Archit. Edu. **71**(1), 57–64 (2017)
30. Adamatzky, A., Ayres, P., Belotti, G., Wösten, H.: Fungal architecture position paper. Int. J. Unconv. Comput. **14** (2019)
31. Chase, J., Ross, P., Wenner, N., Morris, W.: Fungal composites comprising mycelium and an embedded material, Dec. 2019. US Patent App. 16/453,791
32. Meyer, V., Basenko, E.Y., Benz, J.P., Braus, G.H., Caddick, M.X., Csukai, M., de Vries, R.P., Endy, D., Frisvad, J.C., Gunde-Cimerman, N. et al.: Growing a circular economy with fungal biotechnology: a white paper. Fungal Biol. Biotechnol. **7**, 1–23 (2020)
33. Jones, M., Gandia, A., John, S., Bismarck, A.: Leather-like Material Biofabrication using Fungi, Sept. 2020
34. Bahn, Y.-S., Xue, C., Idnurm, A., Rutherford, J.C., Heitman, J., Cardenas, M.E.: Sensing the environment: lessons from fungi. Nat. Rev. Microbiol. **5**(1), 57 (2007)
35. Van Aarle, I.M., Olsson, P.A., Söderström, B.: Arbuscular mycorrhizal fungi respond to the substrate ph of their extraradical mycelium by altered growth and root colonization. New Phytol. **155**(1), 173–182 (2002)
36. Kung, C.: A possible unifying principle for mechanosensation. Nature **436**(7051), 647 (2005)
37. Fomina, M., Ritz, K., Gadd, G.M.: Negative fungal chemotropism to toxic metals. FEMS Microbiol. Lett. **193**(2), 207–211 (2000)
38. Bahn, Y.-S., Mühlschlegel, F.A.: CO_2 sensing in fungi and beyond. Curr. Opin. Microbiol. **9**(6), 572–578 (2006)
39. Howitz, K.T., Sinclair, D.A.: Xenohormesis: sensing the chemical cues of other species. Cell **133**(3), 387–391 (2008)
40. Olsson, S., Hansson, B.S.: Action potential-like activity found in fungal mycelia is sensitive to stimulation. Naturwissenschaften **82**(1), 30–31 (1995)
41. Adamatzky, A.: On spiking behaviour of oyster fungi pleurotus djamor. Sci. Rep. **8**(1), 1–7 (2018)
42. Adamatzky, A., Gandia, A., Chiolerio, A.: Fungal sensing skin (2020). arXiv:2008.09814
43. Beasley, A.E., Powell, A.L., Adamatzky, A.: Fungal photosensors (2020). arXiv:2003.07825
44. Adamatzky, A., Tegelaar, M., Wosten, H.A.B., Powell, A.L., Beasley, A.E., Mayne, R.: On boolean gates in fungal colony. Biosystems **193**, 104138 (2020)
45. Dehshibi, M.M., Adamatzky, A.: Supplementary material for "Electrical activity of fungi: spikes detection and complexity analysis", Aug. 2020. https://doi.org/10.5281/zenodo.3997031. Accessed 24 Aug. 2020
46. Tuteja, N., Mahajan, S.: Calcium signaling network in plants: an overview. Plant Signal. Behav. **2**(2), 79–85 (2007)
47. Chiolerio, A., Adamatzky, A.: Tactile sensing and computing on a random network of conducting fluid channels. Flex. Print. Electron. (2020)
48. Rajan, K., Bocchini, S., Chiappone, A., Roppolo, I., Perrone, D., Castellino, M., Bejtka, K., Lorusso, M., Ricciardi, C., Pirri, C.F., Chiolerio, A.: Worm and bipolar inkjet printed resistive switching devices based on silver nanocomposites. Flex. Print. Electron. **2**, 024002 (2017)

49. Porro, S., Jasmin, A., Bejtka, K., Conti, D., Perrone, D., Guastella, S., Pirri, C.F., Chiolerio, A., Ricciardi, C.: Low-temperature atomic layer deposition of TiO_2 thin layers for the processing of memristive devices. J. Vac. Sci. Technol. A: Vac. Surf. Films **34**, 01A147 (2016)
50. Bevione, M., Chiolerio, A.: Benchmarking of inkjet printing methods for combined throughput and performance. Adv. Eng. Mater. (2021)

On Stimulating Fungi *Pleurotus Ostreatus* with Hydrocortisone

Mohammad Mahdi Dehshibi, Alessandro Chiolerio, Anna Nikolaidou, Richard Mayne, Antoni Gandia, Mona Ashtari-Majlan, and Andrew Adamatzky

Abstract Fungi cells can sense extracellular signals via reception, transduction, and response mechanisms, allowing them to communicate with their host and adapt to their environment. They feature effective regulatory protein expressions that enhance and regulate their response and adaptation to various triggers such as stress, hormones, physical stimuli as light, and host factors. In our recent studies, we have shown that *Pleurotus* oyster fungi generate electrical potential impulses in the form of spike events in response to their exposure to environmental, mechanical and chemical triggers, suggesting that the nature of stimuli may be deduced from the fungal electrical responses. In this study, we explored the communication protocols of fungi as reporters of human chemical secretions such as hormones, addressing whether fungi can sense human signals. We exposed *Pleurotus* oyster fungi to hydrocortisone, which was directly applied to the surface of a fungal-colonised hemp shavings substrate, and recorded the electrical activity of the fungi. Hydrocortisone is a medicinal hormone replacement that is similar to the natural stress hormone cortisol. Changes in cortisol levels released by the body indicate the presence of disease and can have a detrimental effect on physiological process regulation. The response of fungi to hydrocortisone was also explored further using X-ray to reveal changes in

---

M. M. Dehshibi
Department of Computer Science and Engineering, Universidad Carlos III de Madrid, Leganés, Spain

A. Chiolerio
Center for Sustainable Future Technologies, Istituto Italiano di Tecnologia, Torino, Italy

A. Chiolerio · A. Nikolaidou · R. Mayne · A. Adamatzky (✉)
Unconventional Computing Laboratory, UWE, Bristol, UK
e-mail: andrew.adamatzky@uwe.ac.uk

A. Nikolaidou
Department of Architecture, UWE, Bristol, UK

A. Gandia
Institute for Plant Molecular and Cell Biology, CSIC-UPV, Valencia, Spain

M. Ashtari-Majlan
Department of Computer Science, Multimedia, and Telecommunications, Universitat Oberta de Catalunya, Barcelona, Spain

© The Author(s), under exclusive license to Springer Nature Switzerland AG 2023
A. Adamatzky (ed.), *Fungal Machines*, Emergence, Complexity and Computation 47, https://doi.org/10.1007/978-3-031-38336-6_9

the fungi tissue, where receiving hydrocortisone by the substrate can inhibit the flow of calcium and, as a result, reduce its physiological changes. This research could open the way for future studies on adaptive fungal wearables capable of detecting human physiological states and biosensors built of living fungi.

1 Introduction

All living organisms have evolved elaborate communication processes and mechanisms to sense, respond, and adapt to the surrounding environment in order to survive.[1] These processes take place through reception, transduction, and response systems, which enable them to sense and adapt to their surroundings in response to a variety of cues such as nutrients, light, gases, stress, and host factors. Any form of communication requires the existence of three essential elements: a sender, a message, and a receiver. The process begins with a sender releasing a message and ends with a receiver understanding the message [1]. Fungi are composed of eukaryotic cells that report, react, and adapt to external stimuli primarily through signal transduction pathways [2]. They have extracellular and intracellular sensing mechanisms, as well as protein receptors that enable them to detect and respond to a variety of signals. *Pleurotus ostreatus*, a basidiomycete fungi, has effective regulatory protein expression that enhances its adaptation to stress triggers [3].

In our previous studies, we reported that the oyster fungi *Pleurotus djamor* exhibit trains of electrical potential spikes similar to action potential spikes [4–7]. Our initial assumption was that spike trains might reflect increasing mycelium propagation in the substrate, nutrient and metabolite transport, and communication processes within the mycelium network. We investigated the information-theoretic complexity of fungal electrical activity [8–10] to pave the way for additional investigation into sensorial fusion and fungi decision making [11–13]. Later, in a series of laboratory experiments [14, 15], we demonstrated that fungal electrical activity patterns, specifically mycelium bound hemp composite, changes in response to stimuli such as light, mechanical stretching, and attractants and repellents. Our findings demonstrated that fungi are a promising candidate for producing sustainable textiles for use as eco-friendly bio-wearables.

We present an illustrative scoping study in which we investigate the short and long-term dynamics in mycelium of the oyster fungi *Pleurotus ostreatus* in response to stimulation with hydrocortisone. The purpose of this study is to enthuse the scientific community to address the issue of fungi being able to sense animal hormones. The human body's adrenal glands release hormones such as cortisol and adrenaline. Cortisol levels in various bodily fluids can range from 4 to 70 pM [16, 17]. Sweat cortisol levels have a strong correlation with salivary cortisol concentrations [18], and the optimal cortisol level ranges from 0.02 to 0.5 M [16, 17]. Monitoring cortisol

---

[1] The authors would like to thank Vasilis Mitsios and Stamatis Varvanikolakis for coordinating the scans at Bioiatriki, as well as Judith Gómez Cuyàs for her constructive suggestion.

levels in bodily fluids, which can be altered by chronic stress and disease, is critical for maintaining healthy physiological conditions. For this study, not only are the electrical activities investigated, but the substrate is also irradiated with the X-ray beam from multiple angles to produce cross-sectional images of the substrate [19]. This multimodal approach enables us to identify and track the dynamics of changes in the tissue of the mycelium anatomy.

We demonstrated that fungi's electrical responses and reconstructed computed tomography images can be used to detect the presence of stimuli. The findings could lead to the development of biosensing patches for use in organic electronics and bio-electronics, especially with living substrates, which offer a great opportunity to integrate natural systems' parallel sensing and information processing capabilities into future and emerging wearables.

2 Methods

2.1 Experimental Setup

A commercial strain of the fungus *Pleurotus ostreatus* (Mogu's collection code 21–18), preselected for showing a superior fitness growing on different lignocellulosic substrates, was cultured on sterilised hemp shives contained in plastic boxes c. 35×20 cm^2 in darkness at ambient room temperature c. 22 °C. Particles of substrate well colonised by the fungus were spread on rectangular fragments, c. 12×12 cm^2, of moisturised nonwoven hemp fibre mats. When the mats were properly colonised, as visually confirmed by white and homogeneous mycelial growth on the surface, these were used in the experiments. The humidity of the hemp mats ranged from 70 to 80% (MerlinLaser Protimeter, UK). The experiments were carried out in a room with an ambient temperature of c. 21 °C in the absence of light. It is worth mentioning that it takes approximately 25 days for the mats to be properly colonised, depending on the mat size, the ambient room temperature, and the lignocellulosic substrates chosen for the experiments. Figure 1 shows examples of the experimental setups.

Electrical activity of the colonised hemp mats was recorded using pairs of iridium-coated stainless steel sub-dermal needle electrodes (Spes Medica S.r.l., Italy) with twisted cables, via a high-resolution data logger with a 24-bit A/D converter, galvanic isolation, and software-selectable sample rates (Pico Technology, UK). The electrodes were placed in a straight line with a distance of 1–2 cm. To keep the electrodes stable, we put a polyurethane pad underneath the fabric. As a result, the electrode pairs were inserted through the fabric and onto the polyurethane pad, as shown in Fig. 1a.

In each trial, we recorded the electrical activity of seven electrode pairs simultaneously. Each pair of electrodes, referred to as a Channel (Ch), reported a difference in the electrical potential between the electrodes. The electrical activity was recorded

(a) (b)

Fig. 1 Experimental setup. **a** Exemplar locations of electrodes. **b** Electrode pairs and logging setup in the tent

at a rate of one sample per second (1 Hz), with logging times ranging from 60.04 to 93.45 h for the different experiments. Throughout the recording, the logger took as many samples as it could (typically up to 600 per second) and saved the average value. We set the acquisition voltage range to 156 mV with an offset accuracy of 9 µV at 1 Hz to maintain a gain error of 0.1%. Each electrode pair was considered independently with a noise-free resolution of 17 bits and conversion time of 60 ms. We measured and logged a range of resistances 1–1.6 kΩ using Fluke 8846A precision multi-meter, where the test current being 1 ± 0.0013 µA, once per 10 s, 5×10^4 samples per trial [7]. Furthermore, the electrical resistance of the electrodes is always less than 10 Ω, making it minor in comparison to the resistance of the mycelium. Even employing pure gold electrodes would have no effect on the trials. Figure 1b shows one of the recording setup inside a light-proof growing tent.

For stimulation, we used hydrocortisone (Solu-Cortef trademark, 4 mL Act-O-Vial, Pfizer, Athens, Greece). We then applied 2 mL of the resulting solution to the surface of the colonised substrate in the loci surrounding Ch4 and Ch5.

The following was the rationale behind the dosage selection: human patients weighing 80 kg are typically given 20 mg of synthetic cortisol per day for a variety of diseases, such as maintenance therapy for patients with hypopituitarism. As a result, 20 mg divided by 80 kg yields 0.25 mg/kg. Hemp mats colonised with fungi weighed around 100–200 g each, so a 2 mL dose of 250 µg was needed.

2.2 Electrical Activity Analysis

Extracellular signals which surpassed a certain amplitude threshold with depolarisation, repolarisation, and refractory cycles are referred to as spike events (see Fig. 2). Spike events represent the physiological and morphological processes of mycelium in the colonised hemp mat. We proposed a novel method for identifying spiking events, including three main stages as (1) splitting signal into chunks, (2) smoothing the chunk by mapping the constant amplitude to an instantaneous amplitude, and (3) detecting spike events.

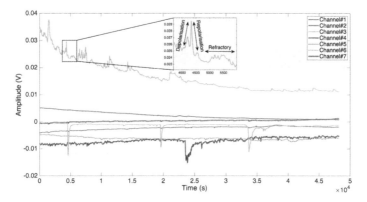

Fig. 2 An example of electrical activity observed in seven channels of colonised hemp mats with fungi over 13.3 h. The inserts are zoomed in on one channel to display a spike event with 'depolarisation', 'repolarisation', and 'refractory' cycles. This spike has a duration of 2258 s and a refractory time of 1426 s. The rates of depolarisation and repolarisation are 54.85 and 45.75 μV/s, respectively

Assume $\mathcal{X} = \{(t_i, x_i)\}_{i=1}^{\mathbb{C}}$ is a recording set of $\mathbb{C}$ channels with the entire length of T seconds and samplings rate of f_s Hz, where x defines the signal's sample value at time t, $1 \leq t \leq T$. Our objective is to detect the set of spike events $\mathcal{S} = \{s_1, s_2, \ldots, s_\eta\}$, where $\eta << T$. We segmented the signal $\mathcal{X}$ using the idea of the variable size sliding window to analyse its dynamics before and after hydrocortisone application in the same intervals. Each segment was then subdivided into chunks of $m = \{1, 2, 4, 8, 16\}$ h. Note that the chunk $m = 16$ in all 16 experiments does not have the same length that resulted in 12–16 h of electrical activity recording.

The electrical activity of the colonised hemp mats with fungi displays diffraction patterns. The presence of these diffraction patterns from multiple slits can cause determining spike events to be distorted. The envelope of an oscillating signal can expand the concept of constant amplitude to instantaneous amplitude and, therefore, bypass multiple slits with a single slit diffraction pattern to outline significant extremes, i.e., spike events [10].

To obtain the signal envelope (ξ), we used the discrete Fourier Transform, as implemented in the Hilbert transform, to detect the analytical signal. Then, inspired by Marple et al. [20], we set the negative frequency in half of each spectral period to zero, resulting in a periodic one-sided spectrum. More formally, using the sampling theorem [21], we convert the input chunk[2], $\mathcal{X}$, into a sequence of values with the sample period of $\tau \triangleq w\frac{T}{f_s}$ (see Eq. 1).

$$L[n] \triangleq \tau \cdot x(n\tau), \ 0 \leq n \leq N - 1 \qquad (1)$$

where N is an even number corresponding to the number of discrete-time analytical signal points. To obtain the N-point one-sided discrete-time analytic signal using

---

[2] Note that here we intentionally drop the m superscript to simplify mathematical notations.

Hilbert transform [20], we need to calculate the discrete-time Fourier transform of $L[n]$, with sampling at τ intervals to prevent aliasing (see Eq. 2). We take the second numerical signal derivation ($L = \frac{\partial^2 \mathcal{X}}{4 \partial t^2}$) to highlight effective signal peaks and neutralise diffraction patterns.

$$F(\omega) = \tau \sum_{n=0}^{N-1} L[n] e^{-i2\pi\omega\tau n} \qquad (2)$$

where $|\omega| \leq \frac{1}{2\tau}$ Hz. To obtain a periodic one-sided spectrum ($Z[k]$), we set the negative frequency in half of each spectral period to zero and calculate the spectrum using Eq. 3.

$$Z[k] = \begin{cases} F[0], & \text{for } k = 0 \\ 2F[k], & \text{for } 1 \leq k \leq \frac{N}{2} - 1 \\ F[\frac{N}{2}], & \text{for } k = \frac{N}{2} \\ 0, & \text{for } \frac{N}{2} + 1 \leq k \leq N - 1. \end{cases} \qquad (3)$$

To obtain the envelope of the original signal $x(t)$, we need to calculate the inverse discrete-time Fourier transform of $F(\omega)$ and $Z[k]$ (see Eq. 4).

$$x_a[n] \triangleq \mathcal{F}^{-1}[F(\omega)], \qquad z_a[n] \triangleq \mathcal{F}^{-1}[Z[k]]. \qquad (4)$$

where $\mathcal{F}^{-1}[\cdot]$ takes the inverse of the Fourier transform, x_a and z_a are the analytical signals, and $z_a[n] = \frac{1}{N\tau} \sum_{k=0}^{N-1} Z[k] e^{\frac{i2\pi kn}{N}}$. By taking the root-mean-square value of the analytical signals as in Eq. 5, we can calculate ξ for the signal $\mathcal{X}$.

$$\begin{aligned} e_a[n] &= x_a[n] + j z_a[n], \\ \xi[n] &= \sqrt{e_a[n] \times \bar{e}_a[n]}. \end{aligned} \qquad (5)$$

where $\bar{e}_a$ is the the complex conjugate of e_a, and j refers to the imaginary part of the analytical signal. We construct an intermediate representation ($\tilde{x}$) of the input signal using the upper and lower envelopes to identify spike events. All amplitude values greater than or equal to the upper envelope and less than or equal to the lower envelope are replaced with the upper and lower envelope values, respectively, while all other amplitude values are preserved. Then, as in Eq. 6, we calculate the absolute differences between the input signal (x) and this intermediate representation ($\tilde{x}$).

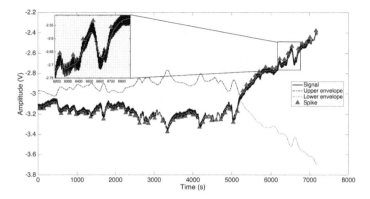

Fig. 3 Construct the intermediate representation of an input signal using Eq. 6, where the upper and lower envelopes are shown in violet and green dash lines, respectively. We zoomed in on one chunk electrical activity to highlight the identified spike events

$$\tilde{x}[n] = \begin{cases} \xi_l[n], & x[n] \leq \xi_l[n] \\ \xi_u[n], & x[n] \geq \xi_u[n] \\ x[n], & \text{otherwise.} \end{cases}$$

$$\Phi[n] = |x[n] - \tilde{x}[n]|. \tag{6}$$

where ξ_l and ξ_u are the lower and upper envelopes, respectively. We locate all local maxima (peaks) of the $\Phi[n]$ where the minimum peak prominence is γ. Here, γ is the 99% of confidence interval calculating using Eq. 7.

$$\gamma = \bar{\Phi}[n] + z^* \frac{\sigma}{\sqrt{N}}. \tag{7}$$

where $z^* = 2.576$ [22], and $\bar{\Phi}[n]$ and σ are the mean and standard deviation of $\Phi[n]$. The proposed spike detection algorithm has true-positive and false-positive rates of 76% and 16%, respectively, implying that we missed 20% of the spike events. However, because this limitation applies to all trials, we report the amortised complexity analysis. Figure 3 shows an example of the proposed method's results.

2.3 CT Images Analysis

To gain a better understanding of the effect of stimuli on the fungal substrate, we prepared two containers. Two hours before being irradiated with an X-Ray beam using the dual-energy Discovery City 750 HD CT scanner, one container received no stimuli, while the other received hydrocortisone. The detector provides for data acquisition with 64 detector rows of 0.625 mm collimation. Together with a z-flying focal spot, this allows simultaneous acquisition of data in 64×0.625 slices. The tubes were operated at 120 kVp. The gantry rotation time of the system is 0.35 s. CT images were reconstructed using Adaptive Statistical Iterative Reconstruction [23]. The image sets were reconstructed at a slice thickness of 1.25 mm, performing 128 slices per rotation (see Fig. 4a). We used two different containers since the fungal colony retains its integrity through the flow of cytoplasm in the mycelium network, where calcium waves [24] and associated electrical potential waves change the propagation coordinate of this flow.

Therefore, we were able to analyse the stimuli spread across the flow of cytoplasm by comparing the CT image of the container that had not received any stimulus with the container that had only one segment exposed (see Fig. 4b, c). When working with images in the spatial domain, we deal with changes in pixel values with respect to the scene. In the frequency domain, however, we are interested in the rate at which the pixel values in the spatial domain change, which provides us with a better understanding of the underlying distribution of changes and complexities. The discrete cosine transform (DCT) was used in this study (see Eq. 8) to transform the input CT image $I_{R \times C}$ from the spatial domain to the frequency domain, allowing us to divide the image into spectral sub-bands of varying significance. This transform can also concentrate the majority of the image's visually important details in just a few DCT coefficients. This property enables us to uncover changes in the substrate that are not apparent throughout visual inspection.

$$DCT_{pq} = \alpha_p \alpha_q \sum_{r=0}^{R-1} \sum_{c=0}^{C-1} I_{rc} \cos \frac{\pi (2r+1) p}{2R} \cos \frac{\pi (2c+1) q}{2C},$$

$$\alpha_p = \begin{cases} 1/\sqrt{R} & p = 0 \\ \sqrt{2/R} & 1 \leq p \leq R-1 \end{cases} \quad \alpha_q = \begin{cases} 1/\sqrt{C} & q = 0 \\ \sqrt{2/C} & 1 \leq q \leq C-1 \end{cases}. \quad (8)$$

where DCT_{pq} is the DCT coefficient in row p and column q of the DCT matrix, $I_{R \times C}$, is the intensity of the pixel in row R and column C. Figure 5 shows three examples of the results obtained by applying DCT to CT images.

2.4 Statistical and Information Theory Metrics

In statistics [25] and information theory [26], the concept of entropy is essential. In statistics, entropy refers to the inference functional for an updating process, and in information theory, it specifies the shortest attainable encoding scheme. Recent advances in complex dynamical systems, on the other hand, have necessitated an extension of the entropy theory beyond the conventional Shannon-Gibbs entropy (H) [27]. In this study, we used the Rényi (R_q) [28] and Tsallis (Tq) [29] additive entropy concepts, which are generalisations of the classical Shannon entropy. Regard-

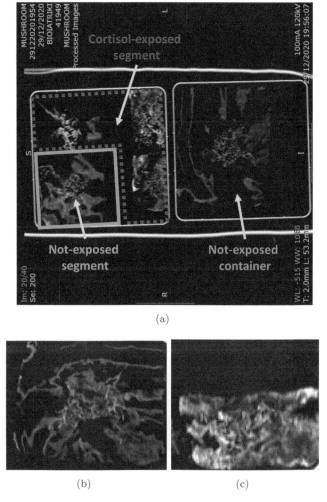

Fig. 4 **a** A CT image of two containers, with the substrate in the left container receiving hydrocortisone two hours before being irradiated with the X-ray beam, and the substrate in the right container receiving no stimulus. **b** The substrate segment in the left container that is not directly exposed to stimuli. **c** The cortisol-exposed segment of the substrate in the left container

less of the generalisation, these two entropy measurements are used in conjunction with the Principle of maximum entropy, with entropy's main application being in statistical estimation theory. Shannon, Tsallis, and Rényi entropy measurements are expressed as in Eq. 9.

$$H(\mathcal{S}) = -\sum_i p(s_i) \log p(s_i),$$

$$T_q(p_i) = \frac{k}{q-1}\left(1 - \sum_i p_i^q\right), \quad R_q(\mathcal{S}) = \frac{1}{1-q}\log\left(\sum_i p_i^q\right). \quad (9)$$

Here, $\mathcal{S}$ is a discrete random variable that represents spike events, with potential outcomes in the set $\mathcal{S} = \{s_1, s_2, \ldots, s_\eta\}$ and corresponding probabilities $p_i \doteq \Pr(\mathcal{S} = s_i)$ for $i = 1, \ldots, \eta$. p_i is a discrete set of probabilities with the condition $\sum_i p_i = 1$, and q is the *entropic-index* or Rényi entropy order with $q \geq 0$ and $q \neq 1$, which in our experiments was set to 2.

We take the logarithm to be in base 2, since we interpreted the entire recording duration T with bits, with '1s' indicating the availability of spike events and '0s' otherwise. To determine spike diversity across all channels, we represent recordings by a binary matrix with a row for each channel. We calculated the Lempel–Ziv complexity (LZc) over channels using the Kolmogorov complexity algorithm [30] to capture both temporal and spatial diversity. We then concatenated the rows of this binary matrix to form a single binary string and normalised LZc by dividing the raw value by the randomly shuffled value obtained for the same binary input sequence to obtain PCIpK. Since the value of PCIpK for a fixed-length binary sequence is

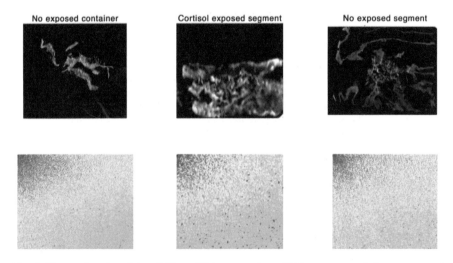

Fig. 5 The results of applying DCT to CT images, where DCT expresses a finite sequence of data points as a sum of cosine functions oscillating at different frequencies. **a** On a non-exposed container, higher energies (red hue) are concentrated in the top left corner, while lower energies (green and blue hues) are concentrated in the bottom right corner. **b** The cortisol-exposed segment has a lower concentration of high energy in the top left corner and a higher dispersion of low energy (blue hue). **c** The no-exposed segment has a moderate energy distribution than the others, where the higher-energy spectral sub-bands are scattered rather than being concentrated in a few DCT coefficients

maximum when it is totally random, the normalised values reflect the degree of signal diversity on a scale of 0 to 1.

Other metrics that were used to quantify the complexity of spike events are Simpson's diversity, Space-filling, and Expressiveness. Simpson's diversity is calculated as $\Gamma = \sum_i (p(s_i))^2$. For $H < 3$, it correlates linearly with Shannon entropy, and the relationship becomes logarithmic for higher values of H. The value of Γ varies from 0 to 1, with 1 representing infinite diversity and 0 representing no diversity. Space-filling (Δ) is the ratio of non-zero entries in the binary representation of $\mathcal{X}$ to the total duration of the recording T. Expressiveness (Υ) is calculated as the Shannon entropy H divided by space-filling ratio Δ, which represents the economy of diversity.

3 Results

Table 1 presents a qualitative description of electrical activity for one of the trials, where all complexity metrics can be compared for the condition in which the substrate was exposed 16 h before hydrocortisone exposure and 1 h after hydrocortisone exposure.

Figure 6 graphically summarises the qualitative description of electrical activity for all trials in a multi-curve layout that can be used to easily identify those metrics that are more influenced by stimulation with hydrocortisone and can thus be used to track the impacts in electrical dynamics complexity. The different panels implicitly represent time, where Fig. 6a refers to 16 h before exposure, Fig. 6b refers to the exposure or trigger event, and Fig. 6c refers to 1 h after exposure. The x-axis, where the channels are aligned, represents space. Those complexity metrics with no variation appear flat (for example after trigger the Shannon entropy for the signal, shown in red in the plots). Those influenced show positive or negative peaks, either to the left (Shannon entropy for the spike event, Simpson's diversity, Space-filling, Expressiveness, Kolmogorov Complexity, and PCIpK) or towards the right of the sample (Tsallis and Rényi entropies). By superimposing complexity metrics on each channel and reconstructing the time evolution of the spiking dynamics the mycelium, we can create a graphical representation of the excitatory or inhibitory state before, during, and after hydrocortisone exposure, demonstrating the formation of an electrical activity fingerprint that corresponds to that specific event (see Fig. 6e).

We found that the PCIpK, Tsallis, and Rényi entropies are the most relevant metrics for system analysis among all complexity measurements considered here (see Fig. 7). The PCIpK measure provides an easy inspection of the substrate's evolution analysis, including its spatio-temporal features. Figure 7a shows that a hydrocortisone stimulus induces a much stronger response in the application locus. Following a stimulus, some excitation events can be seen propagating in the substrate at much lower potentials. Tsallis entropy, as shown in Fig. 7b, is less sensitive but more accurate in tracking the evolution of the reference electrode signal. We can see from Fig. 7c that Rényi entropy helps us to monitor peak evolution over time.

Table 1 Qualitative description of electrical activity (1) 16h prior to hydrocortisone exposure and (2) 1 h after hydrocortisone exposure

Ch#	Spike#	Length (s)	Amplitude (V)	H(signal)	H(spike)	Γ	Δ	Υ	Kolmogorov	PCIpK	T_q	R_q
16 h prior to the trigger event												
1	455	2.91	−1.20	−3.40	242.2	0.99	0.0078	3.07×10^4	0.055	3.86×10^{-03}	−3.230	−0.62
2	449	2.76	−0.32	−3.72	239.6	0.99	0.0077	3.07×10^4	0.052	3.64×10^{-03}	−2.498	−0.29
3	457	2.72	−0.02	−3.88	243.1	0.99	0.0079	3.06×10^4	0.053	3.66×10^{-03}	−0.001	12.52
4	449	3.40	−3.58	−3.74	239.6	0.99	0.0077	3.07×10^4	0.052	3.73×10^{-03}	−46.884	−3.70
5	460	2.23	1.18	−3.43	244.5	0.99	0.0079	3.06×10^4	0.055	3.64×10^{-03}	0.002	12.27
6	446	3.19	−1.09	−3.26	238.3	0.99	0.0077	3.08×10^4	0.052	3.69×10^{-03}	−6.544	−1.49
7	464	2.73	0.26	−3.58	246.2	0.99	0.0080	3.06×10^4	0.054	3.59×10^{-03}	0.102	4.30
1 h after the trigger event												
1	29	2.38	−1.08	−5.72	20.60	0.96	0.0080	2.55×10^3	0.062	0.048	−3.68	−0.78
2	28	2.97	−0.80	−5.49	20.01	0.96	0.0077	2.57×10^3	0.062	0.047	−1.93	0.04
3	30	2.52	−0.14	−5.38	21.18	0.96	0.0083	2.54×10^3	0.065	0.047	−0.12	4.43
4	27	3.15	−6.15	−5.56	19.41	0.96	0.0075	2.58×10^3	0.065	0.051	−47.67	−3.71
5	28	1.41	−0.22	−5.73	20.01	0.96	0.0077	2.57×10^3	0.062	0.045	−0.10	4.64
6	32	2.42	−2.43	−5.70	22.34	0.96	0.0088	2.51×10^3	0.075	0.049	−8.47	−1.79
7	29	2.03	0.25	−5.51	20.60	0.96	0.0080	2.55×10^3	0.068	0.050	0.23	1.83

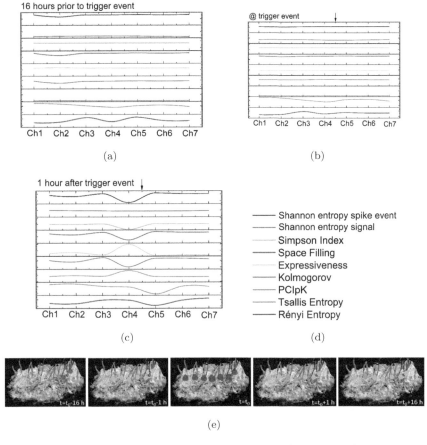

Fig. 6 a Evolution related to a data segment recorded 16 h before hydrocortisone exposure. **b** Cortisol-exposed data segment (trigger event). **c** Evolution related to a data segment recorded 1 h after hydrocortisone exposure. **d** Merit figures of complexity measures performed on recorded potentials, legend. **e** Graphical representation of the living substrate with measurement electrodes and PCIpK complexity defined by a colour code (light blue < -0.5, $-0.5 <$ green $< 0, 0 <$ orange < 0.5, red > 0.5). In the inset, the reference time of each frame is specified

3.1 Internal Inspection of the Fungal Culture

The visual appearance of the substrates colonised by the fungi did not change after exposure to hydrocortisone. However, the impact of hydrocortisone exposure is visible in the distribution of energy levels as a result of applying DCT to CT images, as shown in Fig. 8. The calculated DCT values for no-exposed container, no-exposed

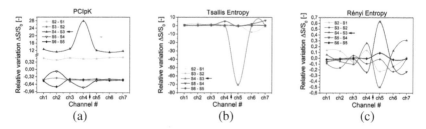

Fig. 7 **a** PCIpK is measured as a function of the measurement segments (which are time-dependent), with the arrows indicating both the spatial locus for hydrocortisone stimulation and the temporal segment corresponding to the trigger. The time interval used to compute the relative variance is denoted by $S(n+1) - S(n)$. **b** Tsallis Entropy. **c** Rényi Entropy

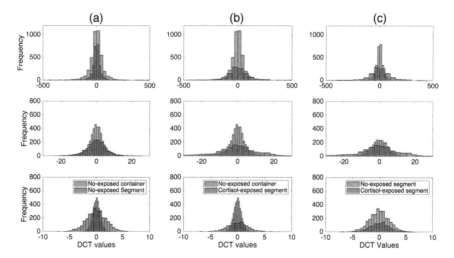

Fig. 8 The comparison of the distribution of energy levels in the calculated DCT for **a** no-exposed container and no-exposed segment, **b** no-exposed container and cortisol-exposed segment, and **c** no-exposed segment and cortisol-exposed segment. The substrate that was not exposed to hydrocortisone had a higher energy frequency than the cortisol-exposed segment and the no-exposed segment adjacent to the cortisol-exposed segment

segment, and cortisol-exposed segment were divided into three parts, including the distribution of values for high, medium, and low energies, which are shown in the first, middle, and last rows, respectively. The substrate that was not exposed to hydrocortisone had a higher energy frequency than the cortisol-exposed segment and the no-exposed segment adjacent to the cortisol-exposed segment.

4 Discussion

The integrity of the fungal colony is preserved by cytoplasm flow in the mycelium network, where calcium waves [24] change the flow's propagation coordination. Cortisol metabolism stimulates the production of receptor activator of nuclear factor-kappa-B ligand (RANKL), a type II membrane protein that regulates bone regeneration and remodelling in mammals [31]. The activity of cells responsible for calcium resorption from bone is inhibited when RANKL is stimulated [32]. The elevated circulating cortisol levels maintain stress levels, triggering physiological changes in the body's regulatory networks. Hog1 is a stress-activated mitogen-activated protein kinase (MAPK) in fungi that is homologous to the p38 MAPK pathways in mammals [33]. When exposed to stress conditions, Hog1 rapidly dephosphorylates and induces the appropriate cellular responses against the offending environmental stimuli [34]. We speculate that receiving cortisol by the substrate can inhibit the flow of calcium and, in turn, reduce its physiological changes.

5 Conclusion

We conclude that the effects of hydrocortisone exposure on a *Pleurotus* ostreatus mycelium can be demonstrated using a well-known combination of techniques in a proper interdisciplinary approach. The computed tomography shows a change in grey tones as a result of hydrocortisone exposure, which is associated with active ion transport triggered in the living mycelium in response to the hormone. Simultaneously, by capturing biopotential signals and extracting their numerical complexity, we could identify consistent metrics for tracing the stimulus and tracking the propagation of ionic waves in response to the hormone. Our research extends our understanding of living fungal, which could be used in future wearable and architectural systems to benefit living species while also maximising sustainability.

References

1. Cottier, F., Mühlschlegel, F.A.: Communication in fungi. Int. J. Microbiol. **2012**(1):1–9 (2012)
2. Alonso-Monge, R., Román, E., Arana, D.M., Pla, J., Nombela, C.: Fungi sensing environmental stress. Clin. Microbiol. Infect. **15**(1), 17–19 (2009)
3. Hou, L., Wang, L., Xiangli, W., Gao, W., Zhang, J., Huang, C.: Expression patterns of two pal genes of pleurotus ostreatus across developmental stages and under heat stress. BMC Microbiol. **19**(1), 1–16 (2019)
4. Adamatzky, A.: On spiking behaviour of oyster fungi pleurotus djamor. Sci. Rep. **8**(1), 1–7 (2018)
5. Beasley, A.E., Powell, A.L., Adamatzky, A.: Capacitive storage in mycelium substrate (2020). arXiv:2003.07816
6. Beasley, A.E., Powell, A.L., Adamatzky, A.: Fungal photosensors (2020). arXiv:2003.07825

7. Adamatzky, A., Gandia, A.: On electrical spiking of ganoderma resinaceum. Biophys. Rev. Lett. 1–9 (2021)
8. Adamatzky, A., Tuszynski, J., Pieper, J., Nicolau, D.V., Rinalndi, R., Sirakoulis, G., Erokhin, V., Schnauss, J., Smith, D.M.: Towards cytoskeleton computers. A proposal. In: Adamatzky, A., Akl, S., Sirakoulis, G. (eds.), From Parallel to Emergent Computing. CRC Group/Taylor & Francis (2019)
9. Adamatzky, A., Gandia, A., Chiolerio, A.: Fungal sensing skin (2020). arXiv:2008.09814
10. Dehshibi, M.M., Adamatzky, A.: Electrical activity of fungi: spikes detection and complexity analysis. Biosystems **203**, 104373 (2021)
11. Adamatzky, A., Ayres, P., Belotti, G., Wösten, H.: Fungal architecture position paper. Int. J. Unconv. Comput. **14** (2019)
12. Adamatzky, A., Tegelaar, M., Wosten, H.A.B., Powell, A.L., Beasley, A.E., Mayne, R.: On boolean gates in fungal colony. Biosystems **193**, 104138 (2020)
13. Goles, E., Tsompanas, M.-A., Adamatzky, A., Tegelaar, M., Wosten, H.A.B., Martínez, G.J.: Computational universality of fungal sandpile automata. Phys. Lett. A 126541 (2020)
14. Adamatzky, A., Gandia, A., Chiolerio, A.: Fungal sensing skin. Fungal Biol. Biotechnol. **8**(1), 1–6 (2021)
15. Adamatzky, A., Nikolaidou, A., Gandia, A., Chiolerio, A., Dehshibi, M.M.: Reactive fungal wearable. Biosystems **199**, 104304 (2021)
16. Jang, H.-J., Lee, T., Song, J., Russell, L., Li, H., Dailey, J., Searson, P.C., Katz, H.E.: Electronic cortisol detection using an antibody-embedded polymer coupled to a field-effect transistor. ACS Appl. Mater. Interfaces **10**(19), 16233–16237 (2018)
17. Kaushik, A., Vasudev, A., Arya, S.K., Pasha, S.K., Bhansali, S.: Recent advances in cortisol sensing technologies for point-of-care application. Biosens. Bioelectron. **53**, 499–512 (2014)
18. Russell, E., Koren, G., Rieder, M., Van Uum, S.H.M.: The detection of cortisol in human sweat: implications for measurement of cortisol in hair. Therap. Drug Monit. **36**(1), 30–34 (2014)
19. Gholami, N., Dehshibi, M.M., Adamatzky, A., Rueda-Toicen, A., Zenil, H., Fazlali, M., Masip, D.: A novel method for reconstructing ct images in gate/geant4 with application in medical imaging: a complexity analysis approach. J. Inf. Proc. **28**, 161–168 (2020)
20. Marple, L.: Computing the discrete-time "analytic" signal via FFT. IEEE Trans. Signal Process. **47**(9), 2600–2603 (1999)
21. Claude Elwood Shannon: Communication in the presence of noise. Proc. IRE **37**(1), 10–21 (1949)
22. Dekking, F.M., Kraaikamp, C., Lopuhaä, H.P., Meester, L.E.: Understanding why and how. Springer Science & Business Media, A Modern Introduction to Probability and Statistics (2005)
23. Barca, P., Giannelli, M., Fantacci, M.E., Caramella, D.: Computed tomography imaging with the adaptive statistical iterative reconstruction (ASIR) algorithm: dependence of image quality on the blending level of reconstruction. Australasian Phys. Eng. Sci. Med. **41**(2), 463–473 (2018)
24. Aramburu, J., Heitman, J., Crabtree, G.R.: Calcineurin: a central controller of signalling in eukaryotes: workshop on the calcium/calcineurin/NFAT pathway: regulation and function. EMBO Rep. 5(4):343–348 (2004)
25. Fan, J., Farmen, M., Gijbels, I.: Local maximum likelihood estimation and inference. J. R. Stat. Soc.: Ser. B (Statistical Methodology) **60**(3), 591–608 (1998)
26. Petz, D.: Quantum Information Theory and Quantum Statistics. Springer Science & Business Media (2007)
27. Shannon, C.E.: A mathematical theory of communication. Bell Syst. Tech. J. **27**(3), 379–423 (1948)
28. Rényi, A.: On measures of entropy and information. In: Proceedings of the Fourth Berkeley Symposium on Mathematical Statistics and Probability, Volume 1: Contributions to the Theory of Statistics, pp. 547–561. University of California Press (1961)
29. Tsallis, C.: Possible generalization of boltzmann-gibbs statistics. J. Stat. Phys. **52**(1), 479–487 (1988)

30. Kaspar, F., Schuster, H.G.: Easily calculable measure for the complexity of spatiotemporal patterns. Phys. Rev. A **36**(2), 842 (1987)
31. Chyun, Y.S., Kream, B.E., Raisz, L.G.: Cortisol decreases bone formation by inhibiting periosteal cell proliferation. Endocrinology **114**(2), 477–480 (1984)
32. Davies, E., Kenyon, C.J., Fraser, R.: The role of calcium ions in the mechanism of ACTH stimulation of cortisol synthesis. Steroids **45**(6), 551–560 (1985)
33. Hohmann, S.: Osmotic stress signaling and osmoadaptation in yeasts. Microbiol. Mol. Biol. Rev. **66**(2), 300–372 (2002)
34. Bahn, Y.-S., Kojima, K., Cox, G.M., Heitman, J.: Specialization of the hog pathway and its impact on differentiation and virulence of cryptococcus neoformans. Mol. Biol. Cell. **16**(5), 2285–2300 (2005)

Fungal Photosensors

Alexander E. Beasley, Michail-Antisthenis Tsompanas, and Andrew Adamatzky

Abstract The rapidly developing research field of organic analogue sensors aims to replace traditional semiconductors with naturally occurring materials. Photosensors, or photodetectors, change their electrical properties in response to the light levels they are exposed to. Organic photosensors can be functionalised to respond to specific wavelengths, from ultra-violet to red light. Performing cyclic voltammetry on fungal mycelium and fruiting bodies under different lighting conditions shows no appreciable response to changes in lighting condition. However, functionalising the specimen using PEDOT:PSS yields in a photosensor that produces large, instantaneous current spikes when the light conditions change. Future works would look at interfacing this organic photosensor with an appropriate digital back-end for interpreting and processing the response.

1 Introduction

The world's drive for sustainability dictates a need to develop 'green architectures', where buildings are made from natural, recyclable materials and incorporate living matter in their constructions [1, 2]. Recent examples include algae facades [3, 4], buildings incorporating bio-reactors [5] and buildings made from pre-fabricated blocks of substrates colonised by fungi [6–8]. Not long ago we proposed growing monolith constructions from live fungal materials, where living mycelium coexist with dry mycelium functionalised with nanoparticles and polymers [9]. In such monolith constructions, fungi could act as optical, tactile and chemical sensors, fuse and process information and make decisions [10].

Functionalisation of living substrates aimed at increasing their sensitivity or conductivity, or imbuing them with novel properties, has been achieved before. Exam-

A. E. Beasley
ARM Ltd, 110 Fulbourn Rd, Cambridge, CB1 9NJ, UK

M.-A. Tsompanas · A. Adamatzky (✉)
Unconventional Computing Laboratory, Department of Computer Science and Creative Technologies, UWE, Bristol, UK
e-mail: andrew.adamatzky@uwe.ac.uk

ples include: tuning electrical properties of plants and slime mould with nanoparticles [11, 12], increasing photosynthetic properties of plants with nanoparticles [13, 14] and hybridizing slime mould with conductive polymers [15, 16]. An important advantage of functionalising living substrates with particles, and/or polymers, is that the substrate will remain functional, as an inanimate electronic device, even when no longer living. Here, in an attempt to advance our ideas of living fungal architectures, sensors and computers, we explore the photosensitive properties of the grey oyster mushroom with aim of designing and prototyping novel organic electronic devices.

The field of organic analogue electronics replaces traditional semi-conductors with organic material. This can be seen in the development of organic thin film transistors [17–19], organic LEDs [20, 21], and sensors, such as photosensors [22–27]. Organic photosensors are a potentially disruptive technology that lend themselves well to being tuned for specific band absorption [28]. Additionally, organic electronics have the potential to self heal as the substrate is a living material [29]. Functionalising organic compounds with polymers can create hybrid components that utilise specific properties from the various individual compounds. For example, increasing the capacitance of a multi-layer capacitor using photosensitivity [30]. Fungi have the capacity to readily transport polymers which can be used to functionalise the specimen. Given these properties, fungi may be a useful substrate for the development of organic photosensors and a digital back-end can be used to interpret and process the signal. The chapter explores the use of the fruiting bodies of the grey oyster fungi as photosensors and present the implications of such an application when used in organic electronic systems.

2 Experimental Method

Samples of mycelium and fruiting bodies of the grey oyster fungi *Pleurotus ostreatus* (Ann Miller's Speciality Mushrooms Ltd, UK), cultivated on damp wood shavings, are explored for their photosensitive properties. Iridium-coated stainless steel subdermal needles with twisted cables (Spes Medica SRL, Italy) were inserted approximately 10 mm apart in samples An LCR meter (BK Precision, model 891) was used to constantly record the DC resistance of the sample. The sample was periodically covered or exposed to intense light (c. 1500 Lux, with a cold light source PL2000, Photonic Optics, USA). Exposure to the high intensity light was in the order of tens of seconds. The container with PEDOT:PSS functionalised fungi was kept in a fume hood for the duration of the experiment.

The I-V characteristics of the fungal substrate were measured using a Keithely source measure unit (SMU) 2450 (Keithley Instruments, USA). Cyclic voltammetry was conducted as using a voltage steps of 0.01 V between −1 and 1 V. The delay between setting each voltage increment was 0.1 s. Cyclic voltammetry was performed under both constant and changing light conditions. Similar tests were performed with a pure resistive load as a control.

In the experiments on functionalised fruit bodies, 10 ml of PEDOT:PSS (poly(3,4-ethylenedioxythiophene) polystyrene sulfonate) (Sigma Aldrich, UK) solution was injected in the stalks the fruit bodies. The fungi were rested for 24 h to allow for translocation of them material towards other parts of the stem and caps of the bodies.

3 Results

Two individual cultures of mycelium and fruiting bodies of the grey oyster mushroom (sample 1 and sample 2) (before functionalisation) were subject to cyclic voltammetry under both constant and changing lighting conditions—Figs. 1 and 2. The cyclic voltammetry gives a mem-fractive response where elements of memristive and mem-capacitive properties are seen [31]. Combining the cyclic voltammetry with instantaneously changing light conditions (runs 3–5 in Fig. 1) showed no notable change when compared to constant light conditions. Cyclic voltammetry of the mycelium specimen under instantaneously changing light conditions (runs 3–5 in Fig. 1) showed no notable change when compared to constant light conditions. Similarly, we saw the same when comparing the cyclic voltammetry of fruiting bodies in Fig. 2. Runs 1 and 2 are constant lighting conditions and run 3 is changing lighting conditions.

As samples of grey oyster fungi are functionalised with PEDOT:PSS they are also dried out. The process of drying the mushrooms reduces their natural conductivity. However the presence of the PEDOT:PSS in the dried fruit bodies . Figure 3 shows the I-V response of the functionalised fruiting body with constant light changes. Again, a mem-fractive response is achieved [31].

In fruiting bodies injected with PEDOT:PSS, rapid changes in lighting conditions produced instantaneous changes in the conducted current. Figures 4 and 5 show the I-V response for the functionalised specimen, under changing light conditions. Figure 4 has the specimen being subject to changes in ambient lab lighting during the positive phase of the cyclic voltammetry between 0 and 0.5 V. Significant spikes, eight fold

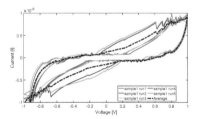

Fig. 1 Cyclic voltammetry of mycelium sample Run 1 shows the response when the sample is in the dark. Run 2 shows the response of the sample under continuous intense illumination. Runs 3–5 shows the response when the sample is subject to sudden changes in light conditions. The average of all runs is shown as a dotted line super imposed over the other traces

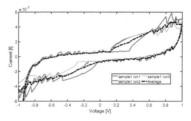

Fig. 2 Cyclic voltammetry of grey oyster mushrooms. Run 1 shows the response when the sample is in the dark. Run 2 shows the response of the sample under continuous intense illumination. Runs 3 shows the response when the sample is subject to sudden changes in light conditions. The average of all runs is shown as a dotted line super imposed over the other traces

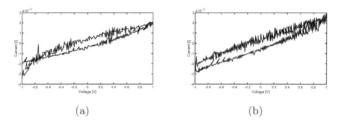

(a) (b)

Fig. 3 Cyclic voltammetry of grey oyster mushrooms functionalised with PEDOT. **a** Specimen kept in darkness. **b** Specimen illuminated with 1500 Lux light

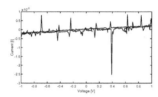

Fig. 4 The fungus functionalised with PEDOT:PSS exposed to instantaneous changes between darkness and ambient lab lighting conditions

increase in conducted current, are seen that correspond precisely with a change in light conditions.

Figure 5 shows the same specimen being subjected to more extreme changes in lighting conditions. The I-V curves continue to exhibit a number of spikes, where the spikes correspond to the point at which the lighting conditions were changed. Figure 5b has greater noise on the I-V curve making distinguishing the response harder (SNR ranges between c. 5.5 and 1.3). The injected specimen was allowed to progressively dry out as cyclic voltammetry was conducted. The drier the specimen becomes, the noisier the recordings can become.

Dried fruit bodies located further, c. 25 cm, from the injection site of the PDOT:PSS were also subjected to cyclic voltammetry. Figure 6 shows the response of the sample under the control conditions. The specimen had very poor conductive

Fungal Photosensors

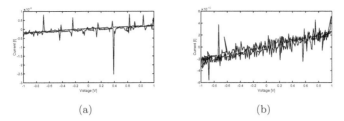

Fig. 5 Cyclic voltammetry of fungi functionalised with PEDOT:PSS. Samples are periodically exposed to intense light changes from darkness to 1500 lux

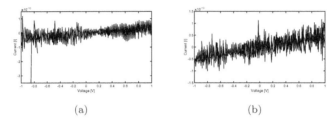

Fig. 6 Cyclic voltammetry of dried sample of grey oyster mushrooms. **a** Specimen kept in darkness. **b** Specimen illuminated with 1500 Lux light

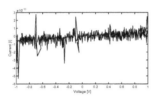

Fig. 7 Cyclic voltammetry of dried sample of grey oyster mushrooms. Samples are periodically exposed to intense light changes from darkness to 1500 lux

properties. Presumably this is due to the fact that the injection site of the PEDOT:PSS was sufficiently far from the recording site that there was, at best, only minimal uptake of the conductive polymer. However, sweeping the voltage did result in some change in conducted current, although always in the pico-Amps range.

Cyclic voltammetry was then performed on the same specimen during intense changes in lighting conditions (Fig. 7). Here, increasing and decreasing the light levels also resulted in an instantaneous spike in the conducted current. However, this specimen was not as responsive as the specimen injected directly with PEDOT:PSS.

4 Conclusions

Fungi are known for their photosensitive nature, with research showing fungi can be more photosensitive than green plants [32, 33]. Fungi are most receptive towards the UV end of the spectrum but exhibit photoresponses across the entire light spectrum. Briefly exposing fungi to light can interrupt their current growth cycle, triggering other responses. The reported photosensitivity of fungi inspired us to check if the fungi can be used as photosensors.

The cyclic voltammetry of the mycelium and fruit bodies specimens shows that exposing the test subject to instantaneously changing light conditions has no discernible impact on conducted current when compared to keeping the light conditions constant. Extreme constant lighting conditions (darkness or extreme illumination) again has little to no bearing on the I–V curves for cyclic voltammetry. Under all conditions, the specimens exhibit characteristic mem-fractance responses [34]. The lack of response to the stimulation with light might be due to the fact that the biochemical responses required to initiate electric responses might take several hours and, therefore, the length of recordings made here was not sufficient to capture such changes. This leads to us to conclude that intact fungi can be not be used as immediate-response photoswitches.

Functionalising the specimen with a conductive polymer, in this case PEDOT:PSS, and re-performing cyclic voltammetry yielded a photosensitive response when the light conditions change instantaneously. Therefore, polyfluorene and PEDOT:PSS photosenesing inks [35], n-Si/PEDOT:PSS solar cells [36], by functionalising mushrooms using the conductive polymer PEDOT:PSS, we were able to produce a photosensor. Furthermore, the spikes in the cyclic voltammetry were significant enough to be detectable by additional hardware. Future work would be to develop the digital back-end that can be used to interpret and process the spiking response of the mushrooms.

References

1. McDonough, W.: Big and Green: Toward Sustainable Architecture in the 21st Century. Princeton Architectural Press (2002)
2. Kibert, C.J.: Sustainable Construction: Green Building Design and Delivery. Wiley (2016)
3. Kim, K.-H.: Beyond green: growing algae facade. In: ARCC Conference Repository (2013)
4. Martokusumo, W., Koerniawan, M.D., Poerbo, H.W., Ardiani, N.A., Krisanti, S.H.: Algae and building façade revisited. a study of façade system for infill design. J. Arch. Urban. **41**(4), 296–304 (2017)
5. Sara Wilkinson, Paul Stoller, Peter Ralph, and Brenton Hamdorf. Feasibility of algae building technology in sydney. *Feasibility of Algae Building Technology in Sydney*, 2016
6. Philip Ross. Your rotten future will be great. *The Routledge Companion to Biology in Art and Architecture*, page 252, 2016
7. Freek VW Appels, Serena Camere, Maurizio Montalti, Elvin Karana, Kaspar MB Jansen, Jan Dijksterhuis, Pauline Krijgsheld, and Han AB Wösten. Fabrication factors influencing

mechanical, moisture-and water-related properties of mycelium-based composites. *Materials & Design*, 161:64–71, 2019
8. Dahmen, J.: Soft matter: Responsive architectural operations. Technoetic Arts **14**(1–2), 113–125 (2016)
9. Adamatzky, A., Ayres, P., Belotti, G., Wösten, H.: Fungal architecture position paper. Int. J. Unconvent. Comput. **14** (2019)
10. Adamatzky, A., Tuszynski, J., Pieper, J., Nicolau, D.V., Rinalndi, R., Sirakoulis, G., Erokhin, V., Schnauss, J., Smith, D.M.: Towards cytoskeleton computers. A proposal. In: Adamatzky, A., Akl, S., Sirakoulis, G. (eds.), From Parallel to Emergent Computing. CRC Group/Taylor & Francis (2019)
11. Gizzie, N., Mayne, R., Patton, D., Kendrick, P., Adamatzky, A.: On hybridising lettuce seedlings with nanoparticles and the resultant effects on the organisms' electrical characteristics. Biosystems **147**, 28–34 (2016)
12. Gizzie, N., Mayne, R., Yitzchaik, S., Ikbal, M., Adamatzky, A.: Living wires-effects of size and coating of gold nanoparticles in altering the electrical properties of physarum polycephalum and lettuce seedlings. Nano LIFE **6**(01), 1650001 (2016)
13. Giraldo, J.P., Landry, M.P., Faltermeier, S.M., McNicholas, T.P., Iverson, N.M., Boghossian, A.A., Reuel, N.F., Hilmer, A.J., Sen, F., Brew, J.A., et al.: Plant nanobionics approach to augment photosynthesis and biochemical sensing. Nat. Mater. **13**(4), 400–408 (2014)
14. Faizan, M., Faraz, A., Yusuf, M., Khan, S.T., Hayat, S.: Zinc oxide nanoparticle-mediated changes in photosynthetic efficiency and antioxidant system of tomato plants. Photosynthetica **56**, 678–686 (2018)
15. Berzina, T., Dimonte, A., Cifarelli, A., Erokhin, V.: Hybrid slime mould-based system for unconventional computing. Int. J. Gen. Syst. **44**(3), 341–353 (2015)
16. Dimonte, A., Battistoni, S., Erokhin, V.: Physarum in hybrid electronic devices. In: Advances in Physarum Machines, pp. 91–107. Springer (2016)
17. Tokito, S.: Flexible printed organic thin-film transistor devices and integrated circuit applications. In: 2018 International Flexible Electronics Technology Conference (IFETC), pp 1–2, Aug. 2018
18. Endoh, H., Toguchi, S., Kudo, K.: High performance vertical-type organic transistors and organic light emitting transistors. In: Polytronic 2007—6th International Conference on Polymers and Adhesives in Microelectronics and Photonics, pp. 139–142, Jan 2007
19. Tang, W., Zhao, J., Li, Q., Guo, X.: Highly sensitive low power ion-sensitive organic thin-film transistors. In: 2018 9th International Conference on Computer Aided Design for Thin-Film Transistors (CAD-TFT), pp. 1, Nov. 2018
20. Mizukami, M., Cho, S., Watanabe, K., Abiko, M., Suzuri, Y., Tokito, S., Kido, J.: Flexible organic light-emitting diode displays driven by inkjet-printed high-mobility organic thin-film transistors. IEEE Electron Dev. Lett. **39**(1), 39–42 (2018)
21. Sano, T., Suzuri, Y., Koden, M., Yuki, T., Nakada, H., Kido, J.: Organic light emitting diodes for lighting applications. In: 2019 26th International Workshop on Active-Matrix Flatpanel Displays and Devices (AM-FPD), vol. 26, pp. 1–4, July 2019
22. Marien, H., Steyaert, M., Heremans, P.: Analog Organic Electronics. Springer (2013)
23. Ocaya, R.O., Al-Sehemi, A.G., Al-Ghamdi, A., El-Tantawy, F., Yakuphanoglu, F.: Organic semiconductor photosensors. J. Alloys Compd. **702**, 520–530 (2017)
24. Dickey, S., Eliasson, B., Moddel, G.: Organic photosensors for ferroelectric liquid crystal spatial light modulators. Organic Thin Films for Photonic Applications, SaE10 (1999)
25. Hamilton, M.C., Kanicki, J.: Organic polymer thin-film transistor photosensors. IEEE J. Sel. Top. Quantum Electron. **10**(4), 840–848 (2004)
26. Manna, E., Xiao, T., Shinar, J., Shinar, R.: Organic photodetectors in analytical applications. Electronics **4**, 688–722 (2015)
27. Zeng, Z., Zhong, Z., Zhong, W., Zhang, J., Ying, L., Gang, Yu., Huang, F., Cao, Y.: High-detectivity organic photodetectors based on a thick-film photoactive layer using a conjugated polymer containing a naphtho[1,2-c:5,6-c]bis[1,2,5]thiadiazole unit. J. Mater. Chem. C **7**, 6070–6076 (2019)

28. Natali, D., Caironi, M.: Organic Photodetectors (2016)
29. Cicoira, F.: Flexible, strechable and healable electronics. In: 2018 International Flexible Electronics Technology Conference (IFETC), p. 1, Aug 2018
30. Lee, H., Kim, J., Kim, H., Kim, Y.: Strong photo-amplification effects in flexible organic capacitors with small molecular solid-state electrolyte layers sandwiched between photo-sensitive conjugated polymer nanolayers. Sci. Rep. **6**, 19527 (2016)
31. Beasley, A.E., Abdelouahab, M.-S., Lozi, R., Powell, A.L., Adamatzky, A.: On memfractance of plants and fungi (2020). arXiv:2005.10500
32. Carlile, M.J.: The Photobiology of Fungi. Ann. Rev. **16** (1965)
33. Furuya, M.: Photobiology of fungi. In: Kendrick, R.E., Kronenberg, G.H.M. (eds.) Photomorphogenesis in Plants. Springer, Dordrecht (1986)
34. Beasley, A.E., Powell, A.L., Adamatzky, A.: Memristive properties of mushrooms (2020). arXiv:2002.06413
35. Lavery, L.L., Whiting, G.L., Arias, A.C.: All ink-jet printed polyfluorene photosensor for high illuminance detection. Organ. Electron. **12**(4), 682–685 (2011)
36. Pietsch, M., Jäckle, S., Christiansen, S.: Interface investigation of planar hybrid n-si/pedot: Pss solar cells with open circuit voltages up to 645 mv and efficiencies of 12.6%. Appl. Phys. A **115**, 1109–1113 (2014)

Reactive Fungal Insoles

Anna Nikolaidou, Neil Phillips, Michail-Antisthenis Tsompanas, and Andrew Adamatzky

Abstract Mycelium bound composites are promising materials for a diverse range of applications including wearables and building elements. Their functionality surpasses some of the capabilities of traditionally passive materials, such as synthetic fibres, reconstituted cellulose fibres and natural fibres. Thereby, creating novel propositions including augmented functionality (sensory) and aesthetic (personal fashion). Biomaterials can offer multiple modal sensing capability such as mechanical loading (compressive and tensile) and moisture content. To assess the sensing potential of fungal insoles we undertook laboratory experiments on electrical response of bespoke insoles made from capillary matting colonised with oyster fungi *Pleurotus ostreatus* to compressive stress which mimics human loading when standing and walking. We have shown changes in electrical activity with compressive loading. The results advance the development of intelligent sensing insoles which are a building block towards more generic reactive fungal wearables. Using FitzhHugh-Nagumo model we numerically illustrated how excitation wave-fronts behave in a mycelium network colonising an insole and shown that it may be possible to discern pressure points from the mycelium electrical activity.

1 Introduction

In-shoe sensor technologies have been widely used in the clinical domain for disease detection, diagnostics and therapeutic use. Smart insoles can detect impairments in balance, gait, posture, muscle strength and cognition, providing valuable information about the user's physical and mental health. They most commonly integrate pressure

---

A. Nikolaidou · N. Phillips · M.-A. Tsompanas · A. Adamatzky (✉)
Unconventional Computing Laboratory, UWE, Bristol, UK
e-mail: andrew.adamatzky@uwe.ac.uk

N. Phillips
e-mail: neil.phillips@uwe.ac.uk

A. Nikolaidou
Department of Architecture, UWE, Bristol, UK

© The Author(s), under exclusive license to Springer Nature Switzerland AG 2023
A. Adamatzky (ed.), *Fungal Machines*, Emergence, Complexity and Computation 47,
https://doi.org/10.1007/978-3-031-38336-6_11

or optical sensor technologies in relevant locations for monitoring the foot-ground interaction force [1], providing complex functions and exhibiting broad sensing range when exposed to mechanical stimuli. Smart insoles can be divided into three subgroups: (1) passive smart insoles able to sense parameters including weight loading of user, local terrain topology, volatile organic compounds, (2) active or reactive smart insoles able to sense and react performing some actions, by integrating an actuator, (3) advanced smart insoles able to sense, react and tailor their behaviour to specific operating circumstances.

Integrating sensor technologies into insoles, patterns and strategies for executing different functional tasks can be assessed, capturing data that can form the basis for use in areas such as rehabilitation, prehabilitation, monitoring elderly people who have mobility problems, mitigating slipping and falling as well as assessing long-term chronic conditions such as dementia, Parkinson's disease and stroke [2]. For example, smart insoles have been used for the measurement of the weight pressure distribution that a patient exerts on each foot, in addition to the gait time, swing time, and stance time of each leg while walking to diagnose several medical conditions [3].

Smart insoles have lately found new applications outside the clinical domain. The proliferation of consumer-grade smart wearables has further propelled the development of commercial in-shoe devices to assess health and wellness-related mobility parameters in activities of daily living [4–7] e.g., pressure mapping of insoles in footwear based on conventional sensor technology (e.g. dielectric layer) have been reported [8, 9] and insoles with diagnostic capabilities are commercially available, for example NURVV Run Smart Insoles [10]. While the use of smart insoles in non-medical applications has recently attracted significant interest, there are important challenges to overcome.

Cost represents a limiting factor. The devices require a number of sensors and actuators driven by electrical circuit components, usually supplied by a battery, making them not only expensive to fabricate but also significantly contributing to the depletion of natural resources. Insoles based on conventional sensor technology have short battery life (e.g. 5 h [11]) and are available in a limited number of insole sizes/shapes (e.g. 6 sizes [12]). Moreover, the performance of smart insoles is directly linked to the number and distribution of the sensors integrated as well as the identification of optimal sensor locations that match areas subject to the highest plantar pressures during gait or standing. These locations can vary depending on the foot sizes and gait patterns of different users and can therefore alter or limit the gait or standing event recognition accuracy. Finally, high sensor responsivity which is the fundamental indicator to ensure a prompt real-time detection of specific biomechanically-relevant gait and standing events and a timely and synchronous action of the linked wearable device, multiple sensor signals as well as information related to the interaction with the external environment are necessary but not always achievable.

Biomaterials, such as mycelium bound composites, present a promising alternative to conventional smart insoles. They exhibit sensing and responsive capabilities without requiring additional space (for support infrastructure) and external inputs (e.g., electrical power sources) to operate, using its own bioelectric activity. Fungal sensors offer increased biodegradability, they are self- sustainable as they can self-grow, self-repair and self-assemble, they are abundant and offer in situ low technology cultivation. Moreover, they present low capital requirements and are easily scalable for the production of customised insole sizes.

In our previous studies, we demonstrated that living blocks of colonised by filamentous polypore fungus *Ganoderma resinaceum* substrate (MOGU's collection code 19-18, Mogu S.r.l., Inarzo, Italy), showed immediate responses in the form of spikes of electrical potential when subjected to weight application (8 kg and 16 kg), recognising the application or removal of weight [13]. We present an illustrative scoping study in which we research the response of mycelium composite insoles to pressure generated by the feet during gait or standing. The primary objective of the studies reported in this paper is to assess the performance and spiking activity of the fungal insoles when exposed to mechanical stimuli and in particular, weight shifting (shifting of weight from toe to heel). We present a prototype of pressure-sensitive fungal insoles, aiming to open up opportunities for further research and discussions on the novel field of responsive smart insoles from living material, with the objective of enabling real-time applications and providing a sustainable, cost efficient and accurate tool for posture, gait and activity recognition events.

2 Methods

A method of forming insoles colonised with fungi from capillary matting (rather than hemp) was developed through multiple iterations. 200 g slab of mixed (mostly Rye grain) seed substrate well colonised with *Pleurotus ostreatus* (Ann Miller's Speciality Mushrooms, UK, https://www.annforfungi.co.uk/shop/oyster-grain-spawn/) was placed in the bottom of clean plastic container (5l, 280 × 145 × 110 mm, Amazon, UK). Multi-layered, absorbent, capillary action matting, c. 3 mm thickness, made from bonded, non-toxic, wool and acrylic fibres, 450 gm^- (manufactured by Tech-Garden, UK) was manually cut into the shape of insoles (UK ladies size 10) using hemp scissors (manufactured by Pemmiproducts, Germany), see Fig. 1a. The bespoke insole was sprayed with deionized water (15 MΩ cm, model Essential, Millipore, UK) and placed on spawn bed, see Fig. 1b. The plastic container was placed inside a polypropylene bag (type 14A) fitted with 0.5 μm air filter patch (Ann Miller's Speciality Mushrooms, UK) and sealed with food storage clip (model Bevara, IKEA, UK). The insole was kept at ambient room temperature 18 °C to 22 °C inside a growth tent (in darkness). The insole was checked for growth every 3 days and additional moisture added (via manual water spray bottle) as required. After c. 3 weeks, uniform mycelium growth throughout the capillary matting was observed, see Fig. 1b. The colonised insole was carefully lifted off the spawn bed, see Fig. 1c.

A bespoke test rig was developed to apply compressive loading to insoles to replicate the weight of a human when walking and standing. A prosthetic foot (ladies, UK size 10) was 3D printed in acrylonitrile butadiene styrene (with Ultimaker S5, UK), see Fig. 2. The top part of the prosthetic foot was intentionally printed flat. A pivot joint (with integrated locking mechanism) was positioned between the weight

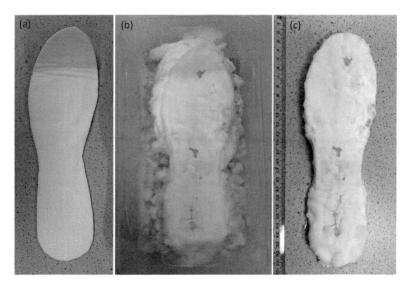

Fig. 1 **a** Capillary matting cut into insole pattern **b** insole on bed of spawn **c** well colonised insole

and aluminium plate on top of the prosthetic foot to provide control over how the compressive load was distributed across the insole, see Fig. 2b. Test rig frame was assembled from plywood sheet (18mm thickness) with plastic pipe (110mm diameter). Mild steel bar (100mm diameter) was free to move vertically inside the pipe to provide compressive loading on the insole at the bottom. For example, 500mm length of steel bar weighs 37 kg which approximates a woman (with size 10 feet) standing on two feet (two 500mm lengths can be used to replicate standing on one foot). The weight(s) could be raised and locked in the retracted position (by manual winch, fitted with a ratchet locking mechanism) to enable insoles to be interchanged and the load varied. To prevent the colonised insole slowly dehydrating over time (which might affect measurements) a sheet of capillary matting was placed under the insole, see Fig. 2a. The end of the capillary matting was left in a tray of de-ionised water which provided a source of moisture.

Three modes of compressive loading were explored (i) toe bias, Fig. 2d (ii) heel bias, Fig. 2c and (iii) uniformly distributed, Fig. 2b.

Electrical activity of the mycelium colonising insoles was recorded using eight pairs of stainless steel sub-dermal needle electrodes (Spes Medica S.r.l., Italy), with twisted cables and ADC-24 (PICO Technology, UK) high-resolution data logger with a 24-bit A/D converter, galvanic isolation and software-selectable sample rates. The pairs of electrodes were pierced through the insole's edge as shown in Fig. 2. We recorded electrical activity one sample per second. During the recording, the logger has been doing as many measurements as possible (typically up to 10 per second) and saving the average value. We set the acquisition voltage range to 156 mV. Each pair

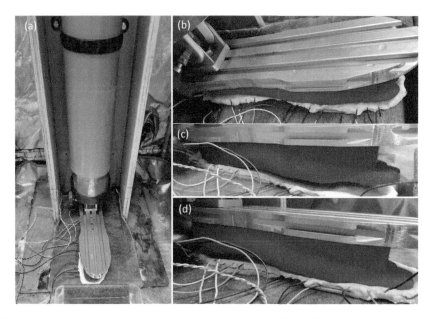

Fig. 2 Bespoke insole test rig **a** setup inside growth tent **b** weight uniformly distributed via pivot joint on prosthetic foot **c** heel bias **d** toes bias

of electrodes, called a channel (Ch), reported a difference of the electrical potential between the electrodes. Distance between electrodes was 1 to 2 cm.

Numerical modelling of the electrical activity was implemented as follows.

We used an artistic image of the mycelium network (Fig. 3a) projected onto a 364×985 nodes grid. The original image $M = (m_{ij})_{1 \leq j \leq n_i, 1 \leq j \leq n_j}, m_{ij} \in \{r_{ij}, g_{ij}, b_{ij}\}$, where $n_i = 364$ and $n_j = 985$, and $1 \leq r, g, b \leq 255$ (Fig. 3a), was converted to a conductive matrix $C = (m_{ij})_{1 \leq i, j \leq n}$ (Fig. 3b) derived from the image as follows: $m_{ij} = 1$ if $r_{ij} > 170$, $g_{ij} > 170$ and $b_{ij} < 200$; a dilution operation was applied to C.

FitzHugh-Nagumo (FHN) equations [14–16] is a qualitative approximation of the Hodgkin-Huxley model [17] of electrical activity of living cells:

$$\frac{\partial v}{\partial t} = c_1 u(u - a)(1 - u) - c_2 uv + I + D_u \nabla^2 \qquad (1)$$

$$\frac{\partial v}{\partial t} = b(u - v), \qquad (2)$$

where u is a value of a trans-membrane potential, v a variable accountable for a total slow ionic current, or a recovery variable responsible for a slow negative feedback, I is a value of an external stimulation current. The current through intra-cellular spaces is approximated by $D_u \nabla^2$, where D_u is a conductance. Detailed explanations of the 'mechanics' of the model are provided in [18], here we shortly repeat some insights.

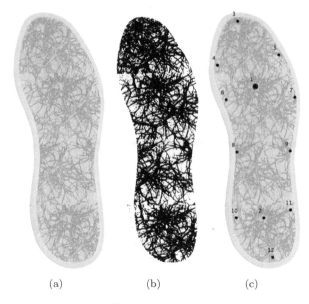

Fig. 3 Image of the fungal colony, 1000 × 960 pixels used as a template conductive for FHN. **a** Original image, mycelium is seen as green pixels. **b** Conductive matrix C, conductive pixels are black. **c** Configuration of electrodes

The term $D_u \nabla^2 u$ governs a passive spread of the current. The terms $c_2 u(u-a)(1-u)$ and $b(u-v)$ describe the ionic currents. The term $u(u-a)(1-u)$ has two stable fixed points $u = 0$ and $u = 1$ and one unstable point $u = a$, where a is a threshold of an excitation.

We integrated the system using the Euler method with the five-node Laplace operator, a time step $\Delta t = 0.015$ and a grid point spacing $\Delta x = 2$, while other parameters were $D_u = 1, a = 0.13, b = 0.013, c_1 = 0.26$. We controlled excitability of the medium by varying c_2 from 0.05 (fully excitable) to 0.015 (non excitable). Boundaries are considered to be impermeable: $\partial u/\partial \mathbf{n} = 0$, where $\mathbf{n}$ is a vector normal to the boundary.

To record dynamics of excitation in the network, as if in laboratory experiments, we simulated electrodes by calculating a potential p_x^t at an electrode location x as $p_x = \sum_{y:|x-y|<2}(u_x - v_x)$. Configuration of electrodes $1, \ldots, 16$ is shown in Fig. 3c.

To imitate a pressure onto insole we perturbed the medium around electrodes E_1 or E_2 or both.

Time-lapse snapshots provided were recorded at every 100th time step, and we display sites with $u > 0.04$; videos and figures were produced by saving a frame of the simulation every 100th step of the numerical integration and assembling the saved frames into the video with a play rate of 30 fps. Videos are available at https://doi.org/10.5281/zenodo.5091807.

3 Results

Electrical activity was recorded for c. 30 min immediately before, c. 30 min during and c. 30 min immediately after 35 kg was evenly distributed across the insole, see Fig. 4.

Electrical activity was recorded for ~24 min immediately before 35kg was evenly distributed across the insole for c. 72 min. The prosthetic foot was then tilted back to bias the weight onto the heel region of the insole c. 24 min. The prosthetic foot was then tilted forward to bias the weight onto the toe region of the insole for c. 24 min. Finally, the weight was removed for c. 24 min. Eight pairs of needle electrodes were distributed along the side of the insole as shown in Fig. 5.

When a small area, c. 10 nodes, is perturbed the excitation starts propagating along the simulated mycelium network (Fig. 6ab). With time, typically after 50K-60K iterations of the integration, the excitation wave fronts span all the mycelium network (Fig. 6cde). Due to inhomogeneity of the network source so the spiral waves are formed. They become sources of oscillatory excitations. Repeated propagation of the spiral waves is reflected in the oscillatory activity levels on the mycelium network (Fig. 7).

Is it possible to discern what loci of the insole a pressure was applied based on electrical activity of the mycelium network? Yes, as we demonstrate further. Let us start with activity. As evidenced in Fig. 7 the overall activity of the mycelium network depends on the site of pressure application (as imitated via exciting areas around electrodes E_1 or E_2 or both). When consider the three possible scenarios of

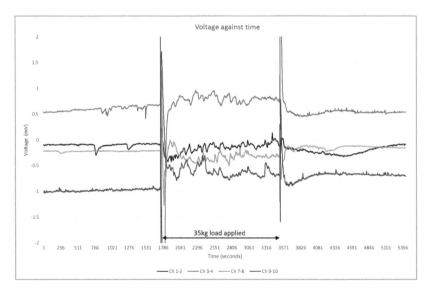

Fig. 4 Electrical activity (recorded on Ch 1–2, Ch 3–4, Ch 7–8, Ch 9–10) before, during and after load (35 kg) was evenly distributed across insole

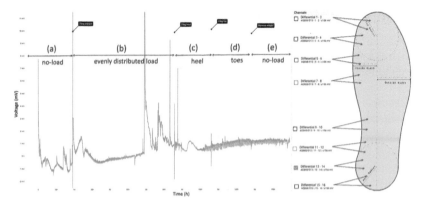

Fig. 5 Exemplar of electrical activity recorded on Ch 13–14 **a** no load **b** load (35 kg) evenly distributed **c** load biased on heel region **d** load biased on toe region **e** no load

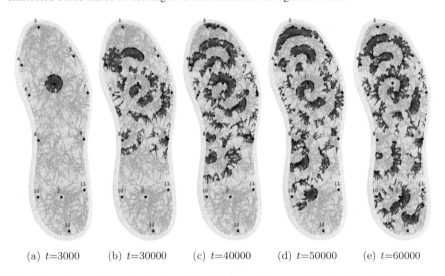

Fig. 6 Snapshots of the excitation dynamics of the mycelium network colonising the insole. Area around electrode E_1 has been excited originally

the stimulation, the overall activity is lowest when area around electrode E_2 is excited ($E_1 = 0$, $E_2 = 1$). This is due to relatively lower number of mycelium strands in that part of the insole. The overall activity increases in the scenario where area round E_1 is excited ($E_1 = 1$, $E_2 = 0$). And the highest level of overall activity is evidenced for the scenario when areas around both electrodes ($E_1 = 1$, $E_2 = 1$).

Coverage frequency could be another indicator to discern geometry of pressure. A coverage frequency at node x is a number of iterations values of excitation variable u_x exceeded 0.1, normalised by maximum coverage frequency amongst the nodes. The coverage frequency is illustrated in Fig. 8. The coverage frequency is maximum around the areas of pressure application and might even reflect a distance, not an

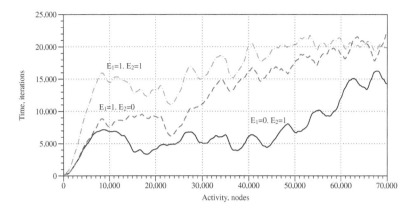

Fig. 7 Activity of the mycelium network, for initial scenarios of excitation around electrode E_2, solid black, E_1, dashed red, E_1 and E_2, dot-dashed blue. The activity is measured in a number of nodes x with $u_x > 0.1$

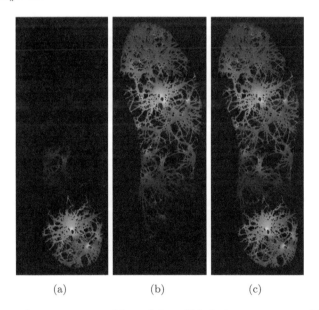

Fig. 8 Coverage frequency, expressed in gradation of black. Areas never covered by excitation wave-fronts are black, areas covered most frequently are white. **a** Area around electrode E_2 have been excited initially. **b** Area around electrode E_1 have been excited initially. **c** Areas around electrodes E_1 and E_2 have been excited initially

Euclidean distance but a distance in the propagation metric of the mycelium networks, form the pressure application site.

A third measure applied to discern geometry of pressure would be spiking activity recorded on the electrodes. Examples of the spiking for all three scenarios of pressure application are shown in Fig. 9. The patterns of spiking activities might give us

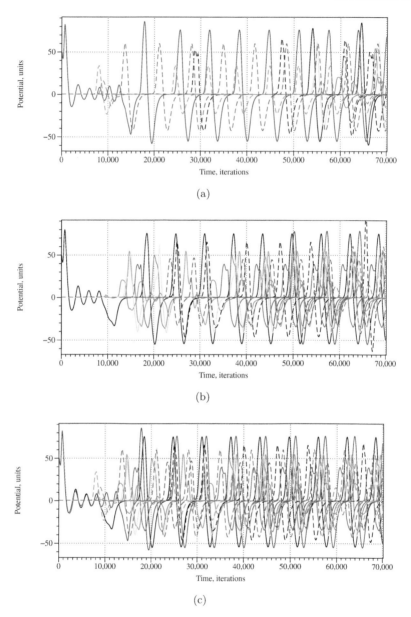

Fig. 9 Potential recorded on electrodes $E_1, \ldots, E_{12}$ for initial scenarios of excitation around electrode **a** E_2, **b** E_1, **c** E_1 and E_2

Table 1 Representation of gates by combinations of spikes. Black lines show the potential when the network was stimulated by input pair (01), red by (10) and green by (11). Adamatzky proposed this representation originally in [19]

Spikes	Gate	Notations
	OR	$x + y$
	SELECT	y
	XOR	$x \oplus y$
	SELECT	x
	NOT-AND	$\overline{x} y$
	AND-NOT	$x \overline{y}$
	AND	xy

unique representations of the geometries of pressure applications. The formal representation of the spiking patterns could be done by distributions of Boolean gates in the spiking activity. This original technique has been developed by us in frameworks of cytoskeleton networks [19], fungal colony [20] and ensemble of proteinoid microspheres [21].

A spiking activity of the mycelium network shown in Fig. 9 in a response to stimulation, i.e. application of inputs $(E_1, E_2) = \{(0, 1), (1, 0), (1, 1)\}$ via impulses at the electrodes E_1 and E_2, recorded from electrodes $E_1, \ldots, E_{12}$. We assume that each spike represents logical TRUE and that spikes occurring within less than $2 \cdot 10^2$ iterations are simultaneous. Then a representation of gates by spikes and their combinations can be implemented as shown in Table 1. By selecting specific intervals of recordings we can realise several gates in a single site of recording. In this particular case we assumed that spikes are separated if their occurrences are more than 10^3 iterations apart. In the simulated scenarios, we found that the following Boolean functions can be implemented on the electrodes $E_1, \ldots, E_{12}$. Three OR gates are realised on electrodes E_3, E_8 and E_{12}. Ten SELECTy, where y=TRUE signifies initial excitation around electrode E_2, are realised on electrodes E_3 and E_{12}. Fifty SELECTx, where x=TRUE signifies initial excitation around electrode E_1, are realised on the electrodes but E_1, E_9 and E_{11}. Five NOT-AND gate, in the form NOT x AND y, are realised on electrodes E_2, E_9 and E_{10}. The implementation of logical functions on the electrodes will allow for logical inference about geometries of pressure applied to insoles.

4 Discussion

Initial testing of insoles made of other materials (not reported here) confirmed the necessity to use a material compatible with the biological organism being utilised. For example, off-the-shelf hemp insoles (sourced from six different manufacturers) showed poor fungal colonisation (possibly due to unknown chemical processes to minimise bacterial growth which could lead to undesirable foot wear odour). Further testing of bespoke insoles cut from non-woven hemp matting (manufactured by Pemmiproducts, Germany) showed improved colonisation by fungi but inconsistent electrical activity (possibly due to inconsistent moisture level) and weak mechanical robustness. Following a process of trial and error with different materials over several months, capillary matting was identified as a strong candidate for the desired functionality.

It was observed that the risk of unwanted bacterial infection during colonisation of the insole could be reduced by keeping the insole inside a sealed local environment (for example, plastic bag fitted with sub-micron air filter) that mitigates airborne infection. Additionally, enclosing in a bag helps to prevent the insole dehydrating by maintaining a high local air humidity. Keeping the insole in darkness or low intensity light encourages its colonisation.

Optionally, forming the insole from two sheets of capillary matting allows a sandwich to be formed with nutrient layer (such as Rye grain seeds) between the top and bottom layer. This can allow the fungi to remain active for longer. Insoles infused with flour paste were prone to infection (even with sterilisation via autoclave).

The large internal volume and porous seals (long fabric zips) on the growth tent containing test rig was prone to low humidity, even with an open container of water present. It was found adding a sheet of capillary matting under the insole, see Fig. 2a, helped to maintain an adequate moisture level in the insole for fungal activity. The end of the sheet was left in a tray of de-ionised water which provided a source of moisture.

Oscillations in plant membrane are already know [22]. The physiological role of such oscillations has been the subject of much speculation. It has been hypothesised these oscillations are links to plants' adaptive response to environmental stresses [23].

The number of spikes (<0.1 mV) recorded over three c. 30 min periods before, during after even compressive load are summarised in Table 2. Further, it was observed that periodicity of electrical spikes changed when the mycelium was under compressive load.

Measurements indicate a layer of mycelium integrated into an insole shows electrical response to mechanical stimulation with change in oscillatory activity. In particular, the number of spikes increases under compressive load. The response to removing weight is different to applying weight.

To examine this response in more detail, the distribution of the weight across the insole was varied and applied in different regions (i.e., toe, heel, whole). That was designed in order to be similar to the way people change their weight distribution from heel to toe while walking. The number of spikes in each time period

Table 2 Electrical response to weight (electrical spikes <0.1 mV)

Differential channels	30 min period before	30 min period during weight	30 min period after weight
Ch 1–2	2	10	1
Ch 3–4	2	6	0
Ch 7–8	0	8	3
Ch 9–10	0	7	2

were automatically counted by utilising SciPy, an open-source collection containing mathematical algorithms and functions built on an extension of Python (https://docs.scipy.org). In specific, the function *find_peaks* was utilised to identify peaks with a prominence of 0.03 *mV* for this application. Then, a custom program was developed in Python to calculate the time difference between two spikes. Histograms of the distribution of time differences between spikes in each insole region under no load, heel biased and toe biased load were produced and they were drawn over the same axes to allow the comparison of the results as shown in Fig. 10. Data visualization software Tableau was used to identify the potential (mV) differences under no load, even load, heel biased and toe biased load for every channel separately as shown in Fig. 11. The amplitude of spikes decreases under even load application compared to no load application. No conclusions could be made regarding the electrical activity under heel biased and toe biased load. It was therefore, not possible to accurately discern the distribution of compressive loading across the insole from analysis of electrical responses.

Experimentation identified the rate of occurrence of spikes is lower than desirable to accurately infer the weight bearing when walking. However, when standing for a period, electrical activity can be collected and analysed to infer weight bearing which can be used for anatomical diagnostics [24, 25]. Continuous monitoring of feet could offer numerous medical/health benefits [26–28]. For example, early detection of health related conditions (such as knee injury) or tiredness. It can also be useful for sports training [29–31].

In the experimental setup described, sensory mapping is limited to '1.5D' (8 pairs of differential electrodes in a row and limited vertical motion) however this could be expanded to '2.5D' by mapping with a 2D distributed array of electrodes across the insole. For example, needle electrodes replaced by thin/flexible wires integrated into the capillary matting such that uninsulated sections of the wires are spatially distributed and electrical connections are realised to the edge of the insole.

Direct conversion of mechanical energy into electricity offers potential as power source for various systems [32–34].

It was observed during the fabrication and testing of a diverse range of prototype wearables (including clothing) that capillary matting offer superior durability over hemp matting (in particular on flexible clothing joints, knees and elbows) and easy of

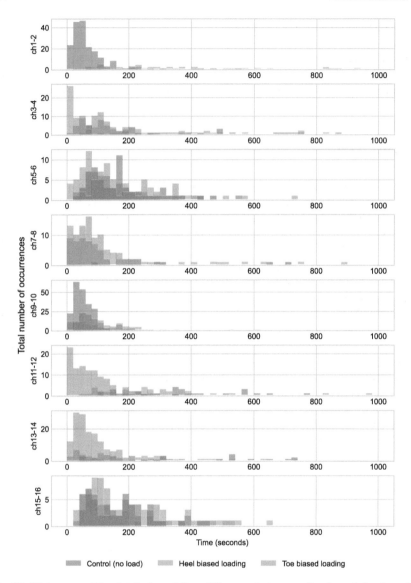

Fig. 10 Histograms of the distribution of time differences between spikes in each insole region under a range of load condition

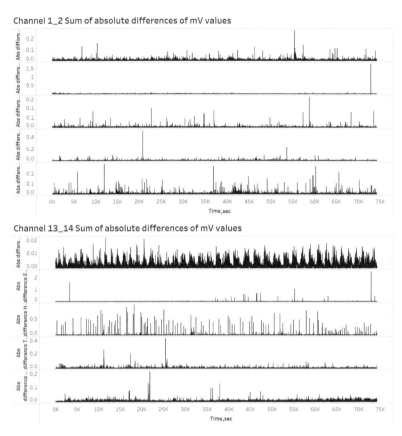

Fig. 11 Graph showing decrease of amplitude when under even load compared to no load application for Channels 1–2 and 13–14

interfacing to conventional fabrics (sewing and gluing). Therefore, capillary matting might be useful substrate for a range of smart fungal wearables.

Smart footwear offers benefits to safety footwear [35]. For example, automatic notification of injury to user and emergency services. Awareness of foot activity might also be beneficial in various environments such as driving (for example, enabling the vehicle to respond before the driver's foot has touched a pedal).

Simulation using FitzhHugh-Nagumo model numerically illustrated how excitation wave-fronts behave in a mycelium network colonising an insole and shown that it is possible to discern pressure points from the mycelium electrical activity.

5 Conclusion

Electrical activity (spiking) was recorded in mycelium bound composites fabricated into insoles. The number and periodicity of electrical spikes change when the mycelium is subjected to compressive loading. We have shown that it might be possible to discern the loading from the electrical response of the fungi to stimuli. The results advance the development of intelligent sensing insoles which are a building block towards more generic reactive fungal wearables. Electrical activity changes in both spatial and temporal domains. Using FitzhHugh-Nagumo model we numerically illustrated how excitation wave-fronts behave in a mycelium network colonising an insole and shown that it might be possible to discern pressure points from the mycelium electrical activity. Fungal based insoles offer augmented functionality (sensory) and aesthetic (personal fashion).

References

1. Martini, E., Fiumalbi, T., Dell'Agnello, F., Ivanić, Z., Munih, M., Vitiello, N., Crea, S.: Pressure-sensitive insoles for real-time gait-related applications. Sensors **20**(5), 1448 (2020)
2. Munoz-Organero, M., Parker, J., Powell, L., Mawson, S.: Assessing walking strategies using insole pressure sensors for stroke survivors. Sensors **16**(10), 1631 (2016)
3. Khoo, I.-H., Marayong, P., Krishnan, V., Balagtas, M.N., Rojas, O.: Design of a biofeedback device for gait rehabilitation in post-stroke patients. In: 2015 IEEE 58th International Midwest Symposium on Circuits and Systems (MWSCAS), pp. 1–4. IEEE (2015)
4. Park, J., Na, Y., Gu, G., Kim, J.: Flexible insole ground reaction force measurement shoes for jumping and running. In: 2016 6th IEEE International Conference on Biomedical Robotics and Biomechatronics (BioRob), pp. 1062–1067. IEEE (2016)
5. Ramirez-Bautista, J.A., Huerta-Ruelas, J.A., Chaparro-Cárdenas, S.L., Hernández-Zavala, A.: A review in detection and monitoring gait disorders using in-shoe plantar measurement systems. IEEE Rev. Biomed. Eng. **10**, 299–309 (2017)
6. Zhang, H., Zanotto, D., Agrawal, S.K.: Estimating cop trajectories and kinematic gait parameters in walking and running using instrumented insoles. IEEE Robot. Autom. Lett. **2**(4), 2159–2165 (2017)
7. Razak, A.H.A., Zayegh, A., Begg, R.K., Wahab, Y.: Foot plantar pressure measurement system: a review. Sensors **12**(7), 9884–9912 (2012)
8. Tao, J., Dong, M., Li, L., Wang, C., Li, J., Liu, Y., Bao, R., Pan, C.: Real-time pressure mapping smart insole system based on a controllable vertical pore dielectric layer. Microsyst. Nanoeng. **6**(1), 62 (2020)
9. Gao, L., et al.: Highly sensitive pseudocapacitive iontronic pressure sensor with broad sensing range. Nano-Micro Lett. **13**(1), 1–14 (2021)
10. Nurvv Ltd. NURVV Run Smart Insoles (2022). https://www.nurvv.com/en-gb/products/nurvv-run-insoles-trackers/. Accessed 18 Sept. 2022
11. Nurvv Ltd. NURVV Run Smart Insoles operating time (2022). https://www.nurvv.com/en-gb/products/nurvv-run-insoles-trackers/. Accessed 18 Sept. 2022
12. Nurvv Ltd. NURVV Run Smart Insoles (2022). https://www.nurvv.com/en-gb/support/sizing/. Accessed 18 Sept. 2022
13. Adamatzky, A., Gandia, A.: Living mycelium composites discern weights via patterns of electrical activity. J. Bioresour. Bioprod. **7**(1), 26–32 (2022)
14. FitzHugh, R.: Impulses and physiological states in theoretical models of nerve membrane. Biophys. J. **1**(6), 445–466 (1961)

15. Nagumo, J., Arimoto, S., Yoshizawa, S.: An active pulse transmission line simulating nerve axon. Proc. IRE **50**(10), 2061–2070 (1962)
16. Pertsov, A.M., Davidenko, J.M., Salomonsz, R., Baxter, W.T., Jalife, J.: Spiral waves of excitation underlie reentrant activity in isolated cardiac muscle. Circ. Res. **72**(3), 631–650 (1993)
17. Beeler, G.W., Reuter, H.: Reconstruction of the action potential of ventricular myocardial fibres. J. Physiol. **268**(1), 177–210 (1977)
18. Rogers, J.M., McCulloch, A.D.: A collocation-Galerkin finite element model of cardiac action potential propagation. IEEE Trans. Biomed. Eng. **41**(8), 743–757 (1994)
19. Adamatzky, A., Huber, F., Schnauß, J.: Computing on actin bundles network. Sci. Rep. **9**(1), 1–10 (2019)
20. Adamatzky, A., Tegelaar, M., Wosten, H.A.B., Powell, A.L., Beasley, A.E., Mayne, R.: On boolean gates in fungal colony. Biosystems **193**, 104138 (2020)
21. Adamatzky, A.: Towards proteinoid computers (2021). arXiv:2106.00883
22. Baluska, D.V.F., Mancuso, S. (eds.): Oscillations in plants. Communication in Plants: Neuronal Aspects of Plant Life, pp. 261–275. Springer, Berlin, Heidelberg (2006)
23. Shabala, S., Shabala, L., Gradmann, D., Chen, Z., Newman, I., Mancuso, S.: Oscillations in plant membrane transport: model predictions, experimental validation, and physiological implications. J. Exp. Botany **57**(1), 171–184, 12 (2005)
24. Ghosh, A.K., Tibarewala, D.N., Mukherjee, P., Chakraborty, S., Dr Ganguli, S.: Preliminary study on static weight distribution under the human foot as a measure of lower extremity disability. Med. Biol. Eng. Comput. **17**, 737–41, 12 (1979)
25. Cavanagh, M.M., Rodgers, P. R.: Pressure Distribution Underneath the Human Foot, pp. 85–95. Springer Netherlands, Dordrecht (1985)
26. Lin, F., Wang, A., Zhuang, Y., Tomita, M.R., Wenyao, X.: Smart insole: a wearable sensor device for unobtrusive gait monitoring in daily life. IEEE Trans. Ind. Inf. **12**(6), 2281–2291 (2016)
27. Xu, W., Huang, M.-C., Amini, N., Liu, J.J., He, L., Sarrafzadeh, M.: Smart insole: a wearable system for gait analysis. In: Proceedings of the 5th International Conference on PErvasive Technologies Related to Assistive Environments, PETRA '12, New York, NY, USA. Association for Computing Machinery (2012)
28. Wang, B., Rajput, K.S., Tam, W.-K., Tung, A.K.H., Yang, Z.: Freewalker: a smart insole for longitudinal gait analysis. In: 2015 37th Annual International Conference of the IEEE Engineering in Medicine and Biology Society (EMBC), pp. 3723–3726 (2015)
29. Ziagkas, E., Loukovitis, A., Zekakos, D.X., Chau, T.D.-P., Petrelis, A., Grouios, G.: A novel tool for gait analysis: validation study of the smart insole podosmart®. Sensors **21**(17) (2021)
30. Tan, A.M., Fuss, F.K., Weizman, Y., Troynikov, O.: Development of a smart insole for medical and sports purposes. Procedia Eng. **112**, 152–156 (2015). 'The Impact of Technology on Sport VI' 7th Asia-Pacific Congress on Sports Technology, APCST2015
31. Oks, A., Katashev, A., Zadinans, M., Rancans, M., Litvak, J.: Development of smart sock system for gate analysis and foot pressure control. In: Kyriacou, E., Christofides, S., Pattichis, C.S. (eds.), XIV Mediterranean Conference on Medical and Biological Engineering and Computing 2016, pp. 472–475, Cham. Springer International Publishing (2016)
32. Sun, J., Guo, H., Schädli, G.N., Tu, K., Schär, S., Schwarze, F.W.M.R., Panzarasa, G., Ribera, J., Burgert, I.: Enhanced mechanical energy conversion with selectively decayed wood. Sci. Adv. **7**(11), eabd9138 (2021)
33. de Fazio, R., Perrone, E., Velázquez, R., De Vittorio, M., Visconti, P.: Development of a self-powered piezo-resistive smart insole equipped with low-power BLE connectivity for remote gait monitoring. Sensors **21**(13) (2021)
34. Wang, W., Cao, J., Yu, J., Liu, R., Bowen, C.R., Liao, W.-H.: Self-powered smart insole for monitoring human gait signals. Sensors **19**(24) (2019)
35. Janson, D., Newman, S.T., Dhokia, V.: Next generation safety footwear. Procedia Manuf. 38:1668–1677 (2019). 29th International Conference on Flexible Automation and Intelligent Manufacturing (FAIM 2019), 24–28 June 2019, Limerick, Ireland, Beyond Industry 4.0: Industrial Advances, Engineering Education and Intelligent Manufacturing

Electrical Response of Fungi to Changing Moisture Content

Neil Phillips, Antoni Gandia, and Andrew Adamatzky

Abstract Mycelium-bound composites are potential alternatives to conventional materials for a variety of applications, including thermal and acoustic building panels and product packaging. If the reactions of live mycelium to environmental conditions and stimuli are taken into account, it is possible to create functioning fungal materials. Thus, active building components, sensory wearables, etc. might be created. This research describes the electrical sensitivity of fungus to changes in the moisture content of a mycelium-bound composite. Trains of electrical spikes initiate spontaneously in fresh mycelium-bound composites with a moisture content between $\sim$95 and $\sim$65%, and between $\sim$15 and $\sim$5% when partially dried. When the surfaces of mycelium-bound composites were partially or totally encased with an impermeable layer, increased electrical activity was observed. In fresh mycelium-bound composites, electrical spikes were seen both spontaneously and when induced by water droplets on the surface. Also explored is the link between electrical activity and electrode depth. Future designs of smart buildings, wearables, fungi-based sensors, and unconventional computer systems may benefit from fungi configurations and biofabrication flexibility.

1 Introduction

Mycelium-bound composites—masses of organic substrates colonised by fungi—are considered environmentally friendly biodegradable biomaterials [1–4]. These fungal-based materials can be used in thermal insulation wall cladding [5–10], acoustic insulation panels [11–13], packaging materials [14–16], wearables [1, 17–22], art [23, 24], and interior design [24, 25]. Fungi are frugal colonisers and are found in most habitats on Earth wherever there is a minimal moisture availability. Remarkably,

---

N. Phillips (✉) · A. Adamatzky
Unconventional Computing Laboratory, UWE, Bristol, UK
e-mail: neil.phillips@uwe.ac.uk

A. Gandia
Institute for Plant Molecular and Cell Biology, CSIC-UPV, Catalonia, Spain

they can thrive in deserted areas thanks to symbiotic relationships with photosynthetic organisms like algae and plants. To this day, most fungal materials are finished and served in a dehydrated form that stops all biological activity throughout the substrate and avoids the eventual regrowth, sporulation, or further decay of the pieces and derived bioburden [12, 26]. Furthermore, downstream preservation techniques such as paint coatings and plasticisation are commonly used to extend the lifespan of these biomaterials to properly fit their functional and commercial purpose as decorative or architectural elements.

Contrary to common practice, in [27] we proposed to develop a structural substrate by using non-dehydrated living fungal mycelium, functionalise the substrate with nanoparticles and polymers to make mycelium-based electronics [28–31], and implement sensorial fusion and decision making in the mycelium networks [32]. Following that vision, the structural substrate—the living mycelium-bound composites — will be used to grow monolithic buildings from the functionalised fungal substrate [33]. Buildings grown with mycelium-bound composites could provide intelligent sensory capability if some parts of the mycelium remain alive, therefore securing a minimal viable moisture content will be crucial to keep the sensorial network of the fungus electrically active. In this case the fungal materials might be able to detect structural loads (dead loads such as the weight of the structure, live loads such as vehicle traffic, building contents, etc, and environmental loads such as wind, snow, etc) [34], illumination [31], temperature, and air pollution.

As part of our research into the sensing characteristics of fungus, we demonstrate how mycelium-bound composites respond to variations in moisture content by modifying their electrical activity. We chose electrical activity as an indicator of fungal response because fungi have been shown to respond to chemical and physical stimuli by changing patterns of electrical activity [35–37] and electrical properties [31].

2 Methods and Materials

Blocks of spawn substrate were bought from commercial suppliers. They were made of rye seeds and millet grain and were well colonised with two types of fungi: *Hericium erinaceus* (supplied by Urban Farm It Ltd, UK, product code M9514) and *Pleurotus ostreatus* (supplied by Mycelia BVBA, BE, product code M2125).

A moisture probe (HOBO EC-5, Tempcon Instrumentation Ltd, UK) was inserted into blocks of colonised substrate and connected to a data logger (HOBO H21 USB Micro Station, Tempcon Instrumentation Ltd, UK). Every ten seconds, the electrical conductivity between the probe's two electrodes was measured and saved (using HOBOware Pro Software from Tempcon Instrumentation Ltd, UK) on a Windows 10 computer for later analysis, see Fig. 1.

The following steps were taken to calibrate the moisture probe: (1) The sample is weighed, and the electrical conductivity between the probe electrodes is measured. (2) The sample is dried in an oven at at 80 °C for ∼48 h. The sample is weighed again, and then the electrical conductivity between the probe electrodes is measured while

Fig. 1 HOBO EC-5 moisture probe and sub-dermal needle electrodes inserted into unwrapped blocks of substrate colonised with *Pleurotus ostreatus*

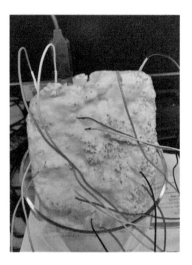

the sample is 'bone dry.' (4) The difference in weight and electrical conductivity between the two situations is calculated.

A freshly unwrapped (∼500 g) block of substrate (rye seeds and millet grain) that had been colonised by *Pleurotus ostreatus* was left to slowly dehydrate at room temperature (18 °C to 22 °C) and ambient humidity (∼30%). A calibrated HOBO EC-5 moisture probe was used to monitore the substrate's moisture content (as previously described). A high-resolution data logger with a 24-bit A/D converter (ADC-24, PICO Technology, UK) and software (PicoLog 6, PICO Technology, UK) with a selectable sample rate and pairs of stainless steel sub-dermal needle electrodes were used to record electrical activity (Spes Medica S.r.l., IT). The sampling period was one second, which was used to record electrical activity. During the recording, the logger took as many measurements as it could (usually up to 600 per second) and saved the average value. The voltage range for acquisition was set to ±39 mV. Each pair of electrodes, called a channel (Ch), reported a difference in the electrical potential between them. The electrodes were pierced through the mycelium on the substrate's surface. The distance between electrodes was 1–2 cm, see Fig. 1.

In a different experiment, a fresh block (∼500 g) of substrate (rye grain seeds and millet grain) that had been colonised by *Pleurotus ostreatus* was left partly wrapped inside the plastic bag it was supplied in. The top of the bag was left open so that the substrate could slowly lose its moisture content. Subdermal needle electrodes with a length of 18 mm length were pushed through the plastic bag and ∼15 mm into the body of the substrate.

5 l bag of substrate (rye and millet grain seed) well colonised with *Ganoderma lucidum* (manufactured by Mycelia BVBA, product code M9726[1]) was divided into the following ten samples (plus waste):

---

[1] https://mycelia.be/shop/m9726-ganoderma-resinaceum/

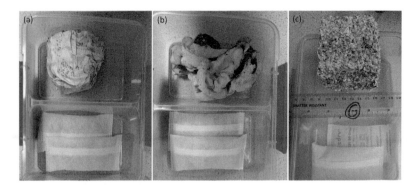

Fig. 2 Drying containers with **a** block of colonised substrate with mycelium surface **b** fruiting bodies **c** block of colonised substrate with bare surface

- 4 sub-blocks with mycelium on exposed surface (80g each), samples 'A', 'B', 'C', 'D', see Fig. 2a.
- 1 collection of fruiting bodies (50g), sample 'E', see Fig. 2b.
- 4 sub-blocks with substrate exposed surface (80 g each), samples 'F', 'G', 'H', 'I', see Fig. 2c.
- 1 collection of substrate fragmented into loose seeds (80 g), sample 'J'

Each portion was put in a 2l, plastic container with a removable airtight lid (model 1720ZS KLIP IT, Sistema, NZ). The divider in the middle allows air flow between the two sides. On the other side, two 50 g sachets of dry silica gel (model WD-1, Viola Technology Ltd, UK) that could hold more than 30 g of water together were put, as shown in the bottom half of Fig. 2. The weight of silica gel was recorded daily for 41 days.

Three configurations of electrodes were used to record electrical activity at different depths in mycelium-bound composite:

1. unmodified sub-dermal needle electrodes (18 mm length) inserted ∼15 mm depth into body of spawn substrate, see Fig. 3a.
2. unmodified sub-dermal needle electrodes inserted through 20 mm thick foam spacer so inserted ∼3 mm depth into mycelium, see Fig. 3b.
3. sub-dermal needle electrodes partly electrically insulated (with 16 mm of heat shrink tubing) inserted ∼18 mm into body of spawn substrate to make electrical contact ∼16 to 18 mm below surface, see Fig. 3c.

Samples of spawn substrate blocks (with a myceliated surface) were taken out of their dehydration containers and the electrical activity was measured with stainless steel sub-dermal needle electrodes, see Fig. 4b, and a data logger (ADC-24, PICO Technology Ltd, UK). For experimental flexibility and consistency, recordings were made inside a custom-made environmental chamber, as shown in Fig. 4a. Temperature control (e.g. $10 \pm 1^{o}C$) was done with a digital thermostat that controlled how the cooling compressor worked. The temperature and humidity inside the chamber were

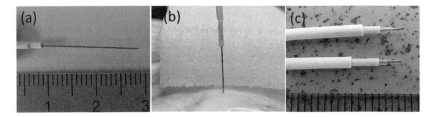

Fig. 3 Three electrode configurations **a** unmodified sub-dermal needle inserted directly **b** unmodified sub-dermal needle inserted through foam spacer **c** partly insulated sub-dermal needles so only tips exposed

Fig. 4 **a** Bespoke environmental chamber **b** Electrodes inserted into fungi

measured with a digital thermo-hygrometer (76114, Trixie Ltd, UK). The humidity in the environmental chamber was raised to ∼75% using an ultrasonic humidifier (3 l Silent, Hffheer Ltd.) filled with deionised water and activated for 15 min every 3 h, see Figs. 4 and 5.

In a separate experiment, a fresh block of substrate (1l bag) that had been well colonised with *Pleurotus ostreatus* was taken out of the bag and left to dry in the air. To record electrical activity, (18 mm length) needle electrodes were inserted ∼15 mm below the skin. After 24 h, the surface of the mycelium was sprayed by hand with de-ionised water. The results are shown in Results shown in Fig. 13.

In a different experiment, to make sure that water on the surface of the mycelium didn't change the electricity conductivity between the electrodes, a block of substrate that had been well colonised with *Pleurotus ostreatus* was left inside a plastic bag with only the top part of the bag open so it could slowly dry out. Eight pairs of subdermal needle electrodes (18 mm length) were inserted (15 mm) through the sides of the

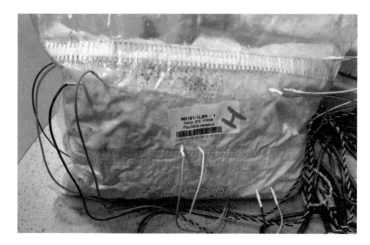

Fig. 5 Water droplets manually sprayed onto mycelium surface inside plastic bag with electrodes inserted from outside

plastic bag into the body of the spawn substrate (see Fig.reffig:Spray top bag). After spontaneous spike trains ceased, de-ionized water was sprayed by hand onto the surface of the mycelium through the open end of the bag, keeping the electrodes away from the sprayed water. Results shown in Fig. 14.

3 Results

3.1 Moisture Content Mapped to Electrical Conductivity

Three blocks of well colonised substrate (two species) were allow to dry under ambient conditions (18 to 22 °C) for 10 d to 12 d, see Table 1.

Table 1 Moisture content of fresh blocks of spawn substrate (from commercial suppliers)

Species	Weight Block (g)	Drying Period (day)	Moisture Content at start (%)	Moisture Content at end (%)	Moisture Content change (%)
Hericium erinaceus	~750	12	~99	~9	~90
Pleurotus ostreatus 'A'	~500	11	~92	~18	~74
Pleurotus ostreatus 'B'	~500	10	~82	~7	~75

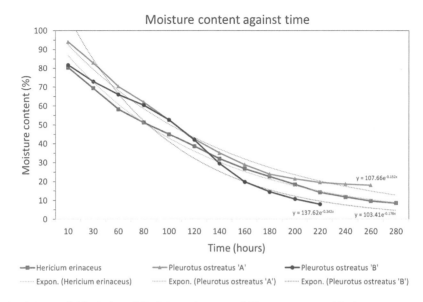

Fig. 6 Rates of dehydration of *Hericium erinaceus* and *Pleurotus ostreatus* blocks

Rate of dehydration inferred from HOBO electrical conductivity probe for *Hericium erinaceus* and *Pleurotus ostreatus* are shown in see Fig. 6.

3.2 Electrical Activity Mapped to Moisture Content

An exemplar of electrical activity against moisture content in unwrapped block of substrate colonised with *Pleurotus ostreatus* is shown in Fig. 7. In this example, spike trains spontaneously initiated after 106 h and ceased after 168 h.

An exemplar of electrical activity against moisture content in partly wrapped block of substrate colonised with *Pleurotus ostreatus* is shown in Fig. 8. In this example, spontaneously spike trains are recorded from the start and ceased after ∼20 h at which time the moisture content has dropped to <∼70%.

3.3 Rate of Water Loss from Substrate Colonised with Fungi

The rate of water loss from substrate colonised with fungi and fruiting bodies in low humidity air against time is shown in Fig. 9. To simplify the comparison, the rate of water loss from fruiting bodies was adjusted pro rota (x1.6) to compensate for a smaller sample mass (50 g rather than 80 g).

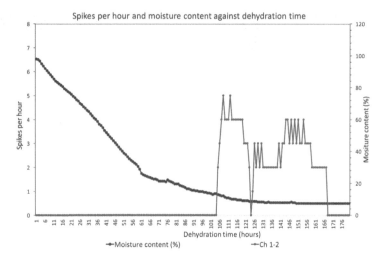

Fig. 7 Electrical activity against moisture content over time of unwrapped block of substrate colonised with *Pleurotus ostreatus*

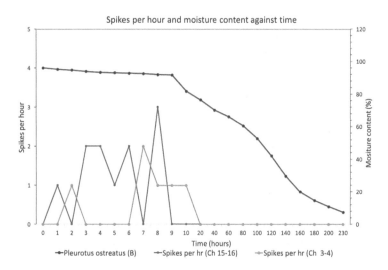

Fig. 8 Electrical activity against moisture content over time in a partly wrapped block of substrate colonised with *Pleurotus ostreatus*

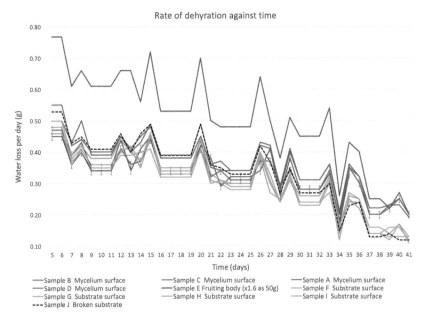

Fig. 9 Rate of water loss from colonised substrate and fruiting bodies against time

3.4 Electrical Activity Mapped to Depth

A comparison of the magnitudes of electrical potentials of trains of electrical spikes recorded with exposed electrodes inserted 3 and 15 mm into the same block of substrate colonised with *Pleurotus ostreatus* is shown in Fig. 10.

Figure 11 shows Ch 1–2 and Ch 3–4 with exposed electrodes inserted 15 mm and Ch 9–10 and Ch 11–12 with exposed electrodes 16–18 mm into the same block of substrate colonised with *Pleurotus ostreatus*.

3.5 Electrical Response to Water Droplets on Mycelium Surface

Multiple trains of electrical spikes were recorded during and immediately after operation of the ultrasonic humidifier, see Fig. 12.

Spike trains triggered both spontaneously and from water droplets sprayed onto the surface of the fresh block of spawn substrate are shown are Fig. 13.

After spontaneous spike trains ceased, electrical activity before, during and after manually spraying water droplets onto the myceliated surface inside the bag is shown in Fig. 14.

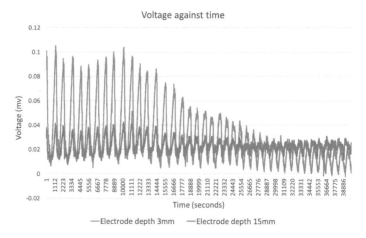

Fig. 10 Exemplar of electrical activity recorded with electrodes inserted 3 and 15 mm

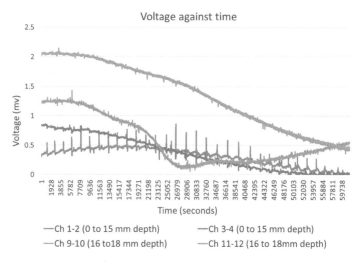

Fig. 11 Exemplar of electrical activity recorded with exposed electrodes inserted 15 and 16 to 18 mm

4 Discussion

Obtaining fresh blocks (~500 g) of electrically active spawn substrate from commercial suppliers at desired times was challenging (e.g. limited stock availability, some blocks didn't show electrical activity). Further, most commercial suppliers were unwilling to provide details of substrate composition (beyond "rye or millet seeds" as considered a 'trade secret'). Therefore, variation in substrate might exist between both batches from the same supplier and different suppliers. Further, the level of colonisation of blocks varied greatly between suppliers and times of

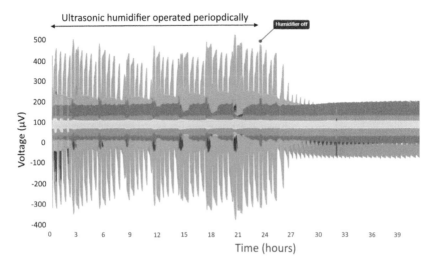

Fig. 12 Exemplar of electrical activity during and after periodic operation of humidifier

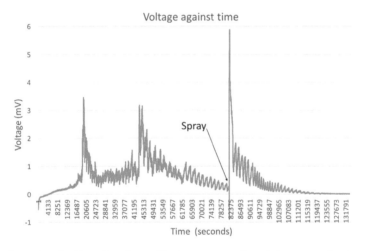

Fig. 13 Exemplar of spike trains triggered both spontaneously and from water droplets sprayed onto the myceliated surface

recordings as the fungi consumed the substrate as a source of nutrients. Additionally, the heterogeneous mixture of substrate and fungi added an additional variable.

Despite an extensive search, no commercial available moisture probe calibrated for substrate colonised with fungi was found. Therefore, it was necessary to calibrate a general-purpose moisture probe for this purpose. Moisture probe HOBO EC-5 and H21 Micro Station from Tempcon Instrumentation Ltd were selected and readings were calibrated by weighting samples before and after oven drying to determine water content. The large physical size of the EC-5 probes meant blocks of spawn

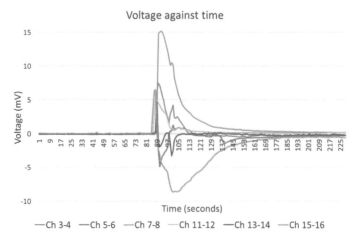

Fig. 14 Exemplar of electrical activity before, during and after manually spraying water droplets onto the surface of mycelium inside the bag (after spontaneous spike trains ceased)

substrate needed to be sufficiently large (e.g. 500 g) to avoid them splitting when the EC-5 probes were inserted into them (which would have interfered with electrical conductivity measurements).

An electrical path from the EC-5 probes via HOBO H21 Micro Station to the data logging computer appeared to interfere with the recording of electrical potentials made with PICO ADC-24. Therefore, it was necessary to unplug the USB connection between HOBO H21 Micro Station and computer during recordings (utilising Micro Station internally batteries for power).

The initial moisture content of fresh spawn substrate blocks (as supplied) is high (typically 80–100%) over time (several days) in low humidity air ($\sim$30% relative humidity) the moisture content reduces significantly (<20%). The drying curves of colonised substrate were found to be similar to organic material (such as seeds and vegetables) dehydrating in air [38–41].

Electrical spikes were not initially recorded in unwrapped fresh blocks of colonised substrate. For example, see Fig. 7. However, as the moisture content dropped (e.g. <20%) the fungi become stressed and spikes spontaneously initiated (e.g. $3 \pm 3 \, h^{-1}$). After a further period of dehydration, the rate spikes decreased and finally ceased (e.g. $\sim$64 h).

Partly wrapped fresh blocks of colonised substrate typically exhibit markedly different electrical characteristics. For example, Fig. 8 shows a partly wrapped block of substrate colonised with *Pleurotus ostreatus* dehydrating. In this example, spontaneously spike trains are recorded from the start and ceased after $\sim$20 h at which time the moisture content has dropped to $<\sim$70%.

The rates of dehydration shown in Figs. 7 and 8 against time are different as the latter is partly enclosed in an (open) plastic bag which slows the rate of water loss from the block.

Oscillations in plant membrane are already known [42].

The rate of water loss from fruiting bodies was considerably higher (40–60%) than substrate both with and without mycelium skin. This suggests that linear cytoplasmic units [43, 44] are drawing water near the surface which increases the rate of evaporation (in an environment with low-humidity air). The rate of water loss from the substrate with a mycelium skin became increasingly higher (10–60%) than substrate without mycelium skin.

Initially, the rate of water loss from the fragmented substrate (mostly loose rye and millet grain seeds) was slightly higher than the average substrate both with and without mycelium skin. However, over time this situation reversed and the rate of water loss become slightly lower than the average substrate both with and without mycelium skin. This suggests that the greater surface area of the fragmented substrate becomes less important as the substrate's moisture content is reduced close to bone dry.

After ~35d the measured rate of 'water loss' levelled off which suggests the remaining measurement of weight gain was water absorbed from the ambient air when the plastic enclosures were opened to remove the sachets of silica gel for weighing. Measuring the weight of the silica gel (rather than directly weighting the sample) provides several advantages including no loss of material or infection of the sample during the weighting process.

Recordings of electrical activity measured just below the surface (0–3 mm) typically contain higher potential differences than those recorded across a broader depth (0 to 15 mm) inside the block, as shown in Fig. 10. This difference is also noticeable in recordings electrodes inserted 0 to 15 mm compared to 16 to 18 mm. For example, Ch 1–2 and 3–4 compared to Ch 9–10 and 11–12 in Fig. 11. This suggests that electrical potential is discharged by electrodes making electrical connections to others parts of the block which are less electrically active. In other words, the high electrical conductivity of the metal electrodes discharges voltage differential as current through other low conductivity parts of the block.

Inserting sub-dermal needles at a significant depth (e.g. 15 mm) into the colonised substrate provides mechanical support to hold the electrodes in position. If sub-dermal needles are only inserted a shallow depth (e.g. 3 mm) then any movement of the flying leads can disturb the connection between the electrode and mycelium. To overcome this issue, a foam spacer was first glued to the surface of the plastic bag containing the spawn substrate. Unmodified sub-dermal needles were then inserted through 20 mm thick foam spacer to securely positioned them ~3 mm depth into mycelium. Other methods of securing the electrodes' positions (e.g. holding the top of the electrodes with a frame rigid relative to the block, electrically insulating the bottom part of the needle and inserting further into the substrate, etc) are possible.

Electrical potential was also observed to vary with electrode separation. A distance of ~20 mm between centres of electrodes was found to be effective for monitoring electrical activity, as was evidenced by the identification of significantly more and larger spikes in the recordings. This suggests that there is an optimum spacing for the electrodes in any environment.

Optimising the relative physical positions of electrodes in colonised substrate (in terms of both depth and spacing) is important to maximising the sensitivity of monitoring and interconnections to other systems.

Figure 12 shows water droplets condensing on the surface of mycelium from the high humidity air (in this example from the ultrasonic humidifier) can trigger trains of electrical spikes. Spikes of diminishing electrical potential continue to occur for ~2h after the humidifier is switched off.

In fresh spawn substrate, spikes can initiate both spontaneously and/or be triggered by water droplets on the surface in mycelium. For example, in Fig. 13, trains of spikes (~3 mV peak) trigger spontaneously ~10 h period. Spraying with de-ionised water triggers a spike train of twice the voltage potential (~6 mV peak). Electrical potential in fruiting bodies with precipitation has been reported [45]. The influence of environmental conditions on the electrical activity of fruiting bodies has been reported [46].

If the spawn substrate is allowed to partly dehydrate, spontaneous spike trains cease. However, electrical response (non-spike train) to water droplets still occurs. Exemplar Fig. 14 shows significant electrical pulses (~15 mV peak), most pulses contain two peaks, the initial one is larger followed (~15 s later) by a second peak. If the spawn substrate is allowed to completely dehydrate, no electrical response to water droplets on the surface of mycelium occurs.

5 Conclusions

Electrical activity in fresh well colonised substrate is significantly greater if part or all of the surface is enclosed with an impermeable layer (e.g. flexible plastic bag or rigid plastic container). For example, trains of electrical spikes initiate spontaneously in fresh spawn substrate blocks with >~65% moisture content inside an (open) plastic bag. If the substrate is unwrapped and allowed to partly dehydrate spikes can spontaneously occur at <~15% moisture content. The rate of change of moisture content in the substrate is affected by how well the substrate's surface is colonised with fungi. In particular, the higher the proportion of the surface covered with fungi the quicker the substrate dehydrates. In fresh spawn, electrical spikes can initiate both spontaneously and/or be triggered by water droplets on the surface in the mycelium. If the spawn is allowed to partly dehydrate, spontaneous spike trains cease. However, electrical response (non-spike train) to water droplets can still occur. The versatility of fungi, in terms of being able to tailor different biofabricated configurations such as composites, flexible tissue, rhizomorphs, and foamy materials, provides a promising opportunity for the development of unconventional computing systems. The suitability of utilising fungi for particular applications needs to be carefully assessed (e.g. quantitative analysis of the selected species) as some fungi form mycotoxins or might become invasive species [47].

6 Funding

This project has received funding from the European Union's Horizon 2020 research and innovation programme FET OPEN "Challenging current thinking" under grant agreement No 858132. The funders played no role in the design of the study and collection, analysis, and interpretation of data.

Acknowledgements We are grateful to Tempcon Instrumentation Ltd for guidance on moisture probe HOBO EC-5 and data logger station HOBO H21.

References

1. Karana, E., Blauwhoff, D., Hultink, E.-J., Camere, S.: When the material grows: A case study on designing (with) mycelium-based materials. Int. J. Des. **12**(2) (2018)
2. Jones, M., Mautner, A., Luenco, S., Bismarck, A., John, S.: Engineered mycelium composite construction materials from fungal biorefineries: a critical review. Mater. Des. **187**, 108397 (2020)
3. Cerimi, K., Akkaya, K.C., Pohl, C., Schmidt, B., Neubauer, P.: Fungi as source for new bio-based materials: a patent review. Fungal Biol. Biotechnol. **6**(1), 1–10 (2019)
4. Javadian, A., Ferrand, H.L., Hebel, D., Saeidi, N.: Application of mycelium-bound composite materials in construction industry: a short review. SOJ Mater. Sci. Eng. **7**, 1–9, 10 (2020)
5. Yang, Z., Zhang, F., Still, B., White, M., Amstislavski, P.: Physical and mechanical properties of fungal mycelium-based biofoam. J. Mater. Civil Eng. **29**(7), 04017030 (2017)
6. Xing, Y., Brewer, M., El-Gharabawy, H., Griffith, G., Jones, P.: Growing and testing mycelium bricks as building insulation materials. In: IOP Conference Series: Earth and Environmental Science, vol. 121, p. 022032. IOP Publishing (2018)
7. Girometta, C., Picco, A.M., Baiguera, R.M., Dondi, D., Babbini, S., Cartabia, M., Pellegrini, M., Savino, E.: Physico-mechanical and thermodynamic properties of mycelium-based bio-composites: a review. Sustainability **11**(1), 281 (2019)
8. Dias, P.P., Jayasinghe, L.B., Waldmann, D.: Investigation of mycelium-miscanthus composites as building insulation material. Res. Mater. **10**, 100189 (2021)
9. Wang, F., LI, H.-G., Kang, S.-S., Bai, Y.-F., Cheng, G.-Z., Zhang, G.-Q.: The experimental study of mycelium/expanded perlite thermal insulation composite material for buildings. Sci. Technol. Eng. **2016**, 20 (2016)
10. Cárdenas-R., J.P.: Thermal insulation biomaterial based on hydrangea macrophylla. In: Bio-Based Materials and Biotechnologies for Eco-Efficient Construction, pp. 187–201. Elsevier (2020)
11. Pelletier, M.G., Holt, G.A., Wanjura, J.D., Bayer, E., McIntyre, G.: An evaluation study of mycelium based acoustic absorbers grown on agricultural by-product substrates. Ind. Crops Prod. **51**, 480–485 (2013)
12. Elsacker, E., Vandelook, S., Van Wylick, A., Ruytinx, J., De Laet, L., Peeters, E.: A comprehensive framework for the production of mycelium-based lignocellulosic composites. Sci. Total Environ. **725**, 138431 (2020)
13. Robertson, O. et al.: Fungal future: a review of mycelium biocomposites as an ecological alternative insulation material. In: DS 101: Proceedings of NordDesign 2020, Lyngby, Denmark, 12th–14th Aug. 2020, pp. 1–13 (2020)
14. Holt, G.A., Mcintyre, G., Flagg, D., Bayer, E., Wanjura, J.D., Pelletier, M.G.: Fungal mycelium and cotton plant materials in the manufacture of biodegradable molded packaging material: evaluation study of select blends of cotton byproducts. J. Biobased Mater. Bioenergy **6**(4), 431–439 (2012)

15. Sivaprasad, S., Byju, S.K., Prajith, C., Shaju, J., Rejeesh, C.R.: Development of a novel mycelium bio-composite material to substitute for polystyrene in packaging applications. Mater. Today: Proc. (2021)
16. Mojumdar, A., Behera, H.T., Ray, L.: Mushroom mycelia-based material: an environmental friendly alternative to synthetic packaging. Microbial Polymers, pp. 131–141 (2021)
17. Nikolaidou, A., Phllips, N., Tsompanas, M.-A., Adamatzky, A.: Reactive fungal insoles. BioRxiv (2022)
18. Adamatzky, A., Nikolaidou, A., Gandia, A., Chiolerio, A., Dehshibi, M.M.: Reactive fungal wearable. Biosystems **199**, 104304 (2021)
19. Silverman, J., Cao, H., Cobb, K.: Development of mushroom mycelium composites for footwear products. Cloth. Text. Res. J. **38**(2), 119–133 (2020)
20. Appels, F.V.W.: The use of fungal mycelium for the production of bio-based materials. Ph.D. thesis, Universiteit Utrecht (2020)
21. Jones, M., Gandia, A., John, S., Bismarck, A.: Leather-like material biofabrication using fungi. Nat. Sustain. 1–8 (2020)
22. Gandia, A., van den Brandhof, J., Appels, F..V..W., Jones, M.P., Shaping the future: Flexible fungal materials. Trends Biotechnol. **39**, 1321–1331 (2021)
23. Meyer, V.: Merging science and art through fungi (2019)
24. Sydor, M., Bonenberg, A., Doczekalska, B., Cofta, G.: Mycelium-based composites in art, architecture, and interior design: a review. Polymers **14**(1), 145 (2022)
25. Ivanova, N.: Fungi for material futures: the role of design. In: Fungal Biopolymers and Biocomposites: Prospects and Avenues, pp. 209–251. Springer (2022)
26. van den Brandhof, J.G., Wösten, H.A.B.: Risk assessment of fungal materials. Fungal Biol. Biotechnol. **9**, 3 (2022)
27. Adamatzky, A., Ayres, P., Belotti, G., Wösten, H.: Fungal architecture position paper. Int. J. Unconven. Comput. **14** (2019)
28. Roberts, N., Adamatzky, A.: Mining logical circuits in fungi. Sci. Rep. **12**, 09 (2022)
29. Beasley, A.E., Powell, A.L., Adamatzky, A.: Capacitive storage in mycelium substrate (2020). arXiv:2003.07816
30. Beasley, A.E., Abdelouahab, M.-S., Lozi, R., Powell, A.L., Adamatzky, A.: Mem-fractive properties of mushrooms (2020). arXiv:2002.06413
31. Beasley, AE., Powell, A.L., Adamatzky: Fungal Photosensors (2020). arXiv:2003.07825
32. Adamatzky, A., Tegelaar, M., Wosten, H.A.B., Powell, A.L., Beasley, A.E., Mayne, R.: On boolean gates in fungal colony. Biosystems **193**, 104138 (2020)
33. Adamatzky, A., Gandia, A., Ayres, P., Wösten, H., Tegelaar, M.: Adaptive fungal architectures. LINKs-Series **5**, 66–77
34. Adamatzky, A., Gandia, A.: Living mycelium composites discern weights via patterns of electrical activity. J. Bioresour. Bioprod. **7**(1), 26–32 (2022)
35. Olsson, S., Hansson, B.S.: Action potential-like activity found in fungal mycelia is sensitive to stimulation. Naturwissenschaften **82**(1), 30–31 (1995)
36. Adamatzky, A.: On spiking behaviour of oyster fungi pleurotus djamor. Sci. Rep. **8**(1), 1–7 (2018)
37. Adamatzky, A., Tuszynski, J., Pieper, J., Nicolau, D.V., Rinalndi, R., Sirakoulis, G., Erokhin, V., Schnauss, J., Smith, D.M.: Towards cytoskeleton computers. A proposal. In: Adamatzky, A., Akl, S., Sirakoulis, G. (eds.), From Parallel to Emergent Computing. CRC Group/Taylor & Francis (2019)
38. Cano-Chauca, M., Ramos, A.M., Stringheta, P.C., Pereira, J.A.M.: Drying curves and water activity evaluation of dried banana. In: Drying 2004-Proceedings of the 14th International Drying Symposium (IDS 2004), São Paulo, Brazil, pp. 22–25 (2004)
39. Sharma, B., Sharma, K.: Studies of drying curves for different vegetables in cabinet dryer. Int. J. Chem. Stud. **9**, 523–527 (2021)
40. Villela, F., Silva, W.R.: Drying curve of corn seeds by the intermittent method. Scientia Agricola **49**, 145–153 (1991)

41. Hustrulid, A., Flikke, A.M.: Theoretical drying curve for shelled corn. Trans. ASAE **2**, 112–114 (1959)
42. Shabala, S., Shabala, L., Gradmann, D., Chen, Z., Newman, I., Mancuso, S.: Oscillations in plant membrane transport: model predictions, experimental validation, and physiological implications. J. Exp. Botany **57**(1), 171–184 (2005)
43. Allen, M.: Mycorrhizal fungi: highways for water and nutrients in arid soils. Vadose Zone J. **6**, 291–297 (2007)
44. Garcia-Rubio, R., Oliveira, H., Rivera, J., Trevijano-Contador, N.: The fungal cell wall: Candida, cryptococcus, and aspergillus species. Front. Microbiol. **10**, 2993 (2020)
45. Fukasawa, Y., Akai, D., Ushio, M., Takehi, T.: Mushroom's electrical conversation after the rain. SSRN 4091460 (2022)
46. Oguntoyinbo, B., Ozawa, T., Kawabata, K., Hirama, J., Yanagibashi, H., Matsui, Y., Kurahashi, A., Shimoda, T., Taniguchi, M., Nishibori, K.: SMA (speaking mushroom approach) environmental control system development: automated cultivation control system characterization, vol. 53, Mar. 2012
47. Sydor, M., Cofta, G., Doczekalska, B., Bonenberg, A.: Fungi in mycelium-based composites: usage and recommendations. Materials **15**(18), 6283 (2022)

Fungal Electronics

Electrical Resistive Spiking of Fungi

Andrew Adamatzky, Alessandro Chiolerio, and Georgios Sirakoulis

Abstract We study long-term electrical resistance dynamics in mycelium and fruit bodies of oyster fungi *P. ostreatus*. A nearly homogeneous sheet of mycelium on the surface of a growth substrate exhibits trains of resistance spikes. The average width of spikes is c. 23 min and the average amplitude is c. 1 kΩ. The distance between neighbouring spikes in a train of spikes is c. 30 min. Typically there are 4–6 spikes in a train of spikes. Two types of electrical resistance spikes trains are found in fruit bodies: low frequency and high amplitude (28 min spike width, 1.6 kΩ amplitude, 57 min distance between spikes) and high frequency and low amplitude (10 min width, 0.6 kΩ amplitude, 44 min distance between spikes). The findings could be applied in monitoring of physiological states of fungi and future development of living electronic devices and sensors.

1 Introduction

Electrical resistance of living substrates is used to identify their morphological and physiological state [1–5]. Examples include determination of states of organs [6], detection of decaying wood in living trees [7, 8], estimation of roots vigour [9], study of freeze-thaw injuries of plants [10], as well as classification of breast tissue [11]. The aim of this study is two-fold.

First aim is to study the dynamics of the fungal electrical resistance during long-term (up to two days of intermittent measurements). Whilst resistive properties of plants and mammals tissue have been studied extensively, results on electrical resistance of fungi are absent. This gap should be properly filled because the fungi is

---

A. Adamatzky (✉) · A. Chiolerio
Unconventional Computing Laboratory, UWE, Bristol, UK
e-mail: andrew.adamatzky@uwe.ac.uk

A. Chiolerio
Center for Sustainable Future Technologies, Istituto Italiano di Tecnologia, Torino, Italy

G. Sirakoulis
Department of Electrical and Computer Engineering, Democritus University of Thrace, Xanthi, Greece

the largest, widely distributed and the oldest group of living organisms [12]. Fungi "possess almost all the senses used by humans" [13]: they can sense light, chemicals, gases, gravity and electric fields. Fungi show a pronounced response to changes in a substrate pH [14], demonstrate mechanosensing [15] and sensing of toxic metals [16], CO_2 [17], and chemical cues, especially stress hormones, from other species [18]. Thus mycelium networks can be used as large-scale distributed sensors. To prototype fungal sensing networks we should know their electrical features and resistance is definitely one of these characteristics.

Second aim is to assess whether fungi can be employed as electronic oscillators. The application domain of the fungal electronic oscillators could be the field of unconventional computing [19], especially in the framework of organic electronics, living sensor and living computing wetware. Feasibility studies with plants [20, 21], slime mould [22–25] and fungi [26] have shown that it is possible to develop electrical analog computing circuits either based or with these living creatures. Biological molecules such as suine microtubules have been shown to enable very fast oscillations, in the tents of MHz range [27]. However, to have a full functional analog computer, we probably need an oscillator. As Horowitz and Hill reported—"A device without an oscillator either doesn't do anything or expects to be driven by something else (which probably contains an oscillator)" [28]. Thus, we envisage that the resistive spiking can be utilised to produce fungal electronic oscillators.

2 Methods

Oyster fungi *P. ostreatus* have been cultivated on hemp substrate in plastic containers in darkness and at ambient temperature 20–23 °C. We used the substrate after it was nearly fully colonised by mycelium, which was indicated by an almost everywhere white colour and white film of nearly homogeneous mycelium, sometimes called 'skin' on surface of the substrate that was formed. The electrical resistance of the skin was measured as follows. We used iridium-coated stainless steel sub-dermal needle electrodes (Spes Medica S.r.l., Italy), with twisted cables. The pairs of electrodes were inserted in fungal skin, while the distance between electrodes was kept 1 cm (Fig. 1a). Twelve trials of measurements were undertaken with fungal skin. In six trials, we also undertook recordings of the fruit body's resistance, where electrodes were inserted in stalks of the bodies (Fig. 1b). The resistance was measured using the Kelvin, or 4-wire, resistance measurement method (Fig. 1c) and logged using Fluke 8846A precision multimeter, the test current being 1 ± 0.0013 µA (each measurement taken in less than 10 ms) once per 10 s, $5 \cdot 10^4$ samples per trial. When characterising trains of spikes, we measured spike average width w, average amplitude a and average distance between spikes d. To check if there are potential oscillations of voltage applied to the fungi, we applied direct current voltage with GW Instek GPS-1850D laboratory DC power supply and measured voltage with Fluke 8846A (Fig. 1d).

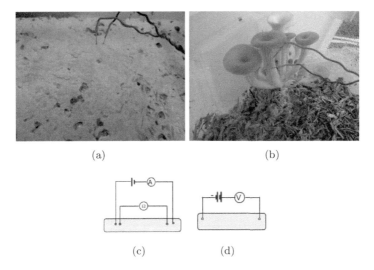

Fig. 1 Experimental setup. **a** Electrodes are inserted in a hemp substrate nearly fully colonised by *P. ostreatus*. **b** Electrodes are inserted in stalk of a fruit body of *P. ostreatus*. **c** Scheme of electrical resistance measurement. **d** Scheme of measuring fungal electrical potential under DC applied

3 Results

The resistance of fungal mycelium exhibits very slow, $1.5 \cdot 10^4$–$3 \cdot 10^4$ s, disordered changes of the resistance values with trains of spikes, of increased resistance, emerging. An example of the long-term recording is shown in Fig. 2a and a train of spikes in Fig. 2b. In over 16 trials we inferred the following parameters of spikes: $w = 1380$ s (median 1190 s, $\sigma = 77$), $a = 1{,}036$ Ω (median 815 Ω, $\sigma = 674$), $d = 1830$ s (median 175 s, $\sigma = 87$). Spike width versus amplitude distribution is shown in Fig. 2c.

Two trials of the resistance recording from substrate, colonised by fungi, have shown outstanding phenomena (although these have not been explicitly included in the above analysis).

More specifically, in one trial (not included in the analyses above) we observed high frequency ($w = 86$ s, median 86 s, $\sigma = 7$ and $d = 283$ s, median 250 s, $\sigma = 166$) and high amplitude ($a = 11{,}448$ Ω, median 10,750 Ω, $\sigma = 3{,}664$). An example of these high amplitude spikes is shown in Fig. 2d.

In another trial we observed very slow variations of resistance (c. $3 \cdot 10^4$ from the start of the ascent to the end of the descent). On tops of these variations there were trains of 5–7 spikes. An example is shown in Fig. 2e. Average widths w of these spike is $1{,}094$ s ($\sigma = 475$), $a = 158$ Ω ($\sigma = 49$) and $d = 1{,}135$ s ($\sigma = 442$).

In fruit bodies, we typically recorded two types of spikes: low frequency ($w = 1{,}690$ s, $\sigma = 32$ and $d = 3{,}450$ s, $\sigma = 161$) and high amplitude ($a = 1{,}632$ Ω, $\sigma = 116$) and high frequency ($w = 580$ s, $\sigma = 16$ and $d = 2{,}630$ s, $\sigma = 188$) and

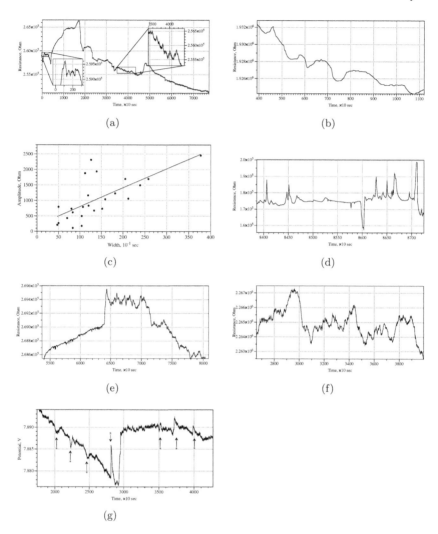

Fig. 2 a Slow variations of resistance with trains of spikes, zoomed in the inserts, are usually observed in long term recordings. **b** Example of a train of 5 spikes. **c** Spike width w versus amplitude a distribution. Line is a linear fit $a = 6w + 195$, $R^2 = 0.49$. **d** Examples of high amplitude and high frequency spikes. **e** Example of a spike train on top of a very slow variation of resistance. **f** Example of resistance recorded on fruit bodies. **g** Oscillation of electrical potential under 10 V DC applied, where spikes analysed are marked by '*'

low amplitude ($a = 611$ Ohm, $\sigma = 266$). Figure 2f shows a typical train of four high frequency spikes followed by a train of low frequency spikes.

To assess feasibility of the living fungal oscillator, we conducted a series of scoping experiments by applying direct voltage to the fungal substrate and measuring output voltage. An example of the electrical potential of a substrate colonised by fungi under 10 V applied is shown in Fig. 2g. Voltage spikes are clearly observed. Spikes with amplitude above 1 mV, marked by '*', except the spike marked by 's' have been analysed. We can see two trains of three spikes each. Average width of the spikes is 1,050 s ($\sigma = 9.2$, median 1,090 s), average amplitude 2.5 mV ($\sigma = 0.68$, median 2.2 mV), while average distance between spikes is 2,318 s ($\sigma = 25.6$, median 2370 s).

4 Discussion

We demonstrated that oyster fungi *P. ostreatus* undergo oscillations of resistance with trains of resistive spiking emerging. Spikes amplitude vary from 1 to 1.6 kΩ and width of spikes from 23 up to 28 min. A distance between spikes in a train varies from 30 to nearly 60 min. The oscillations of resistance have so low frequency that could not be explained using conventional electronics framework (e.g. charging of a mycelium during probing) and resistance sampling was with very low frequency (once per 10 s). Thus the only feasible explanation, we see is the translocation of water and metabolites taking place in the mycelium. This translocation is periodic, and more likely guided by calcium waves. Increase in a liquid in the mycelium loci leads to reduced resistance. When the translocated mass of metabolites leaves the area, the resistance increases. Rates of the translocation, measured by injecting fluorescent dye in hyphae, reported in [29] are 2–6 cm/hour for small specimen and 9–15 cm/hour for large specimen. The distance between electrodes in our experiments was 1 cm, thus the above rate can be translated to the following width of resistive spikes—10–30 min and 4–7 min. The first estimates matches in scale with resistive spikes measured in our experiments. The widths of resistive spikes are twice the widths of electrical potential spikes observed by us previously in fruit bodies of *P. ostreatus* [30]. All the above indicate that the resistive spiking observed is not an artefact but manifestation of physiological processes in fungal mycelium and fruit bodies. Therefore one of the application domains of the proposed methodological setup and delivered results could be in monitoring physiological states of fungi: the physiological states might reflect states of ecosystems inhabited by fungi [31]. In experiments with fungal oscillator we have found that at some stages the fungal skin exhibits oscillations of the electrical potential. A width of a voltage spike is c. 18 min, which is slightly less than an average width of resistive spikes, and an average amplitude c. 2.5 mV (at 10 V DC voltage applied). The amplitude is not as high as would be expected from our previous experience with slime mould electronic oscillators [32]. This may be due to the fact that in the experiments with slime mould, the electrodes were connected by a single protoplasmic tube, so its resistance was crucial, while in the fungal skin the current can also propagate along remnants of the wet hemp substrate. A very low frequency of fungal electronic oscillators does not preclude us from considering inclusion of the oscillators in fully living or hybrid analog circuits embedded into fungal architectures [33] and future specialised circuits and processors made from living and functionalised with nanoparticles fungi as have been illustrated in prototypes of hybrid electronic devices with slime mould [22, 34–37].

References

1. Crile, G.W., Hosmer, H.R., Rowland, A.F.: The electrical conductivity of animal tissues under normal and pathological conditions. Amer. J. Physiol.-Legacy Content **60**(1), 59–106 (1922)
2. Schwan, H.P., Kay, C.F.: Specific resistance of body tissues. Circul. Res. **4**(6), 664–670 (1956)
3. McAdams, E.T., Jossinet, J.: Tissue impedance: a historical overview. Physiol. Meas. **16**(3A), A1 (1995)
4. Héroux, P., Bourdages, M.: Monitoring living tissues by electrical impedance spectroscopy. Ann. Biomed. Eng. **22**(3), 328–337 (1994)
5. Dean, D.A., Ramanathan, T., Machado, D., Sundararajan, R.: Electrical impedance spectroscopy study of biological tissues. J. Electrostat. **66**(3–4), 165–177 (2008)
6. Gersing, E.: Impedance spectroscopy on living tissue for determination of the state of organs. Bioelectrochem. Bioenerg. **45**(2), 145–149 (1998)
7. Skutt, H.R., Shigo, A.L., Lessard, R.A.: Detection of discolored and decayed wood in living trees using a pulsed electric current. Can. J. Fore. Res. **2**(1), 54–56 (1972)
8. SA Al Hagrey: Electrical resistivity imaging of tree trunks. Near Surf. Geophys. **4**(3), 179–187 (2006)
9. Taper, C.D., Ling, R.S.: Estimation of apple rootstock vigor by the electrical resistance of living shoots. Can. J. Bot. **39**(7), 1585–1589 (1961)
10. Zhang, M.I.N., Willison, J.H.M.: Electrical impedance analysis in plant tissues: the effect of freeze-thaw injury on the electrical properties of potato tuber and carrot root tissues. Can. J. Plant Sci. **72**(2), 545–553 (1992)
11. Estrela Da Silva, J., Marques De Sá, J.P., Jossinet, J.: Classification of breast tissue by electrical impedance spectroscopy. Med. Biol. Eng. Comp. **38**(1), 26–30 (2000)
12. Carlile, M.J., Watkinson, S.C., Gooday, G.W.: The Fungi. Gulf Professional Publishing (2001)
13. Bahn, Y.-S., Xue, C., Idnurm, A., Rutherford, J.C., Heitman, J., Cardenas, M.E.: Sensing the environment: lessons from fungi. Nat. Rev. Microbiol. **5**(1), 57 (2007)
14. Van Aarle, I.M., Olsson, P.A., Söderström, B.: Arbuscular mycorrhizal fungi respond to the substrate ph of their extraradical mycelium by altered growth and root colonization. New Phytol. **155**(1), 173–182 (2002)
15. Kung, C.: A possible unifying principle for mechanosensation. Nature **436**(7051), 647 (2005)
16. Fomina, M., Ritz, K., Gadd, G.M.: Negative fungal chemotropism to toxic metals. FEMS Microbiol. Lett. **193**(2), 207–211 (2000)
17. Bahn, Y.-S., Mühlschlegel, F.A.: Co2 sensing in fungi and beyond. Curr. Opin. Microbiol. **9**(6), 572–578 (2006)
18. Howitz, K.T., Sinclair, D.A.: Xenohormesis: sensing the chemical cues of other species. Cell **133**(3), 387–391 (2008)
19. Adamatzky, A.: Advances in Physarum Machines: Sensing and Computing with Slime Mould, vol. 21. Springer (2016)
20. Gizzie, N., Mayne, R., Patton, D., Kendrick, P., Adamatzky, A.: On hybridising lettuce seedlings with nanoparticles and the resultant effects on the organisms' electrical characteristics. Biosystems **147**, 28–34 (2016)

21. Gizzie, N., Mayne, R., Yitzchaik, S., Ikbal, M., Adamatzky, A.: Living wires-effects of size and coating of gold nanoparticles in altering the electrical properties of physarum polycephalum and lettuce seedlings. Nano LIFE **6**(01), 1650001 (2016)
22. Berzina, T., Dimonte, A., Cifarelli, A., Erokhin, V.: Hybrid slime mould-based system for unconventional computing. Int. J. Gen. Syst. **44**(3), 341–353 (2015)
23. Romeo, A., Dimonte, A., Tarabella, G., D'Angelo, P., Erokhin, V., Iannotta, S.: A bio-inspired memory device based on interfacing physarum polycephalum with an organic semiconductor. APL Mater. **3**(1), 014909 (2015)
24. Berzina, T., Dimonte, A., Adamatzky, A., Erokhin, V., Iannotta, S.: Biolithography: Slime mould patterning of polyaniline. Appl. Surf. Sci. **435**, 1344–1350 (2018)
25. Cifarelli, A., Dimonte, A., Berzina, T., Erokhin, V.: On the loading of slime mold physarum polycephalum with microparticles for unconventional computing application. BioNanoScience **4**(1), 92–96 (2014)
26. Beasley, A.E., Powell, A.L., Adamatzky, A.: Capacitive storage in mycelium substrate (2020). arXiv:2003.07816
27. Wei, Z., Hillier, S., Gadd, G.M.: On resistance switching and oscillations in tubulin microtubule droplets. Environ. Microbiol. **14**, 589–595 (2020)
28. Horowitz, P., Hill, W.: The Art of Electronics. Cambridge University Press (1980)
29. Schütte, K.H.: Translocation in the fungi. New Phytologist **55**(2), 164–182 (1956)
30. Adamatzky, A.: On spiking behaviour of oyster fungi pleurotus djamor. Sci. Rep. **8**(1), 1–7 (2018)
31. Kranabetter, J.M., Friesen, J., Gamiet, S., Kroeger, P.: Epigeous fruiting bodies of ectomycorrhizal fungi as indicators of soil fertility and associated nitrogen status of boreal forests. Mycorrhiza **19**(8), 535–548 (2009)
32. Adamatzky, A., Schubert, T.: Slime mold microfluidic logical gates. Mater. Today **17**(2), 86–91 (2014)
33. Adamatzky, A., Ayres, P., Belotti, G., Wösten, H.: Fungal architecture position paper. Int. J. Unconv. Comput. **14**, 123 (2019)
34. Whiting, J.G.H., Mayne, R., Moody, N., de Lacy Costello, B., Adamatzky, A.: Practical circuits with physarum wires. Biomed. Eng. Lett. **6**(2), 57–65 (2016)
35. Walter, X.A., Horsfield, I., Mayne, R., Ieropoulos, I.A., Adamatzky, A.: On hybrid circuits exploiting thermistive properties of slime mould. Sci. Rep. **6**, 23924 (2016)
36. Ntinas, V., Vourkas, I., Sirakoulis, G.C., Adamatzky, A.I.: Oscillation-based slime mould electronic circuit model for maze-solving computations. IEEE Trans. Circuits Syst. I Reg. Pap. **64**(6), 1552–1563 (2017)
37. Adamatzky, A.: Twenty five uses of slime mould in electronics and computing: survey. Int. J. Unconv. Comput. **11** (2015)

Fungal Capacitors

Konrad Szaciłowski, Alexander E. Beasley, Krzysztof Mech, and Andrew Adamatzky

Abstract The emerging field of living technologies aims to create new functional hybrid materials in which living systems interface and interact with inanimate ones. Combining research into living technologies with emerging developments in computing architecture has enabled the generation of organic electronics from plants and slime mould. Here, we expand on this work by studying capacitive properties of a substrate colonised by mycelium of grey oyster fungi, *Pleurotus ostreatus*. Capacitors play a fundamental role in traditional analogue and digital electronic systems and have a range of uses including sensing, energy storage and filter circuits. Mycelium has the potential to be used as an organic replacement for traditional capacitor technology. Here, wer show that the capacitance of mycelium is in the order of hundreds of picofarads and at the same time a voltage-dependent pseudocapacitance of the order of hundreds of microfarads. We also demonstrate that the charge density of the mycelium 'dielectric' decays rapidly with increasing distance from the source probes. This is important as it indicates that small cells of mycelium could be used as a charge carrier or storage medium, when employed as part of an array with reasonable density.

1 Introduction

The study of novel substrates for sensing, storing and processing information draws on work from the fields of unconventional computing, living technology and organic electronics. The field of unconventional computing aims to define the principles of information processing in living, physical and chemical systems and applies this

---

K. Szaciłowski · K. Mech
Academic Centre for Materials and Nanotechnology, AGH University of Science and Technology, Kraków, Poland

A. E. Beasley
ARM Ltd, 110 Fulbourn Rd, Cambridge CB1 9NJ, UK

A. Adamatzky (✉)
Unconventional Computing Laboratory, UWE, Bristol, UK
e-mail: andrew.adamatzky@uwe.ac.uk

© The Author(s), under exclusive license to Springer Nature Switzerland AG 2023
A. Adamatzky (ed.), *Fungal Machines*, Emergence, Complexity and Computation 47, https://doi.org/10.1007/978-3-031-38336-6_14

knowledge to the development of future computing devices and architectures [1]. Research into living technologies is focused on the co-functional integration of animate and non-organic systems [2]. Finally, organic electronics looks to use naturally occurring materials as analogues to traditional semi-conductor circuits [3], which often requires functionalisation using polymers or metallic compounds to exploit ionic movement [4]. The development of organic electronics promises a technology that is low-cost and has low production temperature requirements [5]. However, difficulties such as relatively low gain of organic transistors (approx. 5), and the behavioural variability inherently means that there is a large amount of research effort being placed into organic thin film transistors [6–9], organic LEDs [10, 11], and organic capacitors [12–14]. The capacitive properties of a device allows it to either store energy or react to AC/DC signals differently. There are a number of applications in which this property may be utilised, such as energy harvesting [15] and memory [16]. Hybrid electronic circuits are a concept that looks to combine traditional silicon, semi-conductor devices with elements found in nature [17, 18].

The capacitive properties of living tissues [19] have a wide range of potential applications, e.g. the estimation of a plan root system size [20, 21], quantifying the DNA content of eukaryotic cells [22], analysing water transport pathways in plants [23], measuring heat injury in plants [24], measuring contents of minerals in bones [25], gauging firmness of apples [26], sugar contents of citrus fruits [27], maturity of avocados [28], estimating depth of epidermal barriers [29], studies of endo- and exocytosis of single cells [30], and approximating mass and morphology of microbial colonies [31–33].

Among various possible biological substrates for electronics, fungal mycelia are one of the most promising materials.

Continuously increasing demand for electronics of new capabilities toward sensing, signal analysis, data acquisition, and information processing pushes scientists for searching for unconventional approaches creating new possibilities that may allow fulfilling the growing requirements of users. The utilisation of fungal mycelium seems to be an interesting approach that may enable the fabrication of a new class of green wearable electronics of the future.

In the present study we focus on the capacitive properties of the mycelium of the grey oyster fungi *Pleurotus ostreatus* for several reasons.

Firstly, research into the capacitive properties of fungi is lacking, despite their huge potential for bioelectronic applications. Fungi are the largest, most widely distributed and oldest group of living organisms on the planet [34]. The smallest fungi are microscopic single cells, the largest, *Armillaria bulbosa*?, occupies 15 hectares and weighs 10 tons [35]. Fungi sense light, chemicals, gases, gravity and electric fields [36] as well as demonstrating mechanosensing behaviour [37]. Thus, their electrical properties can be tuned via various inputs.

Secondly, fungi have the potential to be used as distributed living computing devices, i.e. large-scale networks of mycelium, which collect and analyse information about environment and execute some decision making processes [38].

Finally, there is a growing interest in developing buildings from pre-fabricated blocks of substrates colonised by fungi [39–41]. A recent initiative aims to grow monolithic constructions in which living mycelium coexists with dried mycelium, functionalised with nanoparticles and polymers [42]. In such a case, fungi could be act as optical, tactile and chemical sensors, fuse and process information and perform decision making computations [38].

Providing local charge to areas of mycelium allows the storage of information inside the substrate. Identifying the area around which the induced charge can be detected allows the construction of an array in the substrate where each cell can contain individual bits of information. Determining the capacitive properties of fungi takes a step towards the realisation of fungal analogue circuits—circuits that use fungi to replace traditional semiconductors.

2 Experimental Method

Mycelium of the grey oyster fungi *Pleurotus ostreatus* (Ann Miller's Speciality Mushrooms Ltd, UK) was cultivated on damp wood shavings (Fig. 1). Control samples of the growth medium, wood shavings, were not colonised by mycelium. Iridium-coated stainless steel sub-dermal needles with twisted cables (Spes Medica SRL, Italy) were inserted in the colonised substrate. An LCR meter (BK Precision, model 891) was used to provide a nominal reading of the capacitance of the sample with probes at 10, 20, 40 and 50 mm separation. Cyclic voltammograms in two electrode setup has been recorded with with SP-150 potentiostat (Bio-Logic, France) in the dark. Impedance spectra in the 1 Hz–100 kHz frequency range were recorded with Bode 100 vector network analyser (Omicron Labs, Austria). Voltage dependence of pseudocapacitance was measured using chronocoulometric module of Bio-Logic potentiostat.

The samples were charged using a bench top DC power supply (BK precision 9206) to 50 V. The power supply output was de-activated and the discharge curve was measured using a bench top digital multi-meter (DMM) Fluke 8846A (Fig. 1b). To fully characterise the capacitance of the samples, both the charge and discharge curves were monitored [43] with a number of probe separations (e.g. 10, 20, 40, and 50 mm). Measurements from the DMM were automated through a serial terminal from a host PC. All bench-top equipment was high impedance to limit power lost through leakage in the test equipment. All plots were generated using MATLAB. The sample interval was approximately 0.33 s. Multiple samples and repeated experimental recordings were used to increase the statistical significance of the findings.

Fig. 1 Experimental setup. **a** A sample of the mycelium under test. **b** Voltage discharge measuring set up for mycelium samples

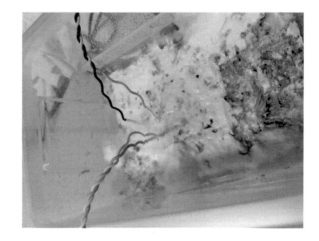

(a)

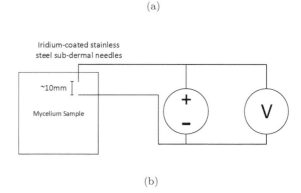

(b)

3 Results

Capacitance Measurement

Samples of the growth medium and mycelium were measured for their capacitance using a standard bench top LCR meter (Table 1). The measured capacitance value of the mycelium substrate was two to four fold greater than that of the growth medium alone. The value recorded for dry wood shavings corresponds to the capacitance of the twisted cable inserted into the sample and served as a background value. Increased humidity of the pristine substrate is higher. The values of capacitance were also effected by the moisture content such that, if the capacitance of the mycelium was measured straight after it is sprayed with water, the capacitance typically increased compared to that of dry substrate.

Table 1 Capacitance of mycelium and growth mediums

Sample	Electrode spacing (mm)	Capacitance (pF)
Dry wood shavings	10	37.6
	20	37.4
Damp wood shavings	10	57
	20	58.6
Mycelium (drying)	10	184
	20	144
	30	146
	40	120
	50	118
Mycelium (freshly watered)	10	193
	20	186
	30	134
	40	125
	50	139

Discharge Characteristics

Discharge characteristics of the growth medium and mycelium samples are shown in Figs. 2 and 3. Discharge curves were produced by setting up the DC power supply and DMM in parallel with each other. The substrate being tested is then charged to 50 V and the power supply output is disabled. The DMM continued to periodically measure the remaining charge in the substrate for a period of time. The sample interval was approximately 0.33 s.

The discharge curves for both the growth medium and the mycelium are very steep—approximated by an exponential Equation (1):

$$f(x) = a \cdot e^{b \cdot x} \tag{1}$$

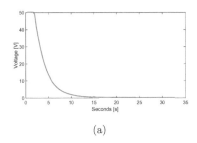
(a)

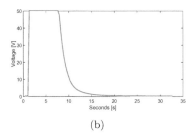
(b)

Fig. 2 Discharge of a substrate after being charged to 50 V with probes separation of 10 mm. **a** Dry wood shavings. **b** Damp wood shavings—shavings are immersed in water for half and hour after which excess water is drained. Data are discrete. Line is for eye guidance only

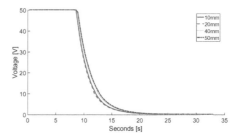

Fig. 3 Mycelium sample is charged to 50 V then allowed to discharge. Electrode spacing is varied (10, 20, 40 and 50 mm). Data are discrete. Line is for eye guidance only

Table 2 Discharge curve fitness approximation coefficients (with 95% confidence bounds)

Sample	a	b
Dry wood shavings	104.9 (102.6, 107.2)	−0.4079 (−0.4151, −0.4007)
Damp wood shavings	6205 (5164, 7245)	−0.6261 (−0.6464, −0.6058)
Mycelium w/10 mm probe separation	2276 (2198, 2354)	−0.4446 (−0.4481, −0.441)
Mycelium w/20 mm probe separation	2688 (2544, 2833)	−0.4639 (−0.4695, −0.4583)
Mycelium w/40 mm probe separation	1569 (1458, 1680)	−0.3948 (−0.4021, −0.3875)
Mycelium w/50 mm probe separation	1413 (1311, 1516)	−0.3817 (−0.3891, −0.3743)

where the parameters for 95% fitness for different mediums can be found in Table 2.

The discharge time is governed by equation $\tau = RC$, where τ is the time constant, R is a resistance, and C is a capacitance.

With capacitance in the order of pico-Farads, and input impedance of the source and measurement equipment in the order of mega-Ohms, it is expected that the discharge will be in the order of seconds.

Comparing the discharge curves of the growth medium to that of the mycelium samples, it is evident that the discharge was not as steep in the mycelium due to the increase in capacitance over the growth medium. However, it was still in the pico-Farad range and, therefore, the majority of charge was lost after just over 5 s. Increasing the separation distance of the probes (Fig. 3) had only a minimal effect on the capacitance of the substrate (shown in Table 1), and therefore minimal effect on the discharge curve.

Observing the charging and discharging behaviour of the sample around the source electrodes helps to build a better picture of the current density of the mycelium substrate. Figure 4 shows how we placed the measurement equipment electrodes 10 mm away from the source electrodes in the mycelium, in three different locations shown in Fig. 5. The sample was then charged for approximately 10 mins before the supply was turned off. The electrodes placed in 'series' (locations (a) and (b) on

Fungal Capacitors

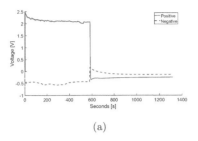
(a)

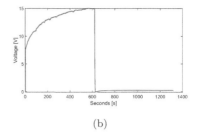
(b)

Fig. 4 Charging mycelium to 50 V with source electrodes 10 mm apart. Substance was charged for approx. 10 min, readings are taken for approx. 22 min in total. **a** Sense electrodes were placed in series, with the source terminals (10 mm away from either positive or negative electrodes). **b** Sense electrodes were in parallel from the source electrodes (10 mm clearance). Data are discrete. Line is for eye guidance only

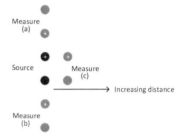

Fig. 5 Measurement probes are arranged around the source probes to examine the charge field in the substrate

Fig. 5) with the charge electrodes (Fig. 4a) show minimum voltage detected beyond the supply electrodes in the horizontal plane, when the supply is active. Placing the measurement probes in 'parallel' (location (c) on Fig. 5) with the supply probes (Fig. 4b) demonstrates the fact that considerably more current is conducted between the supply probes in the vertical plane. When the supply is de-activated, the voltage around the supplies collapses almost instantly.

In order to provide better contact and a more consistent data set, another batch of freshly inoculated substrate was placed in a box with gold-plated silver electrodes (multiply twisted loop, ca 1.5 cm long) with spacing of 2 cm. Changes in electrical properties of growing mycelium samples can be seen using cyclic voltammetry technique as shown in Fig. 6a. These changes, however, does not seem fully systematic and the first (and only one clearly defined) jumps randomly between 0.23 and 0.36 V. This behaviour may be associated with accumulation of more redox active metabolite and indicate that simple voltammetry may not be the best technique for monitoring of growth and evolving properties of mycelium.

Impedance spectroscopy (Fig. 6b) indicates a gradual changes in the spectra, reflecting the mycelium growth and demonstrating that almost fresh (wet substrate) and substrate with well-grown mycelium have completely different electrical properties (Fig. 6). Observed data indicate that both capacity and resistance of the sample change gradually. In first days after inoculation changes are fast, however after day 4 changes are much smaller. Character of recorded impedance spectra indicates a com-

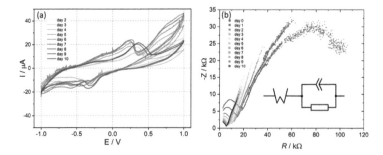

Fig. 6 **a** Cyclic voltammograms recorded during 10 days of mycelium growth. **b** Impedance spectra recorded within 10 days after inoculation of fresh wet substrate with *P. ostreatus* mycelium. Electrode spacing is 20 mm. An inset shown the equivalent circuit used for parameter calculation. Please note that non-ideal constant phase element is used instead of capacitor

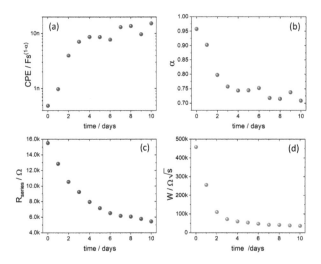

Fig. 7 Time-dependence of the fitting parameters of the equivalent circuit from Fig. 6b

plex electrical behaviour of studied samples—along with capacitive-like behaviour a significant contribution from diffusion-related Warburg-type component can be seen, especially with more matured mycelium. Therefore an equivalent circuit encompassing all these element was suggested.

The changes in the electrical properties of growing mycelium are logical and consistent with expected mycelium topology of percolated network, with percolation increasing with time (Fig. 7). It can be clearly seen as gradually decreasing resistivity of the network, both Ohmic and Warburg components. Interestingly, capacitance of the network stabilised after ca. 4 days, this is also associated with decreasing capacitor quality factor. This is fully justified, as increased percolation of the network created more conductivity pathways of different time constants—proton and ionic conductivity both along the mycelium hyphae and between them. Capacitance

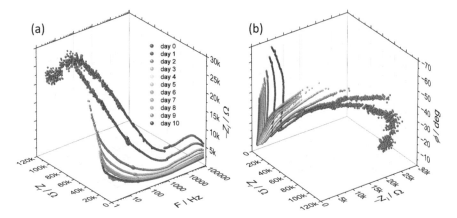

Fig. 8 3D representation of Impedance data: **a** frequency-dependent Nyquis plot and **b** the relation between real impedance (Z_r), imaginary impedance (Z_i) and phase shift angle (ϕ)

can be associated both with electrode/mycelium interface, and contacts between different hyphae. Time course of observed changes suggests the latter as the principal component of observed capacitive behaviour.

This description is a bit simplistic and does not include the whole complexity of electrical properties of mycelium. 3D representation of impedance spectra indicates, much larger complexity of their capacitive response (Fig. 8).

Charge Characteristics

The charge curves in the specimens around the supply electrodes provide an insight into the ability of the substrate to conduct current. The sense electrodes were distanced from the source electrodes by varying amounts. Initially, the growth medium was studied on its own (Fig. 9). The dry wood shavings (Fig. 9a) showed very low voltage across the sense electrodes placed 10 mm away from the source electrodes

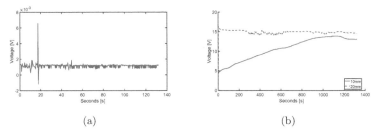

Fig. 9 Charge dynamics of growth substrate with measurement equipment set in parallel with charge electrodes. Electrode pairs were 10 mm apart. **a** Dry wood shavings. **b** Damp wood shavings

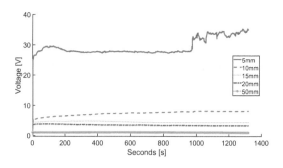

Fig. 10 Charging of mycelium sample with measurement equipment set to measure at different distances from supply (5, 10, 15, 20, and 50 mm). Supply electrodes and measurement electrodes were arranged in parallel with each other

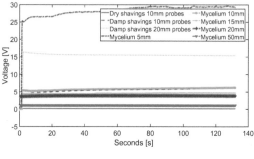

Fig. 11 Charge characteristics of wet and dry growth medium and and mycelium with measurement probes at different distances from source probes

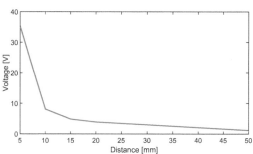

Fig. 12 Maximum measured voltage at increasing distance from supply probes in mycelium. Data are discrete. Line is for eye guidance only

(parallel). The dry shavings essentially acted as an open circuit and the measurement electrodes picked up noise. Damp wood shavings form a more contiguous mass and the introduction of the water helped to conduct current (Fig. 9b). From a 50 V source a maximum voltage of approx. 15 V was reached across the measurement electrodes. Figure 11 shows all charge curves for growth medium and mycelium. From the superimposed plots we are able to observe the maximum voltage recorded from the samples as the distance of the measurement probes are increased away from the charge probes.

Preforming similar experiments with the mycelium substrate (Fig. 10), it was observed that moving the measurement electrodes further from the supply reduced the measured V_{max} over the $\tilde{2}2$ min measurement window. Figure 12 shows that the maximum measured voltage dropped rapidly as the distance from the source elec-

Fig. 13 Current density between the two source electrodes. Current flow is shown in the physical direction rather than convention

trodes increased. At 10 mm away from the source, less than 1/5th of the supply was measured over 22 mins, decreasing to less than 1/10th at 15 mm separation. Figure 13 shows that, for the shortest paths between the two source electrodes, there was a higher current density, indicated by the field lines being closer together. As we move further away from the centre of the two probes, the current density decreased (arrows are shown further apart). Beyond the two probes in the 'y–direction', we would expect to experience very little current flow, however there are still fringing field effects which give rise to the small voltages shown in Fig. 4a. Modern parallel plate capacitors have a large electrode/dielectric surface area to increase the capacitance, and multi-layer capacitors interdigitate the electrode and dielectric to achieve the required surface area in a very small package. Here we have not explored the effect of surface area on capacitance, however, it can be seen from equation (2) that increasing the area of the electrodes, in contact with our mycelium dielectric, would lead to an increase in the capacitance. This was addressed by monitoring the mycelium growth on electrodes implanted into freshly inoculated medium (*vide supra*).

$$C = \frac{\epsilon A}{d} \qquad (2)$$

Where C is the capacitance, ϵ is the relative permittivity, A is the surface area, and d is the separation of the electrodes.

Additionally, the moisture content of the mycelium had an impact on its ability to conduct current. Figure 14 shows that, if the sample of mycelium is continually charged over a period where it is also drying out, the conducted charge can decrease rapidly as the water content vanishes. The measurement electrodes for this sample are 5 mm away from the source, however we see a decrease in measured voltage of almost 15 V over the measurement window.

Application of high potentials (50 V in the case of former experiments) may, however result in electrolysis of mycelial fluid and some irreversible changes. In order to evaluate utility of fungal capacitor charge-discharge tests have been performed with much lower voltages and chronoamperimetric detection. Charging pulses with amplitudes varying from 0.2 to 1.0 V were applied to pairs of electrodes with 2 cm clearance for 30 s, and then discharge current at short circuit conditions was recorded for additional 90 s. Resulted current profiles are shown in Fig. 15a. It can be observed, that effective total capacitance (Fig. 15b) is much higher than results obtained at low amplitude impedance spectroscopy (Fig. 7). It can be justified by a mixed capacitance character with a dominating contribution of electrochemical pseudocapacitance,

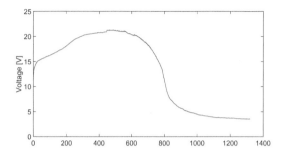

Fig. 14 Mycelium charge measured 5 mm away from 50 V source while mycelium sample starts to dry out. Data are discrete. Line is for eye guidance only

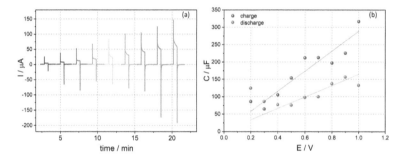

Fig. 15 a Charge/discharge current profiles. **b** Capacitance vs applied voltage for calculated by integration from appropriate parts of the charge/discharge curves

additionally with a significant Faradaic component (please note quasi-reversible peaks in Fig. 6a. We can hypothesise, that ionic processes at the surface of biological membranes are the dominating factor, as the voltage dependence of capacitance is monotonous, within experimental error. This excluded Faradaic component as a dominating one, however its importance should be also noted. It is further supported by observed discrepancy between charge stored during charging pulse and charge collected during discharge. This discrepancy is voltage-independent and indicates significant charge loss, most probably due to ionic diffusion within hyphae.

4 Conclusions

Mycelium exhibits rather unconventional, voltage- and frequency- dependent capacitive characteristics. Depending on the frequency measurement and applied voltage, capacitance of mycelium samples very form hundreds of picofarads to hundreds of microfarads. This enormous discrepancy, spanning six orders of magnitude, is, however not unjustified. In reflects complexity of electrical behaviour of mycelium, which is a consequence of its molecular composition, structure and topology. We cannot expect simple electrical behaviour from wet fibrous hemp-derived substrate, overgrown with living mycelium. Complex topology of the materials, combined with biochemical processes creates a challenging system to study. Simple measurements with benchtop multimeter do not yield high capacitance values, probably due to non-optimal contact of electrodes with mycelium and substrate. High voltage measurements, it turn, may induce some irreversible processes and electrolysis of mycelial cytosol, followed by cell disruption. Not harmful to the mycelium as a whole, these local damages are reflected in rather low charge storage capability due to damaged cell walls. AC measurements also do not give high capacitance values (up to 10 nF), suggesting contribution of diffusive processes (observed as Warburg impedance in impedance experiments) Finally, at low applied voltages, DC experiments demonstrate, however, high charge storage capacity significantly increased up to 150 microfarads. Due to leakage processes and rather low quality factor they cannot be efficiently used for energy storing application, however may be further explored for other application, e.g. sensing, mycelial communication and monitoring of mycelium growth. This shows great potential for the use of mycelium networks to conduct or store charge in local 'hot spots' that are isolated from other areas in the immediate vicinity, especially at low signal amplitudes. However, it is crucial that the moisture content of the mycelium is kept constant since the ability to carry charge is strongly influenced by moisture content.

Any potential analogue circuits implemented with live mycelium will be vulnerable to environmental conditions, especially humidity, availability of nutrients and removal of metabolites. Ideally, the mycelium networks should be stabilised so they continue functioning whilst drying. This stabilisation can be achieved either by coating or priming the mycelium with polyaniline (PANI) or poly(3,4-ethylenedioxythiophene) and polystyrene sulfonate (PEDOT-PSS). This approach has been proven to be successful in experiments with slime mould *P. polycephaum* [44, 44, 45], and thus it is likely that a similar technique may be applied to fungi. Moreover, PABI and PEDOT-PSS incorporated in, or interfaced with, mycelium can bring additional functionality in terms of conductive pathways [46], memory switches [47, 48] and synaptic-like learning [49, 50]. An optional route toward the functional fixation of mycelium would be doping the networks with substances that affect the electrical properties of mycelium, such as carbon nanotubes, graphene oxide, aluminium oxide, calcium phosphate. Similar studies conducted in our laboratory using slime mould and plants have shown that such an approach is feasible [51, 52]. Moreover using a combination of PANI and carbon nanotubes in the mycelium network afford it super-capacitive properties [53, 54]. Another potential direction of future studies would be to increase the capacity of the mycelium as a result of modifying the network geometry by varying nutritional conditions and temperature [55–58], concentration of nutrients [59] or with chemical and physical stimuli [36]. With regards to the impact of our finding for the field of unconventional computing, we believe further research on experimental laboratory implementation of capacitive threshold logic [60, 61], adiabatic capacitive logic [62] and capacitive neuromorphic architectures [63] will yield fruitful insights.

Acknowledgements Authors thank professor Kapela Pilaka for numerous valuable comments and fruitful discussions.

References

1. Adamatzky, A.: Advances in Unconventional Computing. Springer (2016)
2. Bedau, M.A., McCaskill, J.S., Packard, N.H., Rasmussen, S.: Living technology: exploiting life's principles in technology. Artif. Life **16**(1), 89–97 (2010)
3. Stavrinidou, E., Gabrielsson, R., Gomez, E., Crispin, X., Nilson, O., Simon, D.T., Berggren, M.: Electronic plants. Sci. Adv. **1**(10), 1–8 (2015)
4. Leger, J.M.: Organic electronics: the ions have it. Adv. Mater. **20**(4), 837–841 (2008)
5. Marien, H., Steyaert, M., Heremans, P.: Analog Organic Electronics. Springer (2013)
6. Zschieschang, U., Klauk, H.: Organic transistors on paper: a brief review. J. Mater. Chem. C **7**, 5522–5533 (2019)
7. Tokito, S.: Flexible printed organic thin-film transistor devices and integrated circuit applications. In: 2018 International Flexible Electronics Technology Conference (IFETC), pp. 1–2, Aug. 2018
8. Endoh, H., Toguchi, S., Kudo, K.: High performance vertical-type organic transistors and organic light emitting transistors. In: Polytronic 2007—6th International Conference on Polymers and Adhesives in Microelectronics and Photonics, pp. 139–142, Jan. 2007
9. Tang, W., Zhao, J., Li, Q., Guo, X.: Highly sensitive low power ion-sensitive organic thin-film transistors. In: 2018 9th International Conference on Computer Aided Design for Thin-Film Transistors (CAD-TFT), pp. 1–1, Nov. 2018
10. Sano, T., Suzuri, Y., Koden, M., Yuki, T., Nakada, H., Kido, J.: Organic light emitting diodes for lighting applications. In: 2019 26th International Workshop on Active-Matrix Flatpanel Displays and Devices (AM-FPD), vol. 26th, pp. 1–4, July 2019
11. Mizukami, M., Cho, S., Watanabe, K., Abiko, M., Suzuri, Y., Tokito, S., Kido, J.: Flexible organic light-emitting diode displays driven by inkjet-printed high-mobility organic thin-film transistors. IEEE Electron Device Lett. **39**(1), 39–42 (2018)
12. Li, W.H., Ding, K., Tian, H.R., Yao, M.S., Nath, B., Deng, W.H., Wang, Y., Xu, G.: Conductive metal-organic framework nanowire array electrodes for high performance solid state supercapacitors. Adv. Funct. Mater **27**, 1702067 (2017)
13. Sangermano, M., Vitale, A., Razza, N., Favetto, A., Paleari, M., Ariano, P.: Multilayer UV-cured organic capacitors. Polymer **56**, 131–134 (2015)
14. Morimoto, T., Tsushima, M., Suhara, M., Hiratsuka, K., Sanada, Y., Kawasato, T.: Electric double- layer capacitor using organic electrolyte. MRS Proc. **496**, 627 (1997)
15. Beasley, A.E., Bowen, C.R., Zabek, D.A., Clarke, C.T.: Use it or lose it: the influence of second order effects of practical components on storing energy harvested by pyroelectric effects. Tech. Mess. **85**(9), 522–540 (2017)
16. Vadasz, L.L., Chua, H.T., Grove, A.S.: Semiconductor random-access memories. IEEE Spectr. **8**(5), 40–48 (1971)
17. Lu, W., Lieber, C.M.: Nanoelectronics from the bottom up. Nanosci. Technol.: A Collect. Rev. Nat. J. 137–146. World Scientific (2010)
18. Beausoleil, R.G., Kuekes, P.J., Snider, G.S., Wang, S.-Y., Williams, R.S.: Nanoelectronic and nanophotonic interconnect. Proc. IEEE **96**(2), 230–247 (2008)
19. McAdams, E.T., Jossinet, J.: Tissue impedance: a historical overview. Physiol. Meas. **16**(3A), A1 (1995)

20. Chloupek, O.: Evaluation of the size of a plant's root system using its electrical capacitance. Plant Soil **48**(2), 525–532 (1977)
21. Rajkai, K., Végh, K.R., Nacsa, T.: Electrical capacitance as the indicator of root size and activity. Agrokémia és Talajtan **51**(1–2), 89–98 (2002)
22. Sohn, L.L., Saleh, O.A., Facer, G.R., Beavis, A.J., Allan, R.S., Notterman, D.A.: Capacitance cytometry: measuring biological cells one by one. Proc. Nat. Acad. Sci. **97**(20), 10687–10690 (2000)
23. Blackman, C.J., Brodribb, T.J.: Two measures of leaf capacitance: insights into the water transport pathway and hydraulic conductance in leaves. Funct. Plant Biol. **38**(2), 118–126 (2011)
24. Zhang, M.I.N., Willison, J.H.M., Cox, M.A., Hall, S.A.: Measurement of heat injury in plant tissue by using electrical impedance analysis. Can. J. Bot. **71**(12), 1605–1611 (1993)
25. Paul Allen Williams and Subrata Saha: The electrical and dielectric properties of human bone tissue and their relationship with density and bone mineral content. Ann. Biomed. Eng. **24**(2), 222–233 (1996)
26. Bhosale, A.A., Sundaram, K.K.: Firmness prediction of the apple using capacitance measurement. Procedia Technol. **12**, 163–167 (2014)
27. Bhosale, A.A.: Detection of sugar contents in citrus fruits by capacitance method. In: 10th International Conference Interdisciplinarity in Engineering, INTER-ENG 2016, pp. 466–471 (2016)
28. Zachariah, G., Erickson, L.C.: Evaluation of Some Physical Methods for Determining Avocado Maturity, vol. 49. California Avocado Society (1965)
29. Boyce, S.T., Supp, A.P., Harriger, M.D., Pickens, W.L., Wickett, R.R., Hoath, S.B.: Surface electrical capacitance as a noninvasive index of epidermal barrier in cultured skin substitutes in athymic mice. J. Invest. Dermatol. **107**(1), 82–87 (1996)
30. Rituper, B., Guček, A., Jorgačevski, J., Flašker, A., Kreft, M., Zorec, R.: High-resolution membrane capacitance measurements for the study of exocytosis and endocytosis. Nat. Protoc. **8**(6), 1169 (2013)
31. Fehrenbach, R., Comberbach, M., Petre, J.O.: On-line biomass monitoring by capacitance measurement. J. Biotechnol. **23**(3), 303–314 (1992)
32. Neves, A.A., Pereira, D.A., Vieira, L.M., Menezes, J.C.: Real time monitoring biomass concentration in streptomyces clavuligerus cultivations with industrial media using a capacitance probe. J. Biotechnol. **84**(1), 45–52 (2000)
33. Sarra, M., Ison, A.P., Lilly, M.D.: The relationships between biomass concentration, determined by a capacitance-based probe, rheology and morphology of saccharopolyspora erythraea cultures. J. Biotechnol. **51**(2), 157–165 (1996)
34. Carlile, M.J., Watkinson, S.C., Gooday, G.W.: The Fungi. Gulf Professional Publishing (2001)
35. Smith, M.L., Bruhn, J.N., Anderson, J.B.: The fungus Armillaria bulbosa is among the largest and oldest living organisms. Nature **356**(6368), 428 (1992)
36. Bahn, Y.-S., Xue, C., Idnurm, A., Rutherford, J.C., Heitman, J., Cardenas, M.E.: Sensing the environment: lessons from fungi. Nat. Rev. Microbiol. **5**(1), 57 (2007)
37. Kung, C.: A possible unifying principle for mechanosensation. Nature **436**(7051), 647 (2005)
38. Adamatzky, A., Tuszynski, J., Pieper, J., Nicolau, D.V., Rinalndi, R., Sirakoulis, G., Erokhin, V., Schnauss, J., Smith, D.M.: Towards cytoskeleton computers. A proposal. In: Adamatzky, A., Akl, S., Sirakoulis G (eds.) From Parallel to Emergent Computing. CRC Group/Taylor & Francis (2019)
39. Ross, P.: Your rotten future will be great. The Routledge Companion to Biology in Art and Architecture, p 252 (2016)
40. Appels, F.V.W., Camere, S., Montalti, M., Karana, E., Jansen, K.M.B., Dijksterhuis, J., Krijgsheld, P., Wösten, H.A.B.: Fabrication factors influencing mechanical, moisture-and water-related properties of mycelium-based composites. Mater. Des. **161**, 64–71 (2019)
41. Dahmen, J.: Soft matter: responsive architectural operations. Technoetic Arts **14**(1–2), 113–125 (2016)

42. Adamatzky, A., Ayres, P., Belotti, G., Wösten, H.: Fungal architecture position paper. Int. J. Unconvn. Comput. **14**, (2019)
43. Dulik, M., Jurecka, S.: Measuring capacitance of various types of structures. In: 2014 ELEKTRO, pp. 640–644 (2014)
44. Battistoni, S., Dimonte, A., Erokhin, V.: Organic memristor based elements for bio-inspired computing. In: Advances in Unconventional Computing, pp. 469–496. Springer (2017)
45. Cifarelli, A., Berzina, T., Erokhin, V.: Bio-organic memristive device: polyaniline–physarum polycephalum interface. Physica Status Solidi (c) **12**(1–2), 218–221 (2015)
46. Yoo, J.E., Bucholz, T.L., Jung, S., Loo, Y.-L.: Narrowing the size distribution of the polymer acid improves pani conductivity. J. Mater. Chem. 18(26):3129–3135, 2008
47. Howard, G.D., Bull, L., de Lacy Costello, B., Adamatzky, A., Erokhin, V.: A spice model of the peo-pani memristor. Int. J. Bifurc. Chaos **23**(06), 1350112 (2013)
48. Demin, V.A., Erokhin, V.V., Kashkarov, P.K., Kovalchuk, M.V.: Electrochemical model of polyaniline-based memristor with mass transfer step. In: AIP Conference Proceedings, vol. 1648, p. 280005. AIP Publishing LLC (2015)
49. Berzina, T., Smerieri, A., Bernabò, M., Pucci, A., Ruggeri, G., Erokhin, V., Fontana, M.P.: Optimization of an organic memristor as an adaptive memory element. J. Appl. Phys. **105**(12), 124515 (2009)
50. Lapkin, D.A., Emelyanov, A.V., Demin, V.A., Berzina, T.S., Erokhin, V.V.: Spike-timing-dependent plasticity of polyaniline-based memristive element. Microelectron. Eng. **185**, 43–47 (2018)
51. Gizzie, N., Mayne, R., Patton, D., Kendrick, P., Adamatzky, A.: On hybridising lettuce seedlings with nanoparticles and the resultant effects on the organisms' electrical characteristics. Biosystems **147**, 28–34 (2016)
52. Gizzie, N., Mayne, R., Yitzchaik, S., Ikbal, M., Adamatzky, A.: Living wires—effects of size and coating of gold nanoparticles in altering the electrical properties of Physarum polycephalum and lettuce seedlings. Nano LIFE **1**(6), 1650001 (2015)
53. Dong, B., He, B.-L., Cai-Ling, X., Li, H.-L.: Preparation and electrochemical characterization of polyaniline/multi-walled carbon nanotubes composites for supercapacitor. Mater. Sci. Eng. B **143**(1–3), 7–13 (2007)
54. Frackowiak, E., Khomenko, V., Jurewicz, K., Lota, K., Béguin, F.: Supercapacitors based on conducting polymers/nanotubes composites. J. Power Sour. **153**(2), 413–418 (2006)
55. Boddy, L., Wells, J.M., Culshaw, C., Donnelly, D.P.: Fractal analysis in studies of mycelium in soil. Geoderma **88**(3), 301–328 (1999)
56. Ha Thi Hoa and Chun-Li Wang: The effects of temperature and nutritional conditions on mycelium growth of two oyster mushrooms (Pleurotus ostreatus and Pleurotus cystidiosus). Mycobiology **43**(1), 14–23 (2015)
57. Rayner, A.D.M.: The challenge of the individualistic mycelium. Mycologia 48–71 (1991)
58. Regalado, C.M., Crawford, J.W., Ritz, K., Sleeman, B.D.: The origins of spatial heterogeneity in vegetative mycelia: a reaction-diffusion model. Mycol. Res. **100**(12), 1473–1480 (1996)
59. Ritz, K.: Growth responses of some soil fungi to spatially heterogeneous nutrients. FEMS Microbiol. Ecol. **16**(4), 269–279 (1995)
60. Ozdemir, H., Kepkep, A., Pamir, B., Leblebici, Y., Cilingiroglu, U.: A capacitive threshold-logic gate. IEEE J. Solid-State Cir. **31**(8), 1141–1150 (1996)
61. Medina-Santiago, A., Reyes-Barranca, M.A., Algredo-Badillo, I., Cruz, A.M., Gutiérrez, K.A.R., Cortés-Barrón, A.E.: Reconfigurable arithmetic logic unit designed with threshold logic gates. IET Circuits Devices Syst. **13**(1), 21–30 (2018)
62. Pillonnet, G., Fanet, H., Houri, S.: Adiabatic capacitive logic: a paradigm for low-power logic. In: 2017 IEEE International Symposium on Circuits and Systems (ISCAS), pp. 1–4. IEEE
63. Wang, Z., Rao, M., Han, J.-W., Zhang, J., Lin, P., Li, Y., Li, C., Song, W., Asapu, S., Midya, R., et al.: Capacitive neural network with neuro-transistors. Nat. Commun. **9**(1), 1–10 (2018)

Mem-Fractive Properties of Fungi

Alexander E. Beasley, Mohammed-Salah Abdelouahab, René Lozi, Michail-Antisthenis Tsompanas, and Andrew Adamatzky

Abstract Memristors close the loop for I-V characteristics of the traditional, passive, semi-conductor devices. A memristor is a physical realisation of the material implication and thus is a universal logical element. Memristors are getting particular interest in the field of bioelectronics. Electrical properties of living substrates are not binary and there is nearly a continuous transitions from being non-memristive to mem-fractive (exhibiting a combination of passive memory) to ideally memristive. In laboratory experiments we show that living oyster mushrooms *Pleurotus ostreatus* exhibit mem-fractive properties. We offer a piece-wise polynomial approximation of the I-V behaviour of the oyster mushrooms.

1 Introduction

Originally proposed by Chua in 1971 [1], the memristor poses a fourth basic circuit element, whose characteristics differ from that of R, L and C elements. Going through a more general point of view, Chua with coauthors Abdelouahab and Lozi [2], using fractional calculus, published 43 years after his original intuition, a global theory of family of electric elements: the memfractance theory which is the most general theory of such elements with memory, enlarging this family to mem-capacitive and mem-inductive elements of first, second third, etc. order.

A. E. Beasley
ARM Ltd., 110 Fulbourn Rd., Cambridge CB1 9NJ, UK

M.-S. Abdelouahab
Laboratory of Mathematics and their interactions, University Centre Abdelhafid Boussouf, Mila 43000, Algeria

R. Lozi
Université Côte d'Azur, CNRS, LJAD, Nice, France

M.-A. Tsompanas · A. Adamatzky (✉)
Unconventional Computing Laboratory, UWE, Bristol, UK
e-mail: andrew.adamatzky@uwe.ac.uk

© The Author(s), under exclusive license to Springer Nature Switzerland AG 2023
A. Adamatzky (ed.), *Fungal Machines*, Emergence, Complexity and Computation 47,
https://doi.org/10.1007/978-3-031-38336-6_15

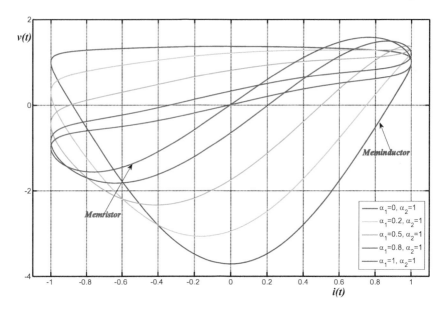

Fig. 1 Ideal plot of voltage-current memfractive elements: from memristor to meminductor [2]

In the memfractance theory the pinch observed in the voltage-current curves of the memristor is only a particular case (Figs. 1 and 2) of this memfractive electric element, allowing more flexibility in modeling.

Memristance has been seen in nano-scale devices where electronic and ionic transport are coupled under an external bias voltage. Strukov et al. [3] posit that the hysteric I-V characteristics observed in thin-film, two-terminal devices can be understood as memristive. However, this is observed behaviour of devices that already have other, large signal behaviours.

The ideal memristor model (Figs. 1 and 2) is shown to display 'lobes' on the I-V characterisation sweeps, indicating that the current resistance is a function of the previous resistance—hence a memristor has memory. For the purposes of analysis, graphs are referred to by their quadrants, starting with quadrant one as the top right and being numbered anti-clockwise.

Similarly, the mem-capacitor and mem-inductor exhibit a change in capacitance/inductance as a function of the applied voltage being swept. The introduction of the mem-capacitor and mem-inductor in [2, 4] complete the non-binary solution space of the mem-fractor that exerts a device may exhibit a combination of memristive, mem-inductor and mem-capacitor elements.

Finding a true memristor is by no means an easy task. Nevertheless, a number of studies have turned to nature to provide the answer, with varying success. Memristive properties of organic polymers were discovered before the 'official' discovery of the memristor was announced in [3]. The first examples of memristors could go back to the singing arc, invented by Duddell in 1900, which was originally used in wireless

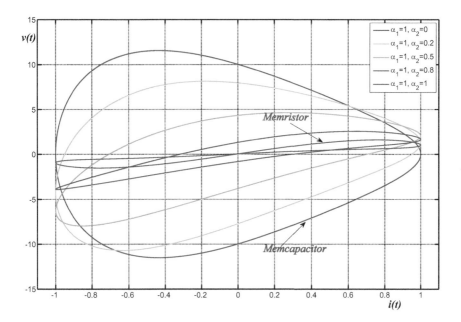

Fig. 2 Ideal plot of voltage-current memfractive elements: from memristor to memcapacitor [2]

telegraphy before the invention of the triode [5]. In addition, memristive properties of organic polymers have been studied since 2005 in experiments focussing on hybrid electronic devices based on the polyaniline-polyethylenoxide junction [6]. Memristive properties of living creatures, their organs and fluids have been demonstrated in skin [7–9], blood [10], plants [11, 12] (including fruits [13, 14]), slime mould [15, 16], tubulin microtubules [17–19]. Most recent results include DNA and melanin based memristive devices [20], biomaterials extracted from plant tissue [21].

We present a study of the I-V characteristics of the fruit bodies of the grey Oyster fungi *Pleurotus ostreatus*. Why fungi? Previously we recorded extracellular electrical potential of Oyster's fruit bodies, basidiocarps [22] and found that the fungi generate action potential like impulses of electrical potential. The impulses can propagate as isolated events, or in trains of similar impulses. Further, we demonstrated, albeit in numerical modelling, that fungi can be used as computing devices, where information is represented by spikes of electrical activity, a computation is implemented in a mycelium network and an interface is realised via fruit bodies [23]. A computation with fungi might not be useful *per se*, because the speed of spike propagation is substantially lower than the clock speed in conventional computers. However, the fungal computation becomes practically feasible when embedded in a slow developing spatial process, e.g. growing architecture structures. Thus, in [24] we discussed how to: produce adaptive building constructions by developing structural substrate using live fungal mycelium, functionalising the substrate with nanoparticles and polymers to make mycelium-based electronics, implementing sensorial fusion and decision making in the fungal electronics.

Why we are looking for mem-fractive properties of fungi? Mem-fractors [2] have combinations of properties exhibited by memristors, mem-capacitors and mem-inductors. A memristor is a material implication [25, 26] and can, therefore, can be used for constructing other logical circuits, stateful logic operations [25], logic operations in passive crossbar arrays of memristors [27], memory aided logic circuits [28], self-programmable logic circuits [29], and memory devices [30]. If strands of fungal mycelium in a mycelium bound composites and the fruit bodies show some mem-fractive properties then we can implement a variety of memory and computing devices embedded directly into architectural building materials made from the fungal substrates [24] and living fungal wearables [31, 32]. The field of living fungal wearables is currently in its infant stage, however it showed undeniably slim shape, good adaptability, and very low energy consumption compared to artificial wearable sensory devices [33]. Mycelium bound composites—masses of organic substrates colonised by fungi—are future environmentally sustainable growing biomaterials [34–36], already they are used in acoustic [37–39] and thermal [40–45] insulation wall cladding and packaging materials [46–48].

In [24] it is proposed to develop a structural substrate by using live fungal mycelium, functionalise the substrate with nanoparticles and polymers to make mycelium-based electronics, implement sensorial fusion and decision making in the mycelium networks [49] and to grow monolithic buildings from the functionalised fungal substrate [50]. Fungal buildings would self-grow, build, and repair themselves subject to substrate supplied, use natural adaptation to the environment, sense all that humans can sense. To implement sensorial integration and make decisions fungal materials will require electronic circuits, the fungal memristors will form essential part of the circuits.

The approach taken has a two-fold novelty component. First, we focus on mem-fractive properties of a substrate, which offers more fuzzy logic like approach of the IV properties of materials. Second, we study electrical IV properties of fungi, which are per se is a novel substrate for future organic electronics.

2 Experimental Set Up

We used grey oyster fungi *Pleurotus ostreatus* (Ann Miller's Speciality Mushrooms Ltd, UK) cultivated on wood shavings. The iridium-coated stainless steel sub-dermal needles with twisted cables (Spes Medica SRL, Italy) were inserted in fruit bodies (Fig. 3) of grey Oyster fungi using two different arrangements: 10 mm apart in the cap of the fungi (cap-to-cap), Fig. 3a, and translocation zones (cap-to-stem), Fig. 3b. I-V sweeps were performed on the fungi samples with Keithley Source Measure Unit (SMU) 2450 (Keithley Instruments, USA) under the following conditions: [−500 mV to 500 mV, −1 V to 1 V] with the samples in ambient lab light (965 Lux). Varying the step size of the voltage sweep allowed testing the I-V characteristics of the subject at

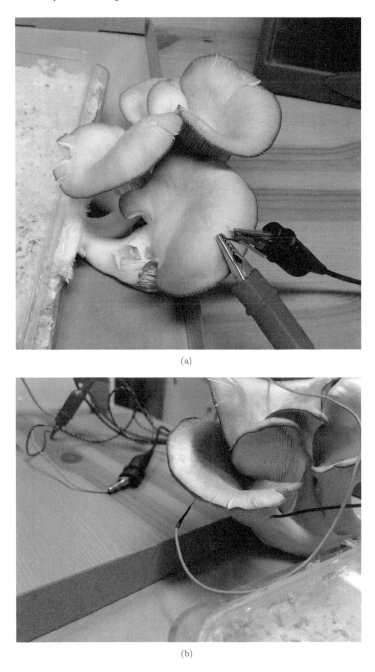

Fig. 3 Positions of electrodes in fruit bodies. **a** Electrodes inserted 10 mm apart in the fruit body cap. **b** One electrode is inserted in the cap with the other in the stem

different frequencies. The voltage ranges are limited so as not to cause the electrolysis of water. Each condition was repeated at least six times over the samples. Voltage sweeps were performed in both directions (cyclic voltammetry) and plots of the I-V characteristics were produced.

MATLAB was used to analyse the frequency and distribution of spiking behaviour observed in the I-V sweeps of the fruiting bodies under test (Sect. 3.1). All histogram plots are binned according to the voltage interval set for the Keithley SMU.

3 Results

Fruit body samples are shown to exhibit memristive properties when subject to a voltage sweep. The ideal memristor model has a crossing point at 0V, where theoretically no current flows. Figures 4 and 5 show the results of cyclic voltammetry of grey oyster fungi with electrodes positioned both in the fungi caps and stems. From Figs. 4 and 5, it can be seen that when 0 V is applied by the source meter, a reading of a nominally small voltage and current is performed. The living membrane is capable of generating potential across the electrodes, and hence a small current is observed. Mem-capacitors produce similar curves to that of an ideal memristor in Fig. 1, when plotting charge (q) against voltage (v) [51]. Additionally, mem-inductors produce similar plots for current (i) against flux (φ). However, the crossing point in the curves observed in quadrant 3 results to a pinched hysteresis loop. That is an indication that the cyclic voltammetry measurement is provided by a device that has mem-fractance properties.

While the sample under test is subjected to a positive voltage (quadrant 1), it can be seen there is nominally a positive current flow. Higher voltages result in a larger current flow. For an increasing voltage sweep there is a larger current flow for the corresponding voltage during a negative sweep.

Similarly, in quadrant 3 where there is a negative potential across the electrodes, the increasing voltage sweep yields a current with smaller magnitude than the magnitude of the current on a negative voltage sweep. Put simply, the fruit body has a resistance that is a function of the previous voltage conditions.

By applying averaging to the performed tests, a clear picture is produced that demonstrates for a given set of conditions, a typical response shape can be expected (Figs. 6 and 7). The stem-to-cap placement of the electrodes in the fruit body yields a tighter range for the response (Figs. 6b and 7b). This can be expected due to the arrangement of the transportation pathways, so-called translocation zone distinct from any vascular hyphae [52, 53], in the fruit body which run from the edge of the cap and down back through the stem to the root structure (mycelium). Cap-to-cap placement of the electrodes applies the potential across a number of the solutes translocation pathways and hence yields a wider range of results. However, for all results, it is observed that the positive phase of the cyclic voltammetry produces a different conduced current than the negative phase. The opening of the hysteresis curve around point zero suggests the fungus is not an ideal mem-ristor, instead it is

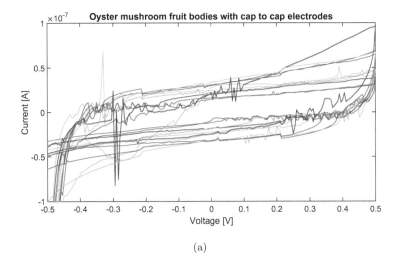

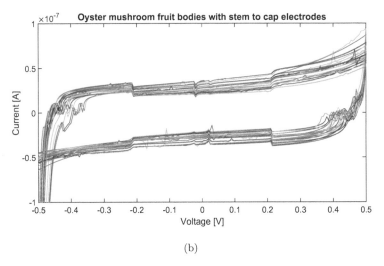

Fig. 4 Raw data from cyclic voltammetry performed over −0.5 V to 0.5 V. **a** Cap-to-cap electrode placement. **b** Stem-to-cap electrode placement

also exhibiting mem-capacitor and mem-inductor effects. The build of charge in the device prevents the curve from closing completely to produce the classic mem-ristor pinching shape.

Reducing the voltage step size (by ten fold, i.e. to 0.001 V) for the I-V characterisation is synonymous to reducing the frequency of the voltage sweep. Decreasing the sweep frequency of the voltage causes the chances of "pinching" in the I-V sweep to increase, as seen in quadrant 1 of Fig. 8. This further reinforces the presence of some mem-capacitor behaviour. Since the charging frequency of the fungus has now been

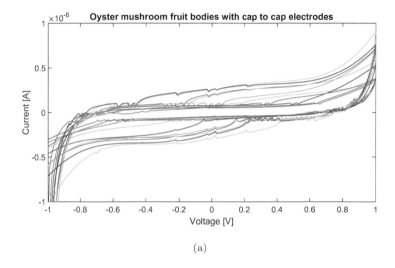

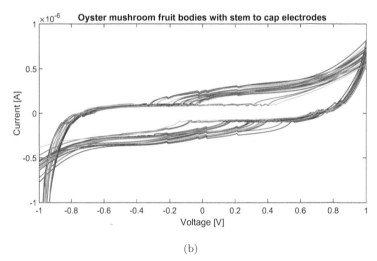

Fig. 5 Raw data from cyclic voltammetry performed over −1 V to 1 V. **a** Cap-to-cap electrode placement. **b** Stem-to-cap electrode placement

reduced there is a greater amount of time for capacitively stored energy to dissipate, thus producing a more 'resistive' plot with a pinch in the hysteresis. However, as indicated by two subsequent runs of voltammetry (Fig. 8), the electrical behaviour of the fungus is heavily altered under these frequencies and, thus, the repeatable observation of similar curves can not be realised (as it was observed in the aforementioned measurements and especially the stem to cap electrode placement). Nonetheless, the production of the curves can be controlled more efficiently by selecting appropriate

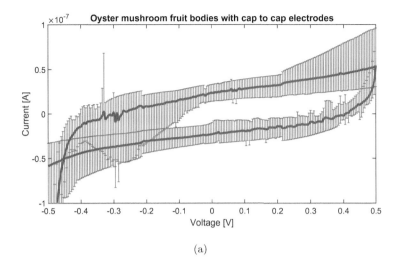

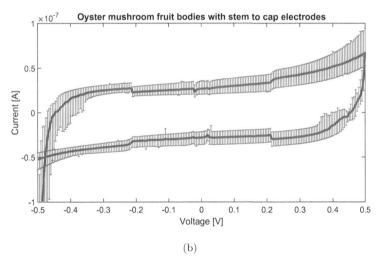

Fig. 6 Average grey oyster fungi fruit bodies I-V characteristics for cyclic voltammetry of −0.5 V to 0.5 V. **a** Cap-to-cap electrode placement. **b** Stem-to-cap electrode placement

frequencies of operation, but less successfully from the fungus substrate part, as this is a living substrate that has inherent stochasticity in the way it metabolises and grows.

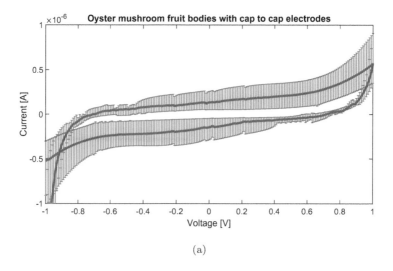

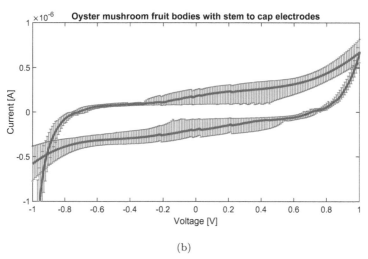

Fig. 7 Average fruit bodies I-V characteristics for cyclic voltammetry of −1 V to 1 V. **a** Cap-to-cap electrode placement. **b** Stem-to-cap electrode placement

3.1 Spiking

It is observed from Figs. 4 and 5 that portions of the cyclic voltammetry result in oscillations in the conduced current, or spiking activity. Oscillations occur most prominently on the positive phase of the cyclic voltammetry as the applied voltage approaches 0V and similarly on the negative phase, again as the applied voltage approaches 0 V. Current oscillations are typically in the order of nano-amps and persist for a greater number of cycles when the electrodes are arranged as a pair

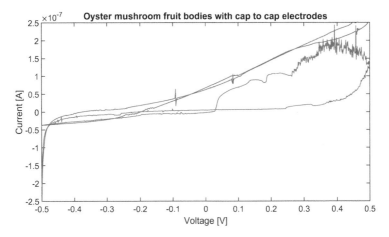

Fig. 8 I-V Characteristics of fungi fruit bodies with the voltage step size set to 0.001 V. The two traces represent repeated runs of the same experiment

on the fruit body cap (between five and ten cycles) compared to the stem-to-cap arrangement (fewer than five repeats).

Figure 9 demonstrates the spiking frequency of a single repeat of the cyclic voltammetry performed between −0.5 V and 0.5 V with the electrodes in a cap-to-cap arrangement. It is shown in the figure that the voltage interval—change in the applied, swept voltage— between spikes in an oscillation period are less than 0.06 V. Figure 10 concatenates the data for all repeats of the cyclic voltammetry performed under four different conditions. It is clearly shown that in cap-to-cap arrangements the voltage interval between spikes is less than when the electrodes are in a translocation arrangement. Any spikes that occur when the voltage interval becomes large can be taken as not occurring during a period of oscillation in the sweep, instead they occur infrequently and randomly during the sweep. Reducing the frequency of the voltage sweep (Fig. 8) also has the effect of removing the current oscillations.

4 Mathematical Model of Mushroom Mem-Fractance

Here we report the I-V characteristics of grey Oyster fungi *Pleurotus ostreatus* fruit bodies. It is evident from the results that grey Oyster fungi display memristive behaviour.

Although the fruit bodies typically do not demonstrate the "pinching" property of an ideal memristor [54], it can be clearly seen that the biological matter exhibits memory properties when the electrical potential across the substrate is swept. A positive sweep yields a higher magnitude current when the applied voltage is positive; and a smaller magnitude current when the applied voltage is negative.

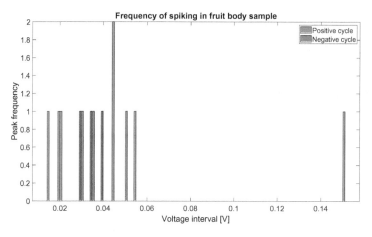

Fig. 9 The voltage interval of spikes in the I-V characteristics of the fruit body for a single run. Note that in this histogram the y-axis refers to frequency of spike occurrence

Fractional Order Memory Elements (FOME) are proposed as a combination of Fractional Order Mem-Capacitors (FOMC) and Fractional Order Mem-Inductors (FOMI) [2]. The FOME (Eq. 1) is based on the generalised Ohm's law and parameterised as follows: α_1, α_2 are arbitrary real numbers—it is proposed that $0 \leq \alpha_1, \alpha_2 \leq 1$ models the solution space by [4], $F_M^{\alpha_1,\alpha_2}$ is the mem-fractance, $q(t)$ is the time dependent charge, $\varphi(t)$ is the time dependent flux. Therefore, the mem-fractance ($F_M^{\alpha_1,\alpha_2}$) is an interpolation between four points: MC—mem-capacitance, R_M—memristor, MI—mem-inductance, and R_{2M}—the second order memristor (Fig. 11). Full derivations for the generalised FOME model are given by [2, 4]. The definition of mem-fractance can be straightforward generalised to any value of α_1, α_2 (see Fig. 27 in [2]).

$$D_t^{\alpha_1}\varphi(t) = F_M^{\alpha_1,\alpha_2}(t)D_t^{\alpha_2}q(t) \tag{1}$$

The appearance of characteristics from various memory elements in the fungal I-V curves supports the assertion that the fungal is a mem-fractor where α_1 and α_2 are both greater than 0 and less than 2.

There is no biological reason for mem-fractance of oyster fungi fruit bodies with stem to cap electrodes, to be a usual closed formula. Therefore, one can get only a mathematical approximation of this function. In the following, we propose two alternatives to obtain the best approximation for mem-fractance in the case of average fruit bodies I-V characteristics for cyclic voltammetry of Fig. 7b.

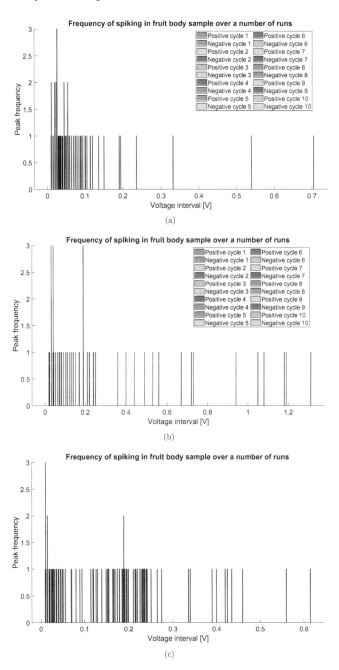

Fig. 10 Concatenations of all spiking data from all data runs for four different test conditions. **a** voltammetry over −0.5 V to 0.5 V, cap-to-cap electrode arrangement. **b** voltammetry over −1 V to 1 V, cap-to-cap electrode arrangement. **c** voltammetry over −0.5 V to 0.5 V, stem-to-cap electrode arrangement. **d** voltammetry over −1 V to 1 V, stem-to-cap electrode arrangement. Legends omitted on **c** and **d** for clarity. Note that in these histograms the y-axes refer to frequency of spike occurrence

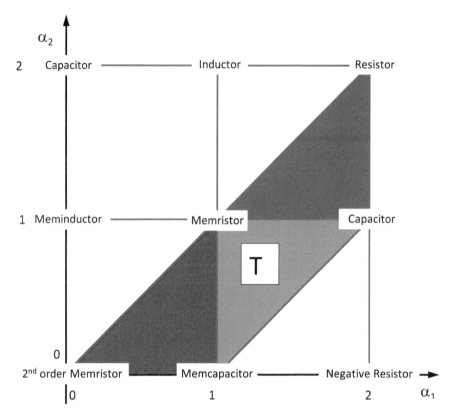

Fig. 11 Principal memfractive elements

4.1 Approximation by Polynomial on the Whole Interval of Voltage

Raw data include the time, voltage and intensity of each reading. There are 171 readings for each run. The process of these data, in order to obtain a mathematical approximation of mem-fractance, in the first alternative, takes 4 steps as follows.

Step 1: approximate $v(t)$ by a twenty-four-degree polynomial (Fig. 12) whose coefficients are given in Table 1.

$$v(t) \approx P(t) = \sum_{j=0}^{j=24} a_j t^j \qquad (2)$$

The polynomial fits very well the experimental voltage curve, as the statistical indexes show in Table 2.

Table 1 Coefficient of P(t)

a_0	−1.047361152400062	a_{13}	1.48292987584698e-16
a_1	0.135299293073760	a_{14}	−8.60157726907686e-19
a_2	−0.0726485498614107	a_{15}	1.59013702626457e-22
a_3	0.0240895989682110	a_{16}	5.80230108481181e-23
a_4	−0.00453232038841485	a_{17}	−7.12198496974121e-25
a_5	0.000531866967507868	a_{18}	5.19611819410190e-27
a_6	−4.19159536470121e-05	a_{19}	−2.64464369703488e-29
a_7	2.33484036114612e-06	a_{20}	9.68672841708898e-32
a_8	−9.51752589043893e-08	a_{21}	−2.52211206380669e-34
a_9	2.90458838155410e-09	a_{22}	4.45025298649318e-37
a_{10}	−6.72265349925510e-11	a_{23}	−4.78342788514078e-40
a_{11}	1.18302125464207e-12	a_{24}	2.36810109946699e-43
a_{12}	−1.56317950862153e-14		

Table 2 Goodness of fit

Sum of squared estimate of errors	$SSE = \sum_{j=1}^{j=n}(v_j - \hat{v}_j)^2$	0.0680517563652170
Sum of squared residuals	$SSR = \sum_{j=1}^{j=n}(\hat{v}_j - \overline{v})^2$	133.688517134422
Sum of square total	SST = SSE + SSR	133.756568890787
Coefficient of determination	$R - \text{square} = \frac{SSR}{SST}$	0.999491226809049

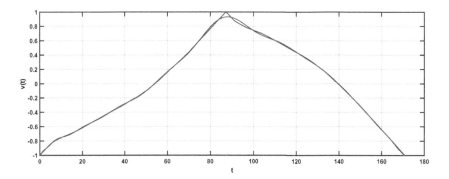

Fig. 12 Voltage versus time and its approximation by a 24-degree polynomial

Table 3 Coefficient of Q(t)

b_0	−2.69478636561017e-06	b_{13}	5.61870303550308e-22
b_1	1.95479195837707e-06	b_{14}	−3.66183256804588e-24
b_2	−7.34738169887512e-07	b_{15}	8.14484000064489e-27
b_3	1.67584032221916e-07	b_{16}	1.36036443304302e-28
b_4	−2.47326661661364e-08	b_{17}	−2.04593370725626e-30
b_5	2.48346182702953e-09	b_{18}	1.59708666114599e-32
b_6	−1.76692818009608e-10	b_{19}	−8.46294727047340e-35
b_7	9.19419585703268e-12	b_{20}	3.19831491989559e-37
b_8	−3.58289124918788e-13	b_{21}	−8.56384614589988e-40
b_9	1.06306849079070e-14	b_{22}	1.55262364796050e-42
b_{10}	−2.42471413376463e-16	b_{23}	−1.71535341852628e-45
b_{11}	4.25821973203331e-18	b_{24}	8.73846352218898e-49
b_{12}	−5.69947824465678e-20		

Table 4 Goodness of fit

Sum of squared estimate of errors	5.84247524503151e-13
Sum of squared residuals	4.07366051979587e-11
Sum of square total	4.13208527224619e-11
Coefficient of determination	0.985860709883522

Step 2: in the same way approximate the current $i(t)$ using a twenty-four-degree polynomial (Fig. 13) whose coefficients are given in Table 3.

$$i(t) \approx Q(t) = \sum_{j=0}^{j=24} b_j t^j \quad (3)$$

Again, the polynomial fits well the experimental intensity curve, as displayed in Table 4.

Step 3: From (Eq. 1) used under the following form $D_t^{\alpha_2} q(t) \neq 0$.

$$F_M^{\alpha_1,\alpha_2}(t) = \frac{D_t^{\alpha_1} \varphi t}{D_t^{\alpha_2} q(t)} \quad (4)$$

and the Rieman-Liouville fractional derivative defined by [55]

$${}_0^{RL} D_t^{\alpha} f(t) = \frac{1}{\Gamma(m-\alpha)} \frac{d^m}{dt^m} \int_0^t (t-s)^{m-\alpha-1} f(s) ds, \, m-1 < \alpha < m \quad (5)$$

together with the formula for the power function

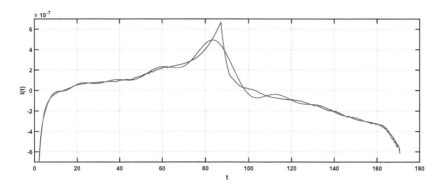

Fig. 13 Current versus time and its approximation by a 24 degree polynomial

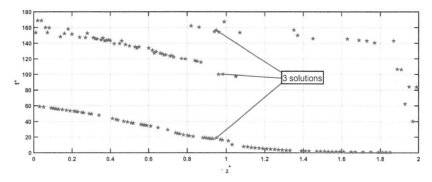

Fig. 14 Zeros $t^*(\alpha_2)$ of the denominator of $F_M^{\alpha_1,\alpha_2}(t)$

$${}_0^{RL}D_t^\alpha\left(at^\beta\right) = \frac{a\Gamma(\beta+1)}{\Gamma(\beta-\alpha+1)}t^{\beta-\alpha}, \beta > -1, \alpha > 0, \quad (6)$$

we obtain the closed formula of $F_M^{\alpha_1,\alpha_2}(t)$, approximation of the true biological memfractance of the Oyster mushroom

$$F_M^{\alpha_1,\alpha_2}(t) = \frac{D_t^{\alpha_1}\varphi(t)}{D_t^{\alpha_2}\varphi(t)} = \frac{{}_0^{RL}D_t^{\alpha_1}\sum_{j=0}^{j=24}\frac{a_j}{j+1}t^{j+1}}{{}_0^{RL}D_t^{\alpha_2}\sum_{j=0}^{j=24}\frac{b_j}{j+1}t^{j+1}} = \frac{\sum_{j=0}^{j=24}\frac{a_j\Gamma(j+1)}{\Gamma(j+2-\alpha_1)}t^{j+1-\alpha_1}}{\sum_{j=0}^{j=24}\frac{b_j\Gamma(j+1)}{\Gamma(j+2-\alpha_2)}t^{j+1-\alpha_2}} \quad (7)$$

Step 4: choice of parameter α_1 and α_2: We are looking for the best value of these parameters in the range $(\alpha_1, \alpha_2) \in [0, 2]^2$. In this goal, we are considering first the singularities of $F_M^{\alpha_1,\alpha_2}(t)$ in order to avoid their existence, using suitable values of the parameters. Secondly, we will choose the most regular approximation. We compute numerically, the values $t^*(\alpha_2)$ which vanish the denominator of $F_M^{\alpha_1,\alpha_2}(t)$ (Fig. 14).

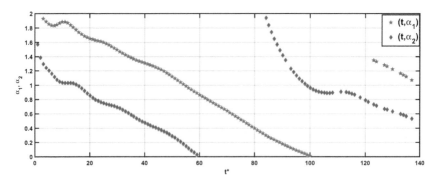

Fig. 15 Zeros $t^*(\alpha_2)$ of $F_M^{\alpha_1,\alpha_2}(t)$ denominator (red dots), and zeros $t^*(\alpha_1)$ of the numerator (blue dots)

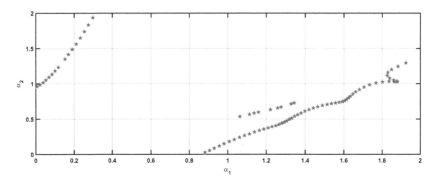

Fig. 16 Values of $(\alpha_1, \alpha_2) \in [0, 2]^2$ for which the zeros $t^*(\alpha_2)$ of denominator of $F_M^{\alpha_1,\alpha_2}(t)$ correspond to the zeros $t^*(\alpha_1)$ of denominator

We observe one, two or three coexisting solutions depending on the value of α_2. Moreover, there is no value of α_2 without zero of the denominator. Therefore, in order to eliminate the singularities, we need to determine the couples $(\alpha_1, \alpha_2) \in [0, 2]^2$, vanishing simultaneously denominator and numerator of $F_M^{\alpha_1,\alpha_2}(t)$ (Figs. 15 and 16).

In the second part of step 4, we choose the most regular approximation. We consider that the most regular approximation is the one for which the function range $(F_M^{\alpha_1,\alpha_2}(t))$ is minimal (Figs. 17 and 18)

$$\text{range}\left(F_M^{\alpha_1,\alpha_2}(t)\right) = \max_{t \in [0,171]}\left(F_M^{\alpha_1,\alpha_2}(t)\right) - \min_{t \in [0,171]}\left(F_M^{\alpha_1,\alpha_2}(t)\right) \quad (8)$$

From the numerical results, the best couple (α_1, α_2) and the minimum range of $F_M^{\alpha_1,\alpha_2}(t)$ are given in Table 5, and the corresponding Mem-fractance is displayed in Fig. 19.

The value of (α_1, α_2) given in Table 5 belongs to the triangle T of Fig. 11, whose vertices are Memristor, Memcapacitor and Capacitor. Which means that

Mem-Fractive Properties of Fungi

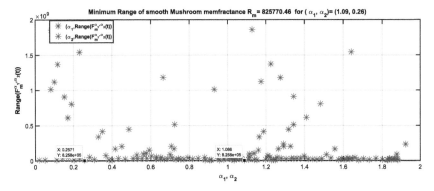

Fig. 17 Values of range $(F_M^{\alpha_1,\alpha_2}(t))$ for $(\alpha_1, \alpha_2) \in [0, 2]^2$

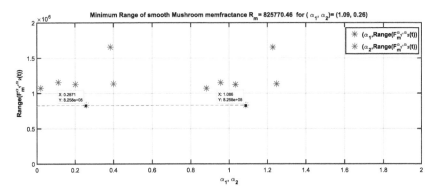

Fig. 18 Magnification of Fig. 17

Table 5 Minimum $F_M^{\alpha_1,\alpha_2}(t)$

α_1	α_2	**Minimum range $F_M^{\alpha_1,\alpha_2}(t)$**
1.08642731	0.25709492	825770.46017259

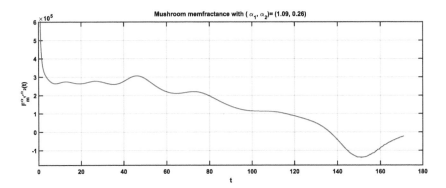

Fig. 19 Mem-fractance for (α_1, α_2) given in Table 5

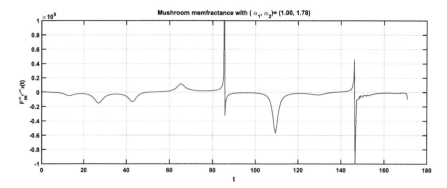

Fig. 20 Mem-fractance with two singularities for $(\alpha_1, \alpha_2) = (1, 1.78348389322388)$

Oyster mushroom fruit bodies with stem to cap electrodes, is like a mix of such basic electronic devices.

As a counter-example of our method for choosing the best possible Mem-fractance, Fig. 20 displays, the Mem-fractance for a non-optimal couple $(\alpha_1, \alpha_2) = (1, 1.78348389322388)$ which presents two singularities.

4.2 Approximate Cycling Voltammetry

From the closed formula of $F_M^{\alpha_1^*, \alpha_2^*}(t)$ it is possible to retrieve the formula of the current function $i(t)$ using (Eq. 1).

$$
\begin{aligned}
i(t) &= D_t^{1-\alpha_2} \left[\frac{D_t^{\alpha_1} \varphi(t)}{F_M^{\alpha_1, \alpha_2}(t)} \right] \\
&= D_t^{1-\alpha_2} \left[\frac{\sum_{j=0}^{j=24} \frac{a_j \Gamma(j+1)}{\Gamma(j+2-\alpha_1)} t^{j+1-\alpha_1}}{\frac{\sum_{j=0}^{j=24} \frac{a_j \Gamma(j+1)}{\Gamma(j+2-\alpha_1)} t^{j+1-\alpha_1}}{\sum_{j=0}^{j=24} \frac{b_j \Gamma(j+1)}{\Gamma(j+2-\alpha_2)} t^{j+1-\alpha_2}}} \right] \\
&= D_t^{1-\alpha_2} \left[\sum_{j=0}^{j=24} \frac{b_j \Gamma(j+1)}{\Gamma(j+2-\alpha_2)} t^{j+1-\alpha_2} \right] \qquad (9) \\
&= \sum_{j=0}^{j=24} \frac{\Gamma(j+2-\alpha_2) b_j \Gamma(j+1)}{\Gamma(j+2-\alpha_2) \Gamma(j+1)} t^{j+1-\alpha_2-(1-\alpha_2)} \\
&= \sum_{j=0}^{j=24} b_j t^j
\end{aligned}
$$

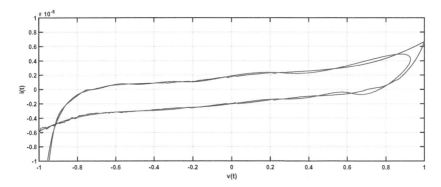

Fig. 21 Comparison between average experimental data of cyclic voltammetry performed over −1 V to 1 V, Stem-to-cap electrode placement, and approximate values of $v(t)$ and $i(t)$

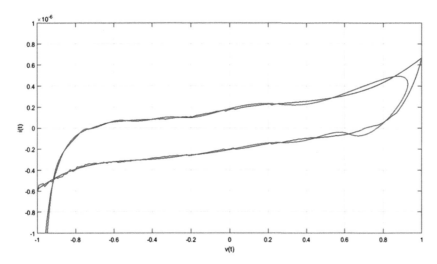

Fig. 22 Both average experimental data curve and the curve computed from closed approximative formula are nested into the histogram of data of all runs

The comparison of average experimental data of cyclic voltammetry performed over −1 V to 1 V, stem-to-cap electrode placement, and closed approximative formula is displayed in Fig. 21, showing a good agreement between both curves except near the maximum value of $v(t)$ and $i(t)$. Figure 22 shows that the curve computed from closed approximative formula belongs to the histogram of data of all runs. The discrepancy between both curves is due to the method of approximation chosen in (2) and (3).

It is possible, as we show in the next subsection to improve the fitting of the approximated curve near the right hand-side vertex, using piecewise polynomial approximation of both $v(t)$ and $i(t)$.

Table 6 Coefficient for i(t)

Coefficient	Value for $0 \leq t \leq T$	Coefficient	Value for $T \leq t < T$
a_0	−0.98299	a'_0	37.16955
a_1	0.02665	a'_1	−1.2986
a_2	−5.91565 E -4	a'_2	0.01826
a_3	1.12211 E -5	a'_3	−1.25146 E -4
a_4	−6.28483 E -8	a'_4	4.12302 E -7
a_5	6.9675 E -11	a'_5	−5.25359 E-19

Table 7 Goodness of fit

Approximation	$t < T$	$t > T$
Coefficient of determination	0.99983	0.9999

4.3 Alternative Approximation of the Cycling Voltammetry

Due to the way of conducting the experiments, the voltage curve presents a vertex, that means that the function $v(t)$ is non-differentiable for $T = 87.23747459$. In fact, the value of T is the average value of the non-differentiable points for the 20 runs.

In this alternative approximation, we follow the same 4 steps as previously, changing the approximation by a twenty-four-degree polynomial to an approximation by a 2-piecewise fifth-degree-polynomial, for both $v(t)$ and $i(t)$.

Step 1: approximation of $v(t)$ by a 2-piecewise fifth-degree-polynomial (Fig. 23) whose coefficients are given in Table 6.

$$v(t) = \begin{cases} P_1(t) = a_0 + a_1 t + a_2 t^2 + a_3 t^3 \\ \quad + a_4 t^4 + a_5 t^5, \text{ for } 0 \leq t \leq T \\ P_2(t) = a'_0 + a'_1 t + a'_2 t^2 + a'_3 t^3 \\ \quad + a'_4 t^4 + a'_5 t^5, \text{ for } T \leq t < 171 \end{cases} \quad (10)$$

The flux is obtained integrating $v(t)$ versus time. The polynomial fits very well the experimental voltage curve, as the statistical indexes show in Table 7.

$$\varphi(t) = \begin{cases} IP_1(t) = a_0 t + \frac{a_1}{2} t^2 + \frac{a_2}{3} t^3 + \frac{a_3}{4} t^4 \\ \quad + \frac{a_4}{5} t^5 + \frac{a_5}{6} t^6, \text{ for } 0 \leq t \leq T \\ IP_2(t) = a'_0 t + \frac{a'_1}{2} t^2 + \frac{a'_2}{3} t^3 + \frac{a'_3}{4} t^4 \\ \quad + \frac{a'_4}{5} t^5 + \frac{a'_5}{6} t^6, \text{ for } T \leq t < 171 \end{cases} \quad (11)$$

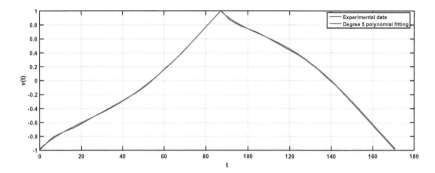

Fig. 23 Voltage versus time and its approximation by 2-piecewise fifth degree polynomial

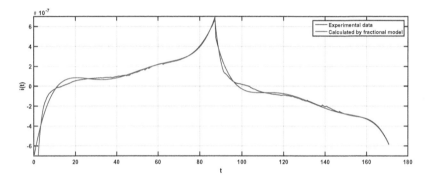

Fig. 24 Current versus time and its approximation by 2-piecewise fifth degree polynomial

Step 2: in the same way, one approximates the current $i(t)$ using a 2-piecewise fifth degree polynomial (Fig. 24) whose coefficients are given in Table 8.

$$i(t) = \begin{cases} P_3(t) = b_0 + b_1 t + b_2 t^2 + b_3 t^3 \\ \qquad + b_4 t^4 + b_5 t^5, \text{ for } 0 \leq t \leq T \\ P_4(t) = b'_0 + b'_1 t + b'_2 t^2 + b'_3 t^3 \\ \qquad + b'_4 t^4 + b'_5 t^5, \text{ for } T \leq t < 171 \end{cases} \quad (12)$$

Again, the polynomial fits very well the experimental voltage curve, as the statistical indexes show in Table 9. Therefore, the charge is given by:

Table 8 Coefficient for i(t)

Coefficient	Value for $0 \le t \le T$	Coefficient	Value for $T \le t < 171$
b_0	−7.21418 E -7	b'_0	2.69466 E -4
b_1	1.11765 E -7	b'_1	−1.05461 E -5
b_2	−6.3792 E -9	b'_2	1.63678 E -7
b_3	1.57327 E -10	b'_3	−1.25915 E -9
b_4	−1.7745 E -12	b'_4	4.80107 E -12
b_5	7.52304 E -15	b'_5	−7.26253 E-15

Table 9 Goodness of fit

Approximation	$t < T$	$t > T$
Coefficient of determination	0.99171	0.98613

$$q(t) = \begin{cases} IP_3(t) = b_0 t + \frac{b_1}{2}t^2 + \frac{b_2}{3}t^3 + \frac{b_3}{4}t^4 \\ \quad + \frac{b_4}{5}t^5 + \frac{b_5}{6}t^6, \text{ for } 0 \le t \le T \\ IP_4(t) = b'_0 t + \frac{b'_1}{2}t^2 + \frac{b'_2}{3}t^3 \\ \quad + \frac{b'_3}{4}t^4 + \frac{b'_4}{5}t^5 + \frac{b'_5}{6}t^6, \text{ for } T \le t < 171 \end{cases} \quad (13)$$

Step 3: Following the same calculus as before with (4), one obtains

$$\text{for } 0 \le t \le T, F_M^{\alpha_1,\alpha_2}(t) = \frac{{}^{RL}_0 D_t^{\alpha_1}\varphi(t)}{{}^{RL}_0 D_t^{\alpha_2}q(t)} = \frac{{}^{RL}_0 D_t^{\alpha_1}[IP_1(t)]}{{}^{RL}_0 D_t^{\alpha_2}[IP_3(t)]}$$
$$= \frac{\sum_{j=1}^{j=5} \frac{a_j \Gamma(j+1)}{\Gamma(j+2-\alpha_1)} t^{j+1-\alpha_1}}{\sum_{j=0}^{j=5} \frac{b_j \Gamma(j+1)}{\Gamma(j+2-\alpha_2)} t^{j+1-\alpha_2}} \quad (14)$$

However, because fractional derivative has memory effect, for $T < t < 171$, the formula is slightly more complicated (depicted in Eq. 15).

Using integration by part repeatedly six times we obtain Eq. 16. In this 2-piece wise approximation, the vertex is non-differentiable, this implies that Eq. 16 expression has a singularity at T (because $(t-T)^{-\alpha_{1,2}} \to \infty$). It could be possible to avoid this singularity, using a 3-piece wise approximation, smoothing the vertex. However, the calculus are very tedious. We will explain, below, what our simpler choice implies. Finally, Eqs. 17 and 18 are obtained.

Mem-Fractive Properties of Fungi

$$F_M^{\alpha_1,\alpha_2}(t) = {}_0^{RL}D_t^{\alpha_1}\varphi(t) \atop {}_0^{RL}D_t^{\alpha_2}q(t)$$

$$= \frac{1}{\Gamma(m_1-\alpha_1)}\frac{d^{m_1}}{dt^{m_1}}\int_0^t(t-s)^{m_1-\alpha_1-1}\varphi(s)ds \atop \frac{1}{\Gamma(m_2-\alpha_2)}\frac{d^{m_2}}{dt^{m_2}}\int_0^t(t-s)^{m_2-\alpha_2-1}q(s)ds, \quad m_1-1<\alpha_1<m_1 \text{ and } m_2-1<\alpha_2<m_2$$

$$= \frac{1}{\Gamma(m_1-\alpha_1)}\frac{d^{m_1}}{dt^{m_1}}\left[\int_0^T(t-s)^{m_1-\alpha_1-1}IP_1(s)ds + \int_T^t(t-s)^{m_1-\alpha_1-1}IP_2(s)ds\right] \atop \frac{1}{\Gamma(m_2-\alpha_2)}\frac{d^{m_2}}{dt^{m_2}}\left[\int_0^T(t-s)^{m_2-\alpha_2-1}IP_3(s)ds + \int_T^t(t-s)^{m_2-\alpha_2-1}IP_4(s)ds\right]$$

$$= \frac{1}{\Gamma(m_1-\alpha_1)}\frac{d^{m_1}}{dt^{m_1}}\sum_{j=0}^{j=5}\left[\frac{a_j}{j+1}\int_0^T(t-s)^{m_1-\alpha_1-1}s^{j+1}ds + \frac{a'_j}{j+1}\int_T^t(t-s)^{m_1-\alpha_1-1}s^{j+1}ds\right] \atop \frac{1}{\Gamma(m_2-\alpha_2)}\frac{d^{m_2}}{dt^{m_2}}\sum_{j=0}^{j=5}\left[\frac{b_j}{j+1}\int_0^T(t-s)^{m_2-\alpha_2-1}s^{j+1}ds + \frac{b'_j}{j+1}\int_T^t(t-s)^{m_2-\alpha_2-1}s^{j+1}ds\right]$$

(15)

$$F_M^{\alpha_1,\alpha_2}(t)$$

$$= \frac{1}{\Gamma(m_1-\alpha_1)}\frac{d^{m_1}}{dt^{m_1}}\sum_{j=0}^{j=5}\left[\frac{a_j}{j+1}\left[\sum_{k=0}^{k=j+1}\frac{-(j+1)!\Gamma(m_1-\alpha_1)(t_T)^{m_1+k-\alpha_1}Tj+1-k}{(j+1-k)!\Gamma(m_1+k+1-\alpha_1)}\right] + \frac{(j+1)!\Gamma(m_1-\alpha_1)t^{m_1+k-\alpha_1}}{\Gamma(m_1+j+1-\alpha_1)}\right] + \frac{a'_j}{j+1}\left[\sum_{k=0}^{k=j+1}\frac{(j+1)!\Gamma(m_1-\alpha_1)(t-T)^{m_1+k-\alpha_1}Tj+1-k}{(j+1-k)!\Gamma(m_1+k+1-\alpha_1)}\right]$$

$$+ \frac{1}{\Gamma(m_2-\alpha_2)}\frac{d^{m_2}}{dt^{m_2}}\sum_{j=0}^{j=5}\left[\frac{b_j}{j+1}\left[\sum_{k=0}^{k=j+1}\frac{-(j+1)!\Gamma(m_2-\alpha_2)(t-T)^{m_2+k-\alpha_2}Tj+1-k}{(j+1-k)!\Gamma(m_2+k+1-\alpha_2)}\right] + \frac{(j+1)!\Gamma(m_2-\alpha_2)t^{m_2+k-\alpha_2}}{\Gamma(m_2+j+1-\alpha_2)}\right] + \frac{b'_j}{j+1}\left[\sum_{k=0}^{k=j+1}\frac{(j+1)!\Gamma(m_2-\alpha_2)(t-T)^{m_2+k-\alpha_2}Tj+1-k}{(j+1-k)!\Gamma(m_2+k+1-\alpha_2)}\right]$$

$$= \frac{1}{\Gamma(m_1-\alpha_1)}\sum_{j=0}^{j=5}(d'_j-a_j)\sum_{k=0}^{k=j+1}\left[\frac{j!\Gamma(m_1-\alpha_1)(t-T)^{m_1+k-\alpha_1}Tj+1-k}{(j+1-k)!\Gamma(m_1+k+1-\alpha_1)}\right] + a_j\frac{j!t^{j+1-\alpha_1}}{\Gamma(j+2-\alpha_1)}$$

$$+ \frac{1}{\Gamma(m_2-\alpha_2)}\sum_{j=0}^{j=5}(b'_j-b_j)\sum_{k=0}^{k=j+1}\left[\frac{j!\Gamma(m_2-\alpha_2)(t-T)^{m_2+k-\alpha_2}Tj+1-k}{(j+1-k)!\Gamma(m_2+k+1-\alpha_2)}\right] + b_j\frac{j!t^{j+1-\alpha_2}}{\Gamma(j+2-\alpha_2)}$$

$$= \sum_{j=0}^{j=5}(a'_j-a_j)\sum_{k=0}^{k=j+1}\left[\frac{j!(t-T)^{k-\alpha_1}Tj+1-k}{(j+1-k)!\Gamma(k+1-\alpha_1)}\right] + a_j\frac{j!t^{j+1-\alpha_1}}{\Gamma(j+2-\alpha_1)}$$

$$\sum_{j=0}^{j=5}(b'_j-b_j)\sum_{k=0}^{k=j+1}\left[\frac{j!(t-T)^{k-\alpha_2}Tj+1-k}{(j+1-k)!\Gamma(k+1-\alpha_2)}\right] + b_j\frac{j!t^{j+1-\alpha_2}}{\Gamma(j+2-\alpha_2)}$$

(16)

$$F_M^{\alpha_1,\alpha_2}(t) = \frac{(t-T)^{-\alpha_1}\left[\sum_{j=0}^{j=5}\left[(a'_j-a_j)\sum_{k=0}^{k=j+1}\left[\frac{j!(t-T)^k T^{j+1-k}}{(j+1-k)!\Gamma(k+1-\alpha_1)}\right]+a_j\frac{j!t^{j+1-\alpha_1}(t-T)^{\alpha_1}}{\Gamma(j+2-\alpha_1)}\right]\right]}{(t-T)^{-\alpha_2}\left[\sum_{j=0}^{j=5}\left[(b'_j-b_j)\sum_{k=0}^{k=j+1}\left[\frac{j!(t-T)^k T^{j+1-k}}{(j+1-k)!\Gamma(k+1-\alpha_2)}\right]+b_j\frac{j!t^{j+1-\alpha_2}(t-T)^{\alpha_2}}{\Gamma(j+2-\alpha_2)}\right]\right]}$$

$$= \sum_{j=0}^{j=5}\left[(a'_j-a_j)\sum_{k=0}^{k=j+1}\left[\frac{j!(t-T)^k T^{j+1-k}}{(j+1-k)!\Gamma(k+1-\alpha_1)}\right]+a_j\frac{j!t^{j+1-\alpha_1}(t-T)^{\alpha_1}}{\Gamma(j+2-\alpha_1)}\right]$$

$$(t-T)^{\alpha_1-\alpha_2}\sum_{j=0}^{j=5}\left[(b'_j-b_j)\sum_{k=0}^{k=j+1}\left[\frac{j!(t-T)^k T^{j+1-k}}{(j+1-k)!\Gamma(k+1-\alpha_2)}\right]+b_j\frac{j!t^{j+1-\alpha_2}(t-T)^{\alpha_2}}{\Gamma(j+2-\alpha_2)}\right]$$

for $0 \leq t \leq T$ (17)

$$F_M^{\alpha_1,\alpha_2}(t) = \begin{cases} \dfrac{\sum_{j=0}^{j=5}\frac{a_j\Gamma(j+1)}{\Gamma(j+2-\alpha_1)}t^{j+1-\alpha_1}}{\sum_{j=0}^{j=5}\frac{b_j\Gamma(j+1)}{\Gamma(j+2-\alpha_2)}t^{j+1-\alpha_2}}, \\ \\ (t-T)^{\alpha_1-\alpha_2}\sum_{j=0}^{j=5}\left[(a'_j-a_j)\sum_{k=0}^{k=j+1}\left[\frac{j!(t-T)^k T^{j+1-k}}{(j+1-k)!\Gamma(k+1-\alpha_1)}\right]+a_j\frac{j!t^{j+1-\alpha_1}(t-T)^{\alpha_1}}{\Gamma(j+2-\alpha_1)}\right] \\ \hline (t-T)^{\alpha_1-\alpha_2}\sum_{j=0}^{j=5}\left[(b'_j-b_j)\sum_{k=0}^{k=j+1}\left[\frac{j!(t-T)^k T^{j+1-k}}{(j+1-k)!\Gamma(k+1-\alpha_2)}\right]+b_j\frac{j!t^{j+1-\alpha_2}(t-T)^{\alpha_2}}{\Gamma(j+2-\alpha_2)}\right] \end{cases}$$

for $T < t < 171$ (18)

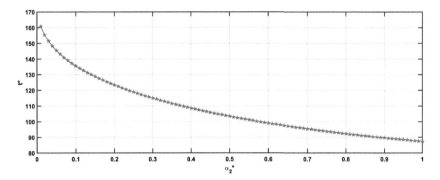

Fig. 25 The first zero $t^*(\alpha_2) \geq T$, of the denominator of $F_M^{\alpha_1,\alpha_2}(t)$, as function of α_2

Step 4: choice of parameter α_1 and α_2: Following the same idea as for the first alternative, we try to avoid singularity for $F_M^{\alpha_1,\alpha_2}(t)$, except of course the singularity near T, which is of mathematical nature (non-differentiability of voltage and intensity at $t = T$). Figure 25 displays the first zero $t^*(\alpha_2) \geq T$, of the denominator of $F_M^{\alpha_1,\alpha_2}(t)$. One can see that $t^*(1) \cong T$.

Figure 26 displays the curves of couples (α_1, α_2) for which the denominator and numerator of $F_M^{\alpha_1,\alpha_2}(t)$ are null simultaneously for $t < T$ and $t > T$. On this figure, the value of α_1, that corresponds to $\alpha_2 = 1$ is $\alpha_1 \approx 1.78348389322388$. The corresponding Mem-fractance is displayed in Fig. 27.

The singularity observed in Figs. 27 and 28 is due to the non-differentiability of both voltage and intensity functions at point T. It is only a mathematical problem of approximation which can be solved using a 3-piecewise polynomial instead of the 2-piecewise polynomial $(P1(t), P2(t))$ and $(P3(t), P4(t))$. The third added piecewise polynomials for $v(t)$ and $i(t)$ being defined on the tiny interval [87.24, 87.90]. However due to more tedious calculus, we do not consider this option in the

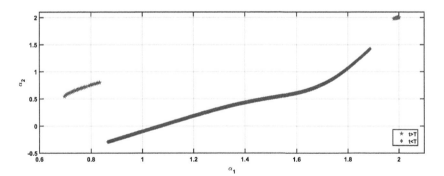

Fig. 26 Couples (α_1, α_2) for which the denominator and numerator of $F_M^{\alpha_1,\alpha_2}(t)$ are null simultaneously for $t < T$ (blue dot) and $t > T$ (red dot)

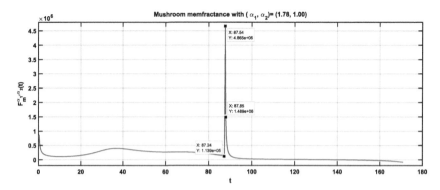

Fig. 27 Mem-fractance for ($\alpha_1 = 1.78, \alpha_2 = 1.00$) given in Table 5

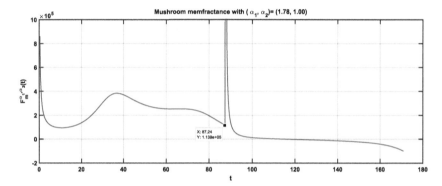

Fig. 28 Magnification of Fig. 27

present study. It is only a math problem, and one can consider that Fig. 28 represents the value of the mem-fractance in the interval $[0, 87.24] \cup [87.90, 171]$.

The value of ($\alpha_1 = 1.78, \alpha_2 = 1.00$) belongs to the line segment of Fig. 11, whose extremities are Memristor, and Capacitor. Which means that Oyster mushroom fruit bodies with stem-to-cap electrodes, is like a mix of such basic electronic devices. The comparison of average experimental data of cyclic voltammetry performed over -1 V to 1 V, stem-to-cap electrode placement, and closed approximative formula is displayed in Fig. 29, showing a very good agreement between both curves.

5 Discussion

Two are the main remarks of this study. First, both approximations used in Sect. 4 converge to Mem-fractance with parameter value (α_1, α_2)—belonging inside or on edge of the triangle T of Fig. 11, whose vertices are Memristor, Memcapacitor and Capacitor. Of course, the value for these approximations are not exactly the same.

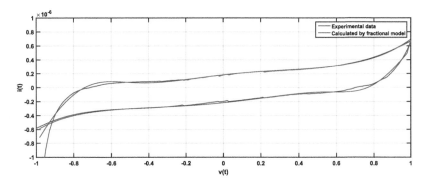

Fig. 29 Comparison between average experimental data of cyclic voltammetry performed over −1 V to 1 V, Stem-to-cap electrode placement, and closed approximative formula

This is in part, due to the fact that we consider that the most regular approximation is the one for which the function range $(F_M^{\alpha_1,\alpha_2}(t))$ is minimal. Other choices based on physiology of Mushroom could be invoked. Moreover, the Mem-fractance is computed on the averaged curve of 20 runs which do not present exactly the same characteristic voltammetry. Oyster mushroom fruit bodies are living substrates. Commonly for living substrates their morphology, i.e. geometry of the translocation zones [52], is changing from one fruit body to another. This high variability prevents exact reproduction of electrical property between the experimental trials.

The second remark is the fact that the use of fractional derivatives to analyze the mem-fractance, is obvious if one considers that fractional derivatives have memory, which allows a perfect modelling of memristive elements. Their handling is however delicate if one wants to avoid any flaw.

Similar I-V characteristics have been experienced for slime mould [15] and apples [13]. The cyclic voltammetry experiments demonstrate that the I-V curve produced from these living substrates is a closed loop where the negative path does not match the positive path. Hence the fungi display the characteristics of a memristor. A similar conclusion is drawn for the microtubule experiments [56]. The microtubule exhibits different resistive properties for the same applied voltage depending on the history of applied voltages.

Additionally, the fruit bodies produce current oscillations during the cyclic voltammetry. This oscillatory effect is only observed on one phase of the voltammetry for a given voltage range which is, again, a behaviour that can be associated to a device whose resistance is a function of its previous resistance. This spiking activity is typical of a device that exhibits memristive behaviours. Firstly, it was reported in experiments with electrochemical devices using graphite reference electrodes, that a temporal dependence of the current of the device—at constant applied voltage— causes charge accumulation and discharge [57]. The spiking is also apparent in some plots, for a large electrode size, in experiments with electrode metal on solution-processed flexible titanium dioxide memristors [58]. A detailed analysis of types of spiking emerging in simulated memristive networks was undertaken in [59]. Most recent research on spiking of memristors demonstrated the rich and complex spiking dynamics of the NbO_2 memristor memristive neurons based on the insulator-metal transition model [60]. Repeatable observations of the spiking behaviour in I-V of the fungi is very important because this opens new pathways for the implementation of neuromorphic computing with fungi. A fruitful theoretical foundation of this field is already well developed [61–66].

6 Conclusion

The fruit bodies of grey oyster fungi *Pleurotus ostreatus* were subjected to I-V characterisation a number of times, from which it was clearly shown that they exhibit mem-fractor properties. Under cyclic voltammetry, the fruit body will conduct differently depending on the phase (positive or negative) of the voltammetry. This behaviour produces the classic "lobes" in the I-V characteristics of a memristor.

However, a biological medium, such as the fruit body of the grey Oyster fungi presented here, will differ from that of the ideal memristor model since the "pinching" behaviour and size of the hysteresis lobes are functions of the frequency of the voltage sweep as well as the previous resistance. Typically, the biological medium generates its own potential across the electrodes, therefore, even when no additional potential is supplied, there is still current flow between the probes. This property of the fungi produces an opening in the I-V curve that is a classic property of the mem-capacitor. Since the fungi are exhibiting properties of both memristors and mem-capacitors, their electrical memory behaviour puts them somewhere in the mem-fractor solution space where $0 < \alpha_1, \alpha_2 < 1$. Hence, it has been shown that fungi act as mem-fractors.

We believe a potential practical implementation of the mem-fractive properties of the fungi would be in the sensorial and computing circuits embedded into mycelium bound composites. In [24] we proposed to develop a structural substrate by using live fungal mycelium, functionalise the substrate with nanoparticles and polymers to make mycelium-based electronics, implement sensorial fusion and decision making in the mycelium networks [49] and to grow monolithic buildings from the functionalised fungal substrate [50]. Fungal buildings would self-grow, build, and repair themselves subject to substrate supplied, use natural adaptation to the environment, sense all that humans can sense. Whilst major parts of a building will be made from dried and cured mycelium composites there is an opportunity to use blocks with living mycelium as embedded living computing elements. Right now we established just some components of the computing fungal architectures. Future challenges will be in implementation of a large scale computing circuits employing mem-fractive properties of the living mycelium and fruit bodies and an integration of living mycelium computers into buildings made from biomaterials.

References

1. Chua, L.: Memristor-the missing circuit element. IEEE Trans. Circuit Theory **18**(5), 507–519 (1971)
2. Abdelouahab, M.-S., Lozi, R., Chua, L.: Memfractance: a mathematical paradigm for circuit elements with memory. Int. J. Bifurc. Chaos **24**(9), 1430023 (2014)
3. Strukov, D.B., Snider, G.S., Stewart, D.R., Stanley Williams, R.: The missing memristor found. Nature **453**(7191), 80–83 (2008)
4. Khalil, N.A., Said, L.A., Radwan, A.G., Soliman, A.M.: General fractional order mem-elements mutators. Microelectron. J. **90**, 211–221 (2019)
5. Ginoux J.-M., Rossetto, B.: The singing arc: the oldest memristor? In: Chaos, CNN, Memristors and Beyond: A Festschrift for Leon Chua With DVD-ROM, composed by Eleonora Bilotta, pp. 494–507. World Scientific (2013)
6. Erokhin, V., Berzina, T., Fontana, M.P.: Hybrid electronic device based on polyaniline-polyethyleneoxide junction. J. Appl. Phys. **97**(6), 064501 (2005)
7. Martinsen, Ø.G., Grimnes, S., Lütken, C.A., Johnsen, G.K.: Memristance in human skin. J. Phys. Conf. Ser. **224**, 012071. IOP Publishing (2010)
8. Pabst, O., Martinsen, Ø.G., Chua, L.: The non-linear electrical properties of human skin make it a generic memristor. Sci. Rep. **8**(1), 1–9 (2018)
9. Pabst, O., Martinsen, Ø.G., Chua, L.: Information can be stored in the human skin memristor which has non-volatile memory. Sci. Rep. **0**(1), 1–10 (2019)
10. Prasad Kosta, S., Kosta, Y.P., Bhatele, M., Dubey, Y.M., Gaur, A., Kosta, S., Gupta, J., Patel, A., Patel, B.: Human blood liquid memristor. Int. J. Med. Eng. Inform. **3**(1), 16–29 (2011)
11. Volkov, A.G., Tucket, C., Reedus, J., Volkova, M.I., Markin, V.S., Chua, L.: Memristors in plants. Plant Signal. Behavior **9**(3), e28152 (2014)
12. Volkov, A.G., Chua, L.: Cyclic voltammetry of volatile memristors in the venus flytrap: short-term memory. Funct. Plant Biol. **48**(6), 567–572 (2021)
13. Volkov, A.G., Markin, V.S.: Electrochemistry of gala apples: memristors in vivo. Russ. J. Electrochem. **53**(9), 1011–1018 (2017)
14. Abdelrahman, D.K., Mohammed, R., Fouda, M.E., Said, L.A., Radwan, A.G.: Memristive bio-impedance modeling of fruits and vegetables. IEEE Access **9**, 21498–21506 (2021)
15. Gale, E., Adamatzky, A., de Lacy Costello, c.B.: BioNanoScience **5**(1), 1–8 (2015)
16. Braund, E., Reck Miranda, E.: On building practical biocomputers for real-world applications: receptacles for culturing slime mould memristors and component standardisation. J. Bionic Eng. **14**(1), 151–162 (2017)
17. del Rocío Cantero, M., Perez, P.L., Scarinci, N., Cantiello, H.F.: Two-dimensional brain microtubule structures behave as memristive devices. Sci. Rep. **9**(1), 1–10 (2019)
18. Chiolerio, A., Draper, T.C., Mayne, R., Adamatzky, A.: On resistance switching and oscillations in tubulin microtubule droplets. J. Colloid Interface Sci. **560**, 589–595 (2020)
19. Tuszynski, J.A., Friesen, D., Freedman, H., Sbitnev, V.I., Kim, H., Santelices, I., Kalra, A.P., Patel, S.D., Shankar, K., Chua, L.O.: Microtubules as sub-cellular memristors. Sci. Rep. **10**(1), 1–11 (2020)
20. More, G.M., Tiwari, A.P., Pawar, K.D., Dongale, T.D., Geun Kim, T.: Bipolar resistive switching in biomaterials: case studies of dna and melanin-based bio-memristive devices. In: Mem-Elements for Neuromorphic Circuits with Artificial Intelligence Applications, pp. 299–323. Elsevier (2021)
21. Sun, B., Guo, T., Zhou, G., Wu, J., Chen, Y., Norman Zhou, Y., Wu, Y.A.: A battery-like self-selecting biomemristor from earth-abundant natural biomaterials. ACS Appl. Bio Mater. **4**(2), 1976–1985 (2021)
22. Adamatzky, A.: On spiking behaviour of oyster fungi pleurotus djamor. Sci. Rep. **8**(1), 1–7 (2018)
23. Adamatzky, A., Tuszynski, J., Pieper, J., Nicolau, D.V., Rinalndi, R., Sirakoulis, G., Erokhin, V., Schnauss, J., Smith, D.M.: Towards cytoskeleton computers. A proposal. In: Adamatzky,

A., Akl, S., Sirakoulis, G. (eds.) From Parallel to Emergent Computing. CRC Group/Taylor & Francis (2019)
24. Adamatzky, A., Ayres, P., Belotti, G., Wösten, H.: Fungal architecture position paper. Int. J. Unconv. Comput. **14** (2019)
25. Borghetti, J., Snider, G.S., Kuekes, P.J., Joshua Yang, J., Stewart, D.R., Stanley Williams, R.: 'memristive' switches enable 'stateful' logic operations via material implication. Nature **464**(7290), 873–876 (2010)
26. Kvatinsky, S., Satat, G., Wald, N., Friedman, E.G., Kolodny, A., Weiser, U.C.: Memristor-based material implication (imply) logic: design principles and methodologies. IEEE Trans. Very Large Scale Integr. (VLSI) Syst. **22**(10), 2054–2066 (2013)
27. Linn, E., Rosezin, R., Tappertzhofen, S., Böttger, U., Waser, R.: Beyond von neumann-logic operations in passive crossbar arrays alongside memory operations. Nanotechnology **23**(30), 305205 (2012)
28. Kvatinsky, S., Belousov, D., Liman, S., Satat, G., Wald, N., Friedman, E.G., Kolodny, A., Weiser, U.C.: Magic-memristor-aided logic. IEEE Trans. Circuits Syst. II: Express Br. **61**(11), 895–899 (2014)
29. Borghetti, J., Li, Z., Straznicky, J., Li, X., Ohlberg, D.A.A., Wu, W., Stewart, D.R., Stanley Williams, R.: A hybrid nanomemristor/transistor logic circuit capable of self-programming. Proc. Natl. Acad. Sci. *106*(6), 1699–1703 (2009)
30. Ho, Y., Huang, G.M., Li, P.: Nonvolatile memristor memory: device characteristics and design implications. In: Proceedings of the 2009 International Conference on Computer-Aided Design, pp. 485–490 (2009)
31. Adamatzky, A., Nikolaidou, A., Gandia, A., Chiolerio, A., Mahdi Dehshibi, M.: Reactive fungal wearable. Biosystems **199**, 104304 (2021)
32. Adamatzky, A., Gandia, A., Chiolerio, A.: Fungal sensing skin. Fungal Biol. Biotechnol. **8**(1), 1–6 (2021)
33. Li, J., Xin, M., Ma, Z., Shi, Y., Pan, L.: Nanomaterials and their applications on bio-inspired wearable electronics. Nanotechnology (2021)
34. Karana, E., Blauwhoff, D., Hultink, E.-J., Camere, S.: When the material grows: a case study on designing (with) mycelium-based materials. Int. J. Des. **12**(2) (2018)
35. Jones, M., Mautner, A., Luenco, S., Bismarck, A., John, S.: Engineered mycelium composite construction materials from fungal biorefineries: a critical review. Mater. Des. **187**, 108397 (2020)
36. Cerimi, K., Can Akkaya, K., Pohl, C., Schmidt, B., Neubauer, P.: Fungi as source for new bio-based materials: a patent review. Fungal Biol. Biotechnol. **6**(1), 1–10 (2019)
37. Pelletier, M.G., Holt, G.A., Wanjura, J.D., Bayer, E., McIntyre, G.: An evaluation study of mycelium based acoustic absorbers grown on agricultural by-product substrates. Ind. Crops Prod. **51**, 480–485 (2013)
38. Elsacker, E., Vandelook, S., Van Wylick, A., Ruytinx, J., De Laet, L., Peeters, E.: A comprehensive framework for the production of mycelium-based lignocellulosic composites. Sci. Total Environ. **725**, 138431 (2020)
39. Robertson, O. et al.: Fungal future: A review of mycelium biocomposites as an ecological alternative insulation material. In: DS 101: Proceedings of NordDesign 2020, Lyngby, Denmark, 12th-14th August 2020, pp. 1–13 (2020)
40. Yang, Z., Zhang, F., Still, B., White, M., Amstislavski, P.: Physical and mechanical properties of fungal mycelium-based biofoam. J. Mater. Civil Eng. **29**(7), 04017030 (2017)
41. Xing, Y., Brewer, M., El-Gharabawy, H., Griffith, G., Jones, P.: Growing and testing mycelium bricks as building insulation materials. In: IOP Conference Series: Earth and Environmental Science, vol. 121, pp. 022032. IOP Publishing (2018)
42. Girometta, C., Maria Picco, A., Michela Baiguera, R., Dondi, D., Babbini, S., Cartabia, M., Pellegrini, M., Savino, E.: Physico-mechanical and thermodynamic properties of mycelium-based biocomposites: a review. Sustainability **11**(1), 281 (2019)
43. Pereira Dias, P., Bhagya Jayasinghe, L., Waldmann, D.: Investigation of mycelium-miscanthus composites as building insulation material. Results Mater. **10**, 100189 (2021)

44. Wang, F., Li, H.-q., Kang, S.-s., Bai, Y.-f., Cheng, G.-z., Zhang, G.-q.: The experimental study of mycelium/expanded perlite thermal insulation composite material for buildings. Sci. Technol. Eng. **2016**, 20 (2016)
45. Pablo Cárdenas-R, J.: Thermal insulation biomaterial based on hydrangea macrophylla. In: Bio-Based Materials and Biotechnologies for Eco-Efficient Construction, pp. 187–201. Elsevier (2020)
46. Holt, G.A., Mcintyre, G., Flagg, D., Bayer, E., Wanjura, J.D., Pelletier, M.G.: Fungal mycelium and cotton plant materials in the manufacture of biodegradable molded packaging material: evaluation study of select blends of cotton byproducts. J. Biobased Mater. Bioenergy **6**(4), 431–439 (2012)
47. Sivaprasad, S., Byju, S.K., Prajith, C., Shaju, J., Rejeesh, C.R.: Development of a novel mycelium bio-composite material to substitute for polystyrene in packaging applications. Mater. Today Proc. (2021)
48. Mojumdar, A., Tanaya Behera, H., Ray, L.: Mushroom mycelia-based material: an environmental friendly alternative to synthetic packaging. Microb. Polym. pp. 131–141 (2021)
49. Adamatzky, A., Tegelaar, M., Wosten, H.A.B., Powell, A.L., Beasley, A.E., Mayne, R.: On boolean gates in fungal colony. Biosystems **193**, 104138 (2020)
50. Adamatzky, A., Gandia, A., Ayres, P., Wösten, H., Tegelaar, M.: Adaptive fungal architectures. LINKs-series **5**, 66–77
51. Yin, Z., Tian, H., Chen, G., Chua, L.O.: What are memristor, memcapacitor, and meminductor? IEEE Trans. Circuits Syst. II: Express Briefs **62**(4), 402–406 (2015)
52. Schütte, K.H.: Translocation in the fungi. New Phytol. **55**(2), 164–182 (1956)
53. Jennings, D.H.: Translocation of solutes in fungi. Biol. Rev. **62**(3), 215–243 (1987)
54. Chua, L.: If it's pinched it's a memristor. Semicond. Sci. Technol. **29**(10), 104001 (2014)
55. Podlubny, I.: Fractional Differential Equations. Academic Press, San Diego (1999)
56. Chiolerio, A., Draper, T.C., Mayne, R., Adamatzky, A.: On resistance switching and oscillations in tubulin microtuble droplets. J. Colloid Interface Sci. **560**, 589–595 (2020)
57. Erokhin, V., Fontana, M.P.: Electrochemically controlled polymeric device: a memristor (and more) found two years ago (2008). arXiv:0807.0333
58. Gale, E., Pearson, D., Kitson, S., Adamatzky, A., de Lacy Costello, B.: The effect of changing electrode metal on solution-processed flexible titanium dioxide memristors. Mater. Chem. Phys. **162**, 20–30 (2015)
59. Gale, E., de Lacy Costello, B., Adamatzky, A.: Emergent spiking in non-ideal memristor networks. Microelectron. J. **45**(11), 1401–1415 (2014)
60. Bo, Y., Zhang, P., Zhang, Y., Song, J., Li, S., Liu, X.: Spiking dynamic behaviors of nbo2 memristive neurons: a model study. J. Appl. Phys. **127**(24), 245101 (2020)
61. Serrano-Gotarredona, T., Prodromakis, T., Linares-Barranco, B.: A proposal for hybrid memristor-CMOS spiking neuromorphic learning systems. IEEE Circuits Syst. Mag. **13**(2), 74–88 (2013)
62. Indiveri, G., Linares-Barranco, B., Legenstein, R., Deligeorgis, G., Prodromakis, T.: Integration of nanoscale memristor synapses in neuromorphic computing architectures. Nanotechnology **24**(38), 384010 (2013)
63. Prezioso, M., Zhong, Y., Gavrilov, D., Merrikh-Bayat, F., Hoskins, B., Adam, G., Likharev, K., Strukov, D.: Spiking neuromorphic networks with metal-oxide memristors. In: 2016 IEEE International Symposium on Circuits and Systems (ISCAS), pp. 177–180. IEEE (2016)
64. Pickett, M.D., Medeiros-Ribeiro, G., Stanley Williams, R.: A scalable neuristor built with mott memristors. Nat. Mater. **12**(2), 114–117 (2013)
65. Linares-Barranco, B., Serrano-Gotarredona, T., Camuñas-Mesa, L.A., Perez-Carrasco, J.A., Zamarreño-Ramos, C., Masquelier, T.: On spike-timing-dependent-plasticity, memristive devices, and building a self-learning visual cortex. Front. Neurosci. **5**, 26 (2011)
66. Indiveri, G., Liu, S.-C.: Memory and information processing in neuromorphic systems. Proc. IEEE **103**(8), 1379–1397 (2015)

Electrical Signal Transfer by Fungi

Neil Phillips, Roshan Weerasekera, Nic Roberts, and Andrew Adamatzky

Abstract Mycelium-bound composites consist of discrete substrate elements joined together by filamentous hypha strands. These composites can be moulded or extruded into custom components of desired shapes. When live fungi are present these composites exhibit electrical conductivity as well as memfractive and capacitive properties. These composites might be used in nonlinear electrical circuits. We investigated the AC conductive properties of mycelium-bound composites and fungal fruit bodies at higher frequencies, spanning three overlapping frequency ranges: 20 Hz to 300 kHz, 10 Hz to 4 MHz, and 50 kHz to 3 GHz, to advance fungal electronics. Our measurements revealed that mycelium-bound composites primarily function as low-pass filters, with an average cut-off frequency of 500 kHz and a roll-off rate of -14 dB/decade. Within the pass band, the average attenuation is less than 1 dB. Fungal fruiting bodies have significantly lower mean cut-off frequencies that range from 5 Khz to 50 Khz depending on the species. Their roll-off range from -20 to -30 decibels per decade, with mean attenuation across the pass band less than 3 decibels. The precise mechanism underlying frequency-dependent attenuation is unclear. However, the high-water content, which is around 80% in mycelium-bound composites and up to 92% in fruiting bodies, is important. Because of the presence of dissolved ionizable solids, this water content is electrically conductive, making it a likely contributing factor. This research looks into the potential applications of mycelium-bound composites and fungal fruiting bodies in analog computing.

N. Phillips (✉) · N. Roberts · A. Adamatzky
Unconventional Computing Laboratory, UWE, Bristol, UK
e-mail: neil.phillips@uwe.ac.u

R. Weerasekera
School of Computer Science, Electrical and Electronic Engineering and Engineering Maths, University of Bristol, Bristol, UK

1 Introduction

Mycelium-bound composites are commonly formed by fungi colonising natural substrates such as Rye grain seeds [1–4]. In a recent paper [5], we proposed a plan to create a functional material out of living fungal mycelium. This entails strengthening the substrate with nanoparticles and polymers in order to construct mycelium-based electronics [6–8]. These composites can be moulded into customised combinations of desired shapes and dimensions using procedures such as moulding [9] or extrusion [10]. There is also the possibility of cultivating mycelium-bound composites for the construction of unified circuits, including constructs such as mycelium networks [11].

At lower frequencies, both mycelium-bound composites and fungal fruit bodies exhibit complicated electrical features, including resistive spikes [12], memfractive behaviour [7, 13], and the ability to control the frequency of external electrical inputs [14]. Moisture concentration affects their electrical characteristics [15]. They are also known for their ability to respond to chemical and physical inputs, which results in changes in their electrical patterns [16–19] and electrical conductivity [8]. In a separate research [13], we developed a unique notion called "direct fungal electronics." This burgeoning field consists of a collection of live electronic devices made of mycelium-bound composites or pure mycelium. These fungal electronic devices may change their impedance and emit bursts of electrical potential in response to external control settings. Fungal electronics can be integrated into fungal materials, worn technologies, or used as standalone sensing and computing tools. Communication between basic fungal-electronic components is critical. Encoding data inside alternating current frequencies is one potential approach for transferring data in a robust manner.

In order to further our investigation into the information-conveying properties of mycelium-bound composites and fungal fruit bodies, we conducted a study to determine how their electrical properties evolve across the higher frequency range [20, 21].

2 Methods and Materials

100 g block of substrate (Rye grain seeds) well colonised with *Pleurotus ostreatus* (Ann Miller's Speciality Mushrooms, UK, [22]) was enclosed in polypropylene bags fitted with $0.5\,\mu$ m air filter patch, see Fig. 2a. The block was kept at ambient room temperature ($18\,°C$ to $22\,°C$) inside a growth tent (in darkness) when not being used in experiments.

Five species of fresh fruiting bodies were sourced from a local supplier (Wholesale Fruit Centre Bristol, UK), see Fig. 1a–e. Two additional species were sourced from a woodland mushroom farm (Livesey Brothers Ltd., Leicestershire, UK), seebreak

Electrical Signal Transfer by Fungi

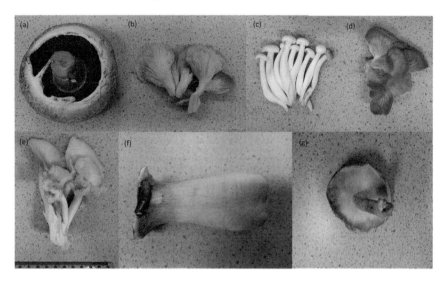

Fig. 1 Fungal fruiting bodies **a** *Agaricus bisporus* (Portobello), **b** *Pleurotus ostreatus* (White Oyster), **c** *Hypsizygus tessellatus* (White Shimeji), **d** *Pleurotus djamor* (Pink Oyster), **e** *Cantharellus cibarius* (Chanterelle), **f** *Pleurotus eryngii* (King Trumpet), **g** *Lentinula edodes* (Shiitake)

Fig. 1f, g. The fungal fruiting bodies were kept at ambient room temperature (18 °C to 22 °C) and initial measurements made within 8 h of purchase.

2.1 Electrical Properties of Mycelium-Bound Composites and Fungal Fruiting Bodies Over the 20 Hz *to* 300 kHz Frequency Range

To make electrical connections to the mycelium-bound composites and fungal fruiting bodies, bespoke electrodes were developed. The copper crimps at the terminal end of the stainless steel sub-dermal needle electrodes (Spes Medica S.r.l., IT) were soldered to the centre conductor of SMA (SubMiniature version A) right angle connectors, see Fig. 2a. The needles were inserted ∼15 mm depth into body of grain spawn and through fungal fruiting bodies of various thicknesses. The distance between the centres of the needle electrodes was maintained at 20 mm by a spacer, see Fig. 2b. 'Radial' measurements were recorded with the electrodes equal distance from the centre of the cap of the fruiting body while 'axial' measurements were recorded with the electrodes perpendicular to centre of the cap. 'Radial' electrodes are therefore across the cap's gills while 'axial' electrodes are aligned with the gills.

Measurements were recorded within a bespoke RF-shielded test chamber, see Fig. 2. The mycelium-bound composite and fruiting bodies were electrically insulated (and physically separated by ∼5 mm) from the inside of diecast aluminium enclosure

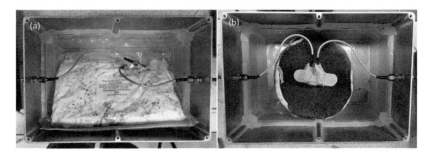

Fig. 2 Needle electrodes inserted into **a** mycelium bound composite in bag **b** stalk of fruiting body (*Agaricus bisporus*) inside RF shielded enclosure via coaxial connections

(model Hammond 1550H, 222 mm × 146 mm × 105 mm) with a polypropylene liner. Coaxial leads and connectors (including through the RF-chamber wall) were SMA type.

Electrical impedance and other circuit parameters were measured using a digital Inductance Capacitance Resistance (LCR) meter (model 891, BK Precision Ltd., UK [23]). The LCR meter was configured to scan across the 20 Hz to 300 kHz frequency range applying 1 V_{rms} sinusoidal voltage waveform through the mycelium bound composite and fruiting bodies. 301 measurements being automatically recorded per sweep with ∼0.05 % accuracy.

2.2 Electrical Properties of Mycelium Bound Composites and Fungal Fruiting Bodies Over the 10 Hz to 4 MHz Frequency Range

The signal propagation was measured using an impedance—amplitude—phase frequency response network analyzer (C60, Cypher Instruments, London, UK [24]). The C60 network analyzer passes 2 V_{pp} sinusoidal voltage waveform through the mycelium bound composite and fruiting bodies at a plurality of frequencies (10 Hz to 4 MHz). The network analyzer was connected to the native CypherGraph (V1.28.0) software package on a Windows computer to control functionality and store measurements. The software evaluates the waveform after it passes through the sample and displays it as a Bode plot, the frequency response was analyzed and measurements stored.

Experimental 'controls' were recorded using the same setup, however, the mycelium-bound composite was replaced with uncolonised substrate (Rye grain seeds in 50 ml glass beaker) and electrically insulating substrate (open cell polyurethane foam with water of different conductivity in 50 ml glass beaker). The 'controls' were subject to the same frequency spectrum of sinusoidal waveforms to explore if part of the signal was being propagated through the fungal hypha rather than the substrate or the instrumentation.

2.3 Electrical Properties of Mycelium Bound Composites and Fungal Fruiting Bodies Over the 50 kHz to 3 GHz Frequency Range

S-parameters S11 and S21 were measured using the previously described setup of electrodes and RF-shielded test chamber. This allowed reflection and transmission measurements. The signal propagation was measured using a Vector Network Analyser (NanoVNA-F V2, Amazon Plc, UK [25]). The instrument's maximum output power depends on the frequency and is between -14dBm and -19dBm. The dynamic range for reflection measurements (S11) is 70 dB or better up to 1.5 GHz and 60 dB or better up to 1.5GHz to 3GHz.

The Vector Network Analyser (VNA) was regularly calibrated to maintain accuracy involving short-circuit, 50 Ω load, and open termination, followed by a direct connection between ports (inside the RF-chamber). VNA was connected to the native NanoVNA-QT VNA Saver (V 0.5.3) software package on a Windows computer to control functionality and store measurements.

3 Results

3.1 Electrical Properties of Mycelium-Bound Composites and Fungal Fruiting Bodies Over the 20 Hz to 300 kHz Frequency Range

The LCR meter estimates a series resistance (R) and a series capacitance (C_s) equivalent model as shown in Fig. 3 for the mycelium substrate.

The electrical impedance $Z(f) = R + \frac{1}{j\omega C_s}$ of the fruiting body (cap and stalk) decreased with increasing frequency over 20 Hz to 300 kHz range. The impedance of mycelium-bound composite also decreased but at a considerably slower rate, see Fig. 4. The stalk of the fruiting body has a higher electrical impedance than the cap. The axial impedance of cap is lower than the radial impedance at lower frequencies

Fig. 3 Equivalent circuit for the mycelium substrate

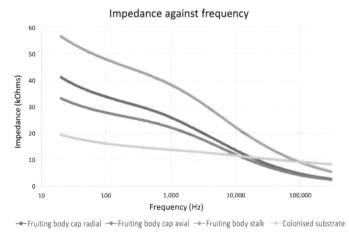

Fig. 4 Impedance against frequency of mycelium-bound composites and fungal fruiting bodies (cap and stalk) over 20 Hz to 300 kHz range

Table 1 Electrical properties of fruiting body and colonised substrate at 300 kHz

	Fruiting body			Colonised substrate
	Cap		Stalk	Body
	Radial	Axial	Radial	
R (kΩ)	1.991	1.782	4.438	8.312
Cs (pF)	348	394	156	551
Z (kΩ)	2.490	2.230	5.580	8.390

becoming similar ~300 kHz. This trend shows that mycelium-bound composites demonstrate lower impedance at higher frequencies.

A summary of other electrical properties of mycelium-bound composites and fungal fruiting bodies is shown in Table 1. The fruiting body's stalk has less than half the capacitance, more than twice the inductance, and lower steady-state DC resistance of the cap (both radial and axial). The steady-state DC resistance of the colonised substrate is higher than the fruiting body.

Using the lump model of the fruiting body and the colonised substrate, we have carried out SPICE S-parameter simulation to understand the frequency dependant characteristics of the mycelium substrate for signal transfer characteristics. Figure 5 depict the S-parameter simulation of the lump model for the return loss (S11) and the gain (S21) for different cases. In very low frequencies up to 100 kHz, a significant amount of the signal is reflected back to the source and the amount of signal transferred through the substrate is very low. However, starting from around 1 MHz onward the return loss becomes negative around −0.1 to −0.5 which is still higher in comparison to a normal conductive substrate.

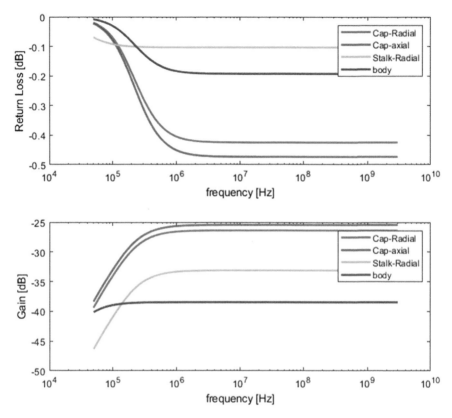

Fig. 5 SPICE S-parameter simulation using the lump equivalent circuit

3.2 Electrical Properties of Mycelium-Bound Composites and Fungal Fruiting Bodies Over the 10 Hz to 4 MHz Frequency Range

The signal propagation passing through mycelium-bound composite was frequency dependent; the lower frequency waveforms passed through with little attenuation while the higher frequencies became increasingly attenuated. The magnitude frequency profile matched that of a low pass filter; the phase response also appeared to correlate with a typical low pass filter. The Bode plot of mycelium-bound composite, see Fig. 6, shows the level of attenuation increased noticeably above 100 kHz (e.g. -8 dB at 1 MHz). The phase decreases noticeably above 10 kHz.

The Bode plots of five species of fungal fruiting bodies are shown in Fig. 7. For measurement consistency, the electrodes (with 20 mm spacing) were normally inserted into the caps of the fruiting bodies. However, the smaller physical size of *Hypsizygus tessellatus* meant that in one recording both electrodes were inserted into the stalk, and for a second recording one electrode was in the stalk and one in

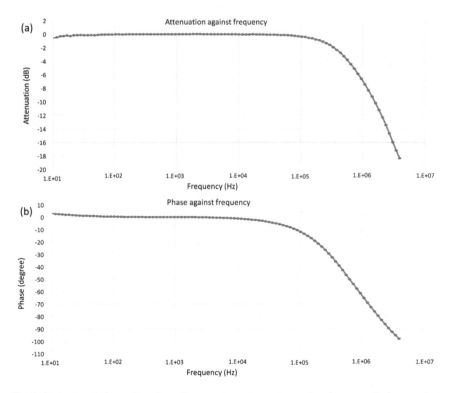

Fig. 6 Bode plot of of mycelium-bound composite **a** attenuation against frequency **b** phase against frequency (10 Hz to 4 MHz)

the cap. The mean cut-off frequency was between 5kHz to 50kHz (depending on species); −20dB/decade to −30dB/decade roll-off, with mean attenuation across the pass band of ∼−3 dB.

To support the analysis of material properties (attenuation and phase against frequency) six configurations were measured: well colonised substrate *Pleurotus ostreatus* on Rye seeds, blocks of 100 g and 750 g), uncolonised substrate (Rye seeds ∼66 % moisture content and Rye seeds with 2 ml of mains water 10 MΩ cm added to increase moisture content to ∼76 %, in 50 ml glass beakers) and open cell polyurethane foam (with 2 ml de-ionised water 10 MΩ cm and 2 ml mains water 0.0025 MΩ cm, in 50 ml glass beakers), see Fig. 8.

It was observed during measurements that the cap and stalk of fruiting bodies can have significantly different electrical properties. By way of example, Fig. 9 shows an exemplar of sequential recording from the same *Agaricus bisporus* fruiting body. It was not possible to measure some species as their stalks were physically too small to accommodate electrodes with 20 mm separation.

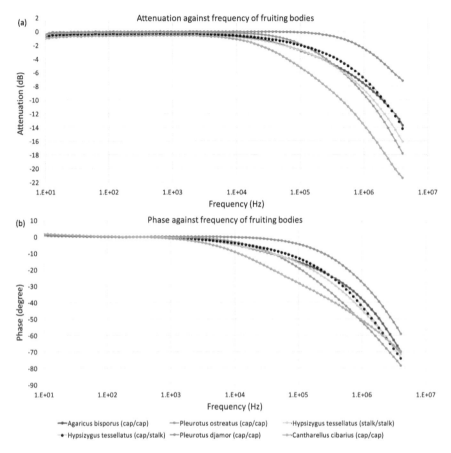

Fig. 7 Bode plots of five species of fungal fruiting bodies **a** attenuation against frequency **b** phase against frequency (10 Hz to 4 MHz)

3.3 Electrical Properties of Mycelium-Bound Composites and Fungal Fruiting Bodies Over the 50 kHz to 3 GHz Frequency Range

Signal propagation in the mycelium-bound composite is frequency-dependent. Figure 10 shows S11 the return loss against frequency of exemplar recording. Resonance peaks are observed around 960 MHz, 1.05 GHz, 1.61 GHz, and 2.04 GHz.

Gain against frequency (S21) of exemplar recording of mycelium-bound composite is shown in Fig. 11. Resonance peaks in gain are observed around 64 MHz, 898 MHz, 1.17 GHz, 1.89 GHz, 1.99 GHz, and 2.70 GHz.

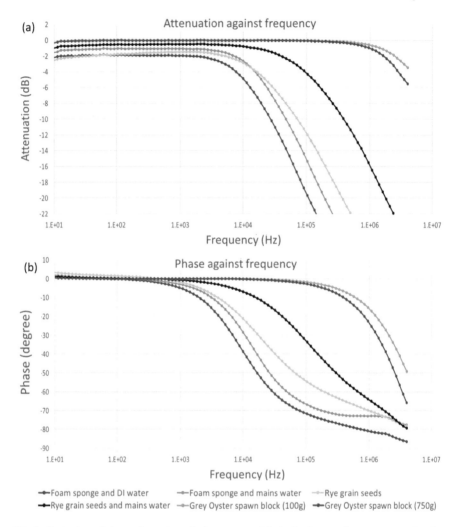

Fig. 8 Bode plots of six configurations; foam sponge and de-ionised water, foam sponge and mains water, Rye grain seeds, Rye grain seeds and mains water, grey oyster spawn block (100 g), Grey Oyster spawn block (750g), **a** attenuation against frequency **b** phase against frequency (10 Hz to 4 MHz)

Return loss against frequency (S11) in cap and stalk of *Agaricus bisporus* (Portobello) fruiting bodies, is shown in Fig. 12. Signal propagation was observed to be frequency-dependent with resonance peaks around 180 MHz, 780 MHz, 1.4 GHz, 2.1 GHz and 2.7 GHz.

Signal gain against frequency (S21) in cap and stalk of *Agaricus bisporus* (Portobello) fruiting bodies is shown in Fig. 13, with peaks in attenuation around 780 MHz, 1.18 GHz, 1.89 GHz, 2.07 GHz, 2.26 GHz, and 2.76 GHz.

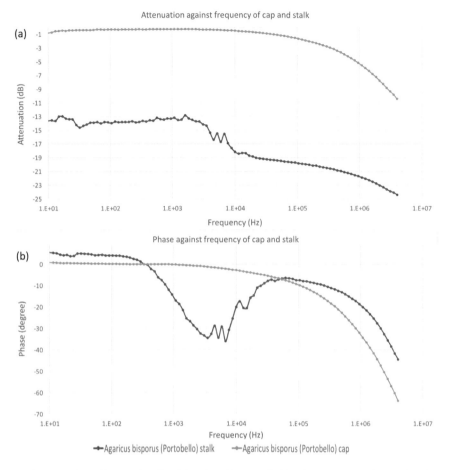

Fig. 9 Bode plots of cap and stalk of *Agaricus bisporus* fruiting body **a** attenuation against frequency **b** phase against frequency (10 Hz to 4 MHz)

The moisture content of mycelium-bound substrates and fruiting bodies was determined by the following procedure: (1) measure the 'wet' weight of the sample (2) dehydrate the sample in an oven at 80 °C for ∼ 48 h (3) measure the 'dry' weight of the sample (4) calculate the difference between 'wet' and 'dry' weights then dividing by the 'dry' weight.

Overall, electrical characteristics were observed to vary with electrode separation. A distance of ∼20 mm between centres of electrodes was found to be effective for most measurements. This suggests that there is an optimum spacing for the electrodes in any environment. Optimising the relative physical positions of electrodes in colonised substrate and fungal fruiting bodies is important to maximising the sensitivity of monitoring and interconnections to other systems.

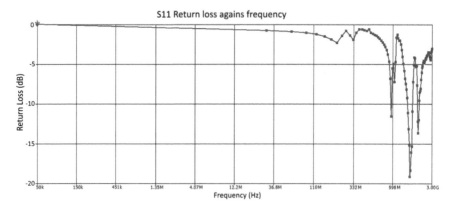

Fig. 10 S11—Return loss against the frequency (50 kHz to 3 GHz) of mycelium bound composite (Rye seeds well colonised with *Pleurotus ostreatus*)

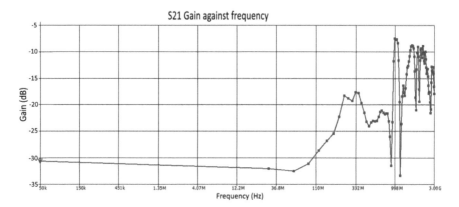

Fig. 11 S21—Gain against frequency (50 kHz to 3 GHz) of mycelium bound composite (Rye seeds well colonised with *Pleurotus ostreatus*)

Obtaining fresh blocks of spawn (e.g. 100 g bags) from commercial suppliers at desired times was challenging (e.g. limited stock availability). Further, most commercial suppliers were unwilling to provide details of substrate composition (beyond "Rye seeds" as considered a 'trade secret'). Therefore, variation in substrate might exist between both batches from the same supplier and different suppliers. The level of colonisation of blocks varied greatly between suppliers and times of recordings as the fungi consumed the substrate as a source of nutrients. Additionally, the heterogeneous mixture of substrate and fungi added an additional variable.

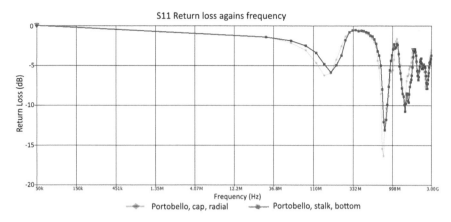

Fig. 12 Return loss against frequency (50 kHz to 3 GHz) of cap and stalk of *Agaricus bisporus* fruiting body

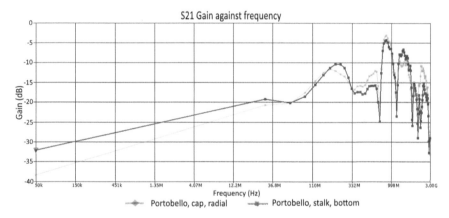

Fig. 13 Gain against frequency (50 kHz to 3 GHz) of cap and stalk of *Agaricus bisporus* fruiting body

4 Discussion

With 80% moisture content, the growth of a strong hypha network through the body of well-colonised substrate, particularly near the surface of fresh spawn, appears to significantly raise the mean cut-off frequency to >500 kHz compared to 10 kHz with bare Rye seeds with 66% moisture content and 50 kHz with Rye seeds moistened with mains water with 76% moisture content.

The lower mean cut-off frequency of fruiting bodies (5 kHz to 50 kHz depending on species) may be useful in a variety of applications, including non-computing electronic circuits (e.g. signal filtering in audio systems). The higher mean cut-off frequency of mycelium-bound composites might be used in high-speed analog computer circuits. 'Switches' based on regulated development of hypha networks, for example, might build ultra-low power consumption signal routers.

Living mycelium might be employed for sensing, sensory fusion, and preprocessing. Furthermore, combining mycelium with silicon components (potential on dried mycelium) may provide innovative analog computing hardware capabilities.

A low pass filter might be used to establish an upper bound on data processing rate in any computational, logic gate circuitry incorporated in or produced from the substrates.

5 Conclusions

The electrical properties of mycelium-bound composites were investigated in order to determine their suitability for signal transmission, particularly in electronic sectors such as bio-hybrid computing systems.

Mycelium-bound composites have low-pass filter properties, with an average cut-off frequency of around 500 kHz, a roll-off rate of around -14 dB/decade, and an average attenuation inside the pass band of less than 1 dB. Fungal fruiting bodies, on the other hand, have substantially lower average cut-off frequencies that range from 5 kHz to 50 kHz depending on the species. Their roll-off rates range from -20 to -30 decibels per decade, and the typical attenuation inside the pass band is less than 3 decibels. While the mechanism underlying frequency-dependent attenuation is unknown, the presence of high-water content, which is electrically conductive due to dissolved ionizable solids (approximately 80% in mycelium-bound composites and up to 92% in fruiting bodies depending on species), is most likely a major contributor.

Mycelium-bound composites and fungal fruiting bodies may be useful components in analog computing systems. They could, for example, serve as frequency filters in oscillator computing and contribute to sensory data processing in edge computing. Living mycelium could be used in sensing, sensory fusion, and preliminary data processing. Furthermore, the integration of mycelium networks with silicon components (possibly with dried mycelium) could lead to novel capabilities in next-generation analogue computing. This research shows that mycelium-bound composites have significant losses in the low-to-high frequency region and have dielectric characteristics. Despite being considered electrical insulators, the moisture content may diminish their electrical breakdown voltage, making them unsuitable for high-voltage applications. While transferring electrical signals through mycelium is difficult, including conductive particles holds promise for future sustainable electronic systems [26].

6 Availability of Data

The raw data required to reproduce these findings are available to download from https://doi.org/10.5281/zenodo.7339710. The processed data required to reproduce these findings are available to download from https://doi.org/10.5281/zenodo.7339728.

References

1. Saez, D., Grizmann, D., Trautz, M., Werner, A.: Exploring the binding capacity of mycelium and wood-based composites for use in construction. Biomimetics **7**(2) (2022)
2. Karana, E., Blauwhoff, D., Hultink, E.-J., Camere, S.: When the material grows: a case study on designing (with) mycelium-based materials. Int. J. Des. **12**(2) (2018)
3. Jones, M., Mautner, A., Luenco, S., Bismarck, A., John, S.: Engineered mycelium composite construction materials from fungal biorefineries: a critical review. Mater. Des. **187**, 108397 (2020)
4. Cerimi, K., Can Akkaya, K., Pohl, C., Schmidt, B., Neubauer, P.: Fungi as source for new bio-based materials: a patent review. Fungal Biol. Biotechnol. **6**(1), 1–10 (2019)
5. Adamatzky, A., Ayres, P., Belotti, G., Wösten, H.: Fungal architecture position paper. Int. J. Unconv. Comput. **14** (2019)
6. Beasley, A.E., Powell, A.L., Adamatzky, A.: Capacitive storage in mycelium substrate (2020). arXiv:2003.07816
7. Beasley, A.E., Abdelouahab, M.-S., Lozi, R., Powell, A.L., Adamatzky, A.: Mem-fractive properties of mushrooms (2020). arXiv:2002.06413
8. Beasley, A.E., Powell, A.L., Adamatzky, A.: Fungal photosensors (2020). arXiv:2003.07825
9. Jones, M., Mautner, A., Luenco, S., Bismarck, A., John, S.: Engineered mycelium composite construction materials from fungal biorefineries: a critical review. Mater. Des. **187**, 108397 (2020)
10. Soh, E., Yong Chew, Z., Saeidi, N., Javadian, A., Hebel, D., Le Ferrand, H.: Development of an extrudable paste to build mycelium-bound composites. Mater. Des. **195**, 109058 (2020)
11. Adamatzky, A., Tegelaar, M., Wosten, H.A.B., Powell, A.L., Beasley, A.E., Mayne, R.: On boolean gates in fungal colony. Biosystems **193**, 104138 (2020)
12. Adamatzky, A., Chiolerio, A., Sirakoulis, G.: On resistive spiking of fungi (2020). arXiv:2009.00292
13. Adamatzky, A., Gandia, A., Chiolerio, A.: Fungal sensing skin. Fungal Biol. Biotechnol. **8**(1), 1–6 (2021)
14. Przyczyna, D., Szacilowski, K., Chiolerio, A., Adamatzky, A.: Electrical frequency discrimination by fungi pleurotus ostreatus (2022). arXiv:2210.01775
15. Phillips, N., Gandia, A., Adamatzky, A.: Electrical response of fungi to changing moisture content. TBC J. (2022)
16. Olsson, S., Hansson, B.S.: Action potential-like activity found in fungal mycelia is sensitive to stimulation. Naturwissenschaften **82**(1), 30–31 (1995)
17. Adamatzky, A.: On spiking behaviour of oyster fungi pleurotus djamor. Sci. Rep. **8**(1), 1–7 (2018)
18. Adamatzky, A., Tuszynski, J., Pieper, J., Nicolau, D.V., Rinalndi, R., Sirakoulis, G., Erokhin, V., Schnauss, J., Smith, D.M.: Towards cytoskeleton computers. A proposal. In: Adamatzky, A., Akl, S., Sirakoulis, G. (eds.) From Parallel to Emergent Computing. CRC Group/Taylor & Francis (2019)
19. Adamatzky, A., Ayres, P., Beasley, A.E., Roberts, N., Tegelaar, M., Tsompanas, M.-A., Wösten, H.A.B.: Logics in fungal mycelium networks (2021). arXiv:2112.07236

20. Shao, B., Weerasekera, R., Tareke Woldegiorgis, A., Zheng, L.-R., Liu, R., Zapka, W.: High frequency characterization and modelling of inkjet printed interconnects on flexible substrate for low-cost rfid applications. In: Electronics System-Integration Technology Conference, pp. 695–700 (2008)
21. Shao, B., Weerasekera, R., Zheng, L.-R., Liu, R., Zapka, W., Lindberg, P.: High frequency characterization of inkjet printed coplanar waveguides. In: IEEE Workshop on Signal Propagation on Interconnects, pp. 1–4 (2008)
22. Miller, A.: Oyster grain spawn (2022). https://www.annforfungi.co.uk/shop/oyster-grain-spawn/. [Online; accessed 18-Sept-2022]
23. Precision, B.K.: Model 891 (2022). https://www.bkprecision.pl/files/891_datasheet.pdf. [Online; accessed 18-Sept-2022]
24. Cypher-Instruments.: C60 network analyzer (2022). http://www.cypherinstruments.co.uk/. [Online; accessed 18-Sept-2022]
25. NanoRFE.: Vector network analyse, model NanoVNA-F V2 (2022). https://nanorfe.com/nanovna-v2.html. [Online; accessed 18-Sept-2022]
26. Danninger, D., Pruckner, R., Holzinger, L., Koeppe, R., Kaltenbrunner, M.: Myceliotronics: fungal mycelium skin for sustainable electronics. Sci. Adv. **8**(45), eadd7118 (2022)
27. Ishtaiwi, M., Hajjyahya, M., Habbash, S.: Electrical properties of dead sea water. J. Appl. Math. Phys. **9**(12), 3094–3101 (2021)
28. Porle, R.R., Ruslan, N.S., Ghani, N.M., Arif, N.A., Ismail, S.R., Parimon, N., Mamat, M.: A survey of filter design for audio noise reduction. J. Adv. Rev. Sci. Res. **12**(1), 26–44 (2015)
29. Williams, A.B.: Analog Filter and Circuit Design Handbook. McGraw-Hill Education (2014)
30. Thiele, N.: Bandpass subwoofer design. J. Audio Eng. Soc. **62**(3), 145–160 (2014)
31. Parhami, B.: Parallel processing with big data (2019)
32. Cao, K., Liu, Y., Meng, G., Sun, Q.: An overview on edge computing research. IEEE Access **8**, 85714–85728 (2020)
33. Varghese, B., Wang, N., Barbhuiya, S., Kilpatrick, P., Nikolopoulos, D.S.: Challenges and opportunities in edge computing. In: 2016 IEEE International Conference on Smart Cloud (SmartCloud), pp. 20–26. IEEE (2016)
34. Krestinskaya, O., Pappachen James, A., Ong Chua, L.: Neuromemristive circuits for edge computing: a review. IEEE Trans. Neural Netw. Learn. Syst. **31**(1), 4–23 (2019)
35. Csaba, G., Raychowdhury, A., Datta, S., Porod, W.: Computing with coupled oscillators: theory, devices, and applications. In: 2018 IEEE International Symposium on Circuits and Systems (ISCAS), pp. 1–5. IEEE (2018)
36. Csaba, G., Porod, W.: Coupled oscillators for computing: a review and perspective. Appl. Phys. Rev. **7**(1), 011302 (2020)
37. Chou, J., Bramhavar, S., Ghosh, S., Herzog, W.: Analog coupled oscillator based weighted ising machine. Sci. Rep. **9**(1), 1–10 (2019)
38. MacKay, D.M.: High-speed electronic-analogue computing techniques. Proc. IEE-Part B: Radio Electron. Eng. **102**(5), 609–620 (1955)
39. Ronaldo da Costa Bento, C., Carlos Gomes Wille, E.: Bio-inspired routing algorithm for manets based on fungi networks. Ad Hoc Netw. **107**, 102248 (2020)
40. Adamatzky, A., Nikolaidou, A., Gandia, A., Chiolerio, A., Mahdi Dehshibi, M.: Reactive fungal wearable. Biosystems **199**, 104304 (2021)

Fungal Computing

Towards Fungal Computer

Andrew Adamatzky

Abstract We propose that fungi *Basidiomycetes* can be used as computing devices: information is represented by spikes of electrical activity, a computation is implemented in a mycelium network and an interface is realised via fruit bodies. In a series of scoping experiments we demonstrate that electrical activity recorded on fruits might act as a reliable indicator of the fungi's response to thermal and chemical stimulation. A stimulation of a fruit is reflected in changes of electrical activity of other fruits of a cluster, i.e. there is distant information transfer between fungal fruit bodies. In an automaton model of a fungal computer we show how to implement computation with fungi and demonstrate that a structure of logical functions computed is determined by mycelium geometry.

1 Introduction

The fungi is a largest, widely distributed and the oldest group of living organisms [1]. Smallest fungi are microscopic single cells. The largest mycelium belongs to *Armillaria bulbosa* which occupies 15 ha and weights 10 tons [2] and the largest fruit body belongs to *Fomitiporia ellipsoidea* which at 20 years old is 11 m long, 80 cm wide, 5 cm thick and has estimated weight of nearly half-a-ton [3]. During last decade we produced nearly forty prototypes of sensing and computing devices from the slime mould *Physarum polycephalum* [4], including shortest path finders, computational geometry processors, hybrid electronic devices, see compilation of latest results in [5]. We found that the slime mould is a convenient substrate for unconventional computing however geometry of the slime mould's protoplasmic networks is continuously changing, thus preventing fabrication of long-living devices, and the slime mould computing devices are confined to experimental laboratory setups. Fungi *Basidiomycetes* are now taxonomically distinct from the slime mould, however their development and behaviour are phenomenologically similar: mycelium

---

A. Adamatzky (✉)
Unconventional Computing Lab, UWE, Bristol, UK
e-mail: andrew.adamatzky@uwe.ac.uk

networks are analogous to the slime mould's protoplasmic networks, and the fruit bodies are analogous to the slime mould's stalks of sporangia. *Basidiomycetes* are less susceptible to infections, when cultured indoors, especially commercially available species, they are larger in size and more convenient to manipulate than slime mould, and they could be easily found and experimented with outdoors. This make the fungi an ideal object for developing future living computing devices. Advancing our recent results on electrical signalling in fungi [6], which in a way is similar to electrical signalling in plants [7], we are exploring computing potential of fungi in present chapter.

2 Mycelium Basis of Fungal Computer

Mycelium propagates by a foraging front and consolidations of mycelial cords behind the front [11]. The foraging front travels outward and produces fruit bodies (Fig. 1ab). The front is also manifested by rings of increased vegetation and 'exhausted' soil (Fig. 1c), see historical overviews in [8, 9]. Propagation/extension of the ring is due to exhaustion of nutrients necessary for fungi growth.

A mycelial growth pattern is determined by nutritional conditions and temperature [11–16], as also demonstrated in computer models in [17, 18]. A complexity of the mycelium network, as estimated by a fractal dimension, is determined by the nutrient availability and the pressure built up between various parts of the mycelial network [19] In domains with high concentration of nutrients mycelia branch, in poor nutrient domains mycelia stop branching [20]. As indicated in [13] optimisation of resources is evidenced by inhibitory effect of contact with baits on the remainder of the colony margin, regression of mycelium originating from the inoculum associated with the renewed growth from the bait, and differences between growth patterns of large and small inocula/baits (Fig. 1d). Optimisation of the mycelial network [15] is quite similar to that of the slime mould *P. polycephalum*, as evidenced in our previous studies, esp. in terms of proximity graphs [21] and transport networks [22]. Exploration of confined spaces by hyphae has been studied in [23–27], and evidences of the efficiency of the exploration provided. All the above indicates that (1) fungal mycelium can solve the same range of computational geometry problems as the slime mould *P. polycephalum* does [5]: shortest path [28–32], Voronoi diagram [33], Delaunay triangulation, proximity graphs and spanning tree, concave hull and, possibly, convex hull, and, with some experimental efforts, travelling salesman problem [34], and (2) by changing environmental conditions we can reprogram a geometry and graph-theoretical structure of the mycelium networks and then utilise electrical activity of fungi [6, 35, 36] to realise computing circuits.

A mycelium is hidden underground, therefore only configurations of fruit bodies can be seen as outputs of a geometric computation implemented by propagating mycelium. Consider the following example of interacting foraging fronts. Propagation of wave-fronts of fungi at large scale was described by Shantz and Piemeisel in Yuma, Colorado, on June 1916 [9]. This is illustrated in (Fig. 1d). There are

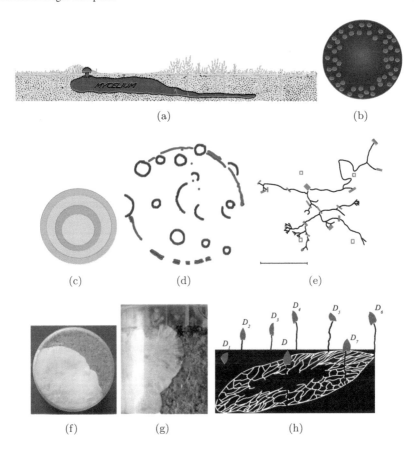

Fig. 1 Development of mycelium in nutrient rich (**abcd**) and nutrient poor (**e**) substrates. **a** A cross-section of a fairy ring produced by *Marasmius Oreades*. Redrawn from [8]. **b** A view from above: the mycelium is dark red, the fruits are red, and the dried fruits are blue. **c** Vegetation profile corresponding to (**b**): outer stimulated (light green) and inner stimulated (dark green) zones of increased vegetation, dead zone (gray) of reduced vegetation and inside zone (yellow) of ambient vegetation. **d** Rings and fragments of rings of *Agaricus campestris* (dark red) inside 65 m ring of *Calvatia ciathyiformi* (fresh fruits are red, dry fruits are blue). Redrawn from [9]. **e** A development pattern of a single mycelial system of *Phanerochaete velutina*. Lines are mycelial cords. Orange/gray rectangles are inoculum blocks, white rectangles decayed inoculum blocks. Scale bar is 1 m. Redrawn from [10]. **f** Photo of mycelium propagating on a nutrient rich cocoa substrate. **g** Zoomed view of the propagating front where branching is articulated. **h** Schematic architecture of a fungal computer. Fruit bodies $D_1, D_2, \ldots$ are I/O interface. Mycelium network **C** is a distributed computing device

two species of fungi *Agaricus campestris* and *Calvatia ciathyiformi*. The ring of *C. ciathyiformi* was nearly 65 m in diameter with 50 fresh fruits. There are several smaller rings of *A. campestris*: in some places they interrupt 'wave-fronts' of *C. ciathyiformi* growth. In theory, such interaction of wave fronts of different species can be used to approximate the Voronoi diagram, as have been done previously with slime mould [33, 37], when planar data points are represented by locations of fungi inoculates.

Also, notice characteristic location of dry fruits of *C. ciathyiformi* (blue in Fig. 1bd), this brings in an analogy with excitable medium: the fresh fruits are analogous to the 'excitation' wave-front and the dried fungi to 'refractory' tails of the excitation waves. Fungi rings can extend up to 200 meters diameter [9]. Analogy between fungi foraging fronts and excitation wave fronts indicate that already algorithms for computing with wave fronts in excitable medium [38, 39] can be realised with foraging mycelium. Said that, solving geometrical problems with mycelium networks does not sound feasible, because the mycelium growth rate is very low, thus a solution of any problem could take weeks and months, if not the years for problems which spatial representation covers hundreds of meters.

In contrast to the slow growth of mycelium, fungi exhibit an electrical response to stimulation in a matter of seconds or minutes [6, 35, 36]. Therefore, a computation using electrical impulses propagating in and modified by the mycelium networks seems to be promising. We propose the following architecture of a fungal computer $\mathcal{A}$ (Fig. 1h): a mycelium **C** is a processor, or rather a network of processors, and fruit bodies $D_1, D_2, \ldots$ comprise I/O interface of the fungal computer. The information is represented by spikes of electrical potential. Thus a state of D_i^t at a time step t could be either binary, depending on whether a spike is present or absent at time t, or multiple valued, depending on a number of spikes in a train, duration of spikes and their amplitudes. Any fruit body can be considered as input and output and the fungal computer $\mathcal{A}$: $\mathbf{D}^{t+w} = \mathcal{F}(\mathbf{D})$, where w is a positive integer. Details on how exactly $\mathcal{A}$ could compute will be analysed in Sect. 4, first let us consider, Sect. 3, few examples from laboratory experiments on endogenous spiking, response of fungi to stimulation and evidences of communication between fruit bodies.

3 Electrical Activity of Fungi

3.1 *Experimental*

We used commercial mushroom growing kits[1] of pearl oyster mushrooms *P. ostreatus*. In the experiments reported seven growing kits have been used. For each kit we recorded electrical activity of the first flush of fruiting bodies only because the first

---

[1] © Espresso Mushroom Company, Brighton, UK.

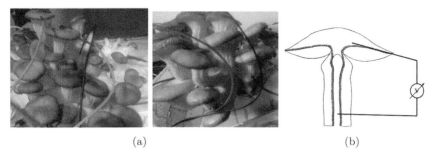

(a) (b)

Fig. 2 Experimental setup. **a** Photographs of fruit bodies with electrodes inserted. **b** Position of electrodes in relation to a translocation zone. Drawing of fruit body is from Schütte [40]; a scheme of electrodes is our

flush is usually provides a maximum yield of fruiting bodies [41] and the growing mycelium in the substrate was less affected by products of fungi metabolism [42, 43]. Each substrate's bag was 22 cm by 10 cm by 10 cm, 800–900 g in weight. The bag was cross-sliced 10 cm vertical and 8 cm horizontal and placed in a cardboard box with 8 cm by 10 cm opening. Experiments were conducted at room temperature in constant (24 h) ambient lighting of 10 lux. Electrical potential of fruit bodies was recorded from the second-third day of their emergence. Resistance between cap and stalk of a fruit body was 1.5 MΩ in average, between any two heads in the cluster 2 MΩ (measured by Fluke 8846A). We recorded electrical potential difference between cap and stalk of the fruit body. We used sub-dermal needle electrodes with twisted cable[2]. Recording electrode was inserted into stalk and reference electrode in the translocation zone, cross-section of a fruit body showing translocation zone, drawing by Schütte [40], of the cap (Fig. 2b); distance between electrodes was 3–5 cm. In each cluster we recorded 4–6 fruit bodies simultaneously (Fig. 2a) for 2–3 days. Electrical activity of fruit bodies was recorded with ADC-24 High Resolution Data Logger[3]. The data logger employs differential inputs, galvanic isolation and software-selectable sample rates—these contribute to a superior noise-free resolution; its 24-bit A/D converter maintains a gain error of 0.1%. Its input impedance is 2 MΩ for differential inputs, and offset error is 36 μV in $\pm$1250 mV range use. We recorded the electrical activity one sample per second; during the recording the logger made as many measurements as possible (typically up 600) per second then saved average value.

---

[2] © SPES MEDICA SRL Via Buccari 21 16153 Genova, Italy.
[3] Pico Technology, St. Neots, Cambridgeshire, UK.

Fig. 3 Co-existence of various types of electrical activity in fruit bodies of the same cluster. **a** Electrical potential recorded for over 16 h on four fruits. **b** Zoomed in area marked 'A' in (**a**). Large amplitude spikes. **c** Zoomed in area marked 'B' in (**a**). Two wave-packets

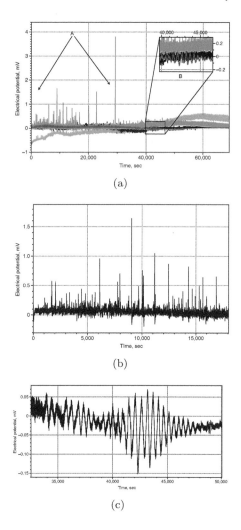

3.2 Endogenous Spiking

As we previously discussed in [6] fruits show a rich family of endogenous, i.e. not caused by purposeful stimulation during experiments, spiking behaviour. Spiking patterns of several types have been observed during simultaneous recording from the different fruits of the same cluster. Recordings of four fruit bodies during nearly 20 h are shown in Fig. 3. Most pronounced patterns are the trains of large amplitude spikes (Fig. 3b) and the wave-packets (Fig. 3c).

Large amplitude spikes (Fig. 3b) have average amplitude 0.77 mV, st. deviation 0.29. The spikes are usually observed in pairs. Average distance between spikes in a pair is 238 s, st. deviation 81 s. Time interval between two largest, over 0.8 mV, spikes varies from 20 min to 48 min. Two wave-packets are shown in Fig. 3c. First wave-packet, roughly 91 min long, consists of ten spikes. Their amplitude varies from 0.05 mV a the beginning to 0.1 mV at the eclipse. Shortest spike is 362 s duration, longest, in the middle of the waveform is 705 s. Second wave-packet, most pronounced, consists of 19 spikes and lasts for 163 min. Amplitudes of the spikes vary from 0.05 mV at the beginning of the wave-packet to 0.2 mV in the middle. Shortest spike is 457 s long, longest spike is 609 s long. Average spike duration is 516 s ($\sigma = 56$), average amplitude is 0.12 mV ($\sigma = 0.06$).

3.3 Signalling Between Fruits

To check if fruits in a cluster would respond to stimulation of their neighbours we conducted experiments illustrated in Fig. 4. Note, fruits which electrical potential recorded were not stimulated (Fig. 4a). Recording on one of the fruiting bodies (Ch3) shows periodic oscillations: average amplitude 0.47 mV ($\sigma = 0.19$), average duration of a spike is 1669 s (st. deviation 570), average period 1819 s (st. deviation 847) (Fig. 4b). Other recorded fruiting bodies also show substantial yet non-periodic changes in the electrical potential with amplitudes up to 1 mV. A thermal stimulation, S1 and S2 in Fig. 4b, leads to a temporal disruption of oscillation of the fruit Ch3, and low amplitude short period spikes in other recorded fruits Ch1, Ch2, Ch4–Ch9. The response of an intact fruit to stimulation of another fruit with an open flame consists of a depolarisation c. 0.02 mV amplitude, c. 6 s duration, followed by a repolarisation c. 0.2 mV amplitude, c. 9 s duration. The depolarisation starts c. 3 s after start of stimulation. This might indicate that it is caused by action potential-like fast dynamical changes. High-amplitude repolarisation takes place at c. 13 s after start of stimulation, when a substantial loci of a fruit cap becomes thermally damaged. Application of ethanol (S3, Fig. 4) and salt (S4, Fig. 4) leads to 0.15–0.45 mV drop in electrical potential, recovery occurs in c. 1200 s.

In experiment illustrated in Fig. 5 we stimulated fruits with open flame, salt and sugar, and recorded electrical responses from non-stimulated neighbours. Application of sugar did not cause any response, and thus can be seen as a control experiment on mechanical stimulation. There were no responses to a short-term (c. 1–2 s) mechanical stimulation recorded. Fruits respond to thermal stimulation of a member of their cluster by a couple of action-potential like impulses (Fig. 5b). The amplitude of the response differ from fruit to fruit, and more likely depends not only on the distance between the recorded fruit and the stimulated fruit but also on the position of electrodes. The fruits respond to saline stimulation of their neighbour in a more uniform manner (Fig. 5c). In 12–15 s after application of salt, electrical potential of the recorded fungi drops by c. 0.2–0.8 mV. The potential recovers in c. 30 s.

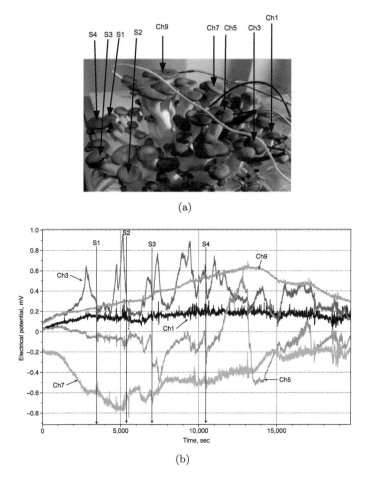

Fig. 4 Stimulation of fruits. **a** Setup of recording, sites of stimulation and location of electrode pairs corresponding to channels Ch1–Ch9. **b** Electrical potential recording on five mushrooms. Channel Ch1 is shown by black, Ch3 red, Ch5 blue, Ch7 green and Ch9 orange. The following stimuli have been applied to fruiting bodies. (S1) 3450 s: start open flame stimulation for 20 s. (S2) 5310 s: start open flame stimulation for 60 s. (S3) 7000 s: ethanol drop is placed on a cap of the fruit. (S4) 10440 s: 15 mg of table salt is placed on a cap of one of fruiting bodies

3.4 Signalling About State of Growth Substrate

To test a response of fruits to environmental changes in the growth substrate we injected 150 mL of sodium chloride (6 mg/mL) in the substrate and recorded electrical potential of a fruit. The moment of injection is reflected in the spike of electrical potential with amplitude 9 mV. This spike might be caused by mechanical stimulation of mycelium (Fig. 6). Four hours after injection, the recorded fruit exhibited trains of spiking activity. Amplitude of spikes vary from 0.29 to 12.3 mV, average 4.1 mV

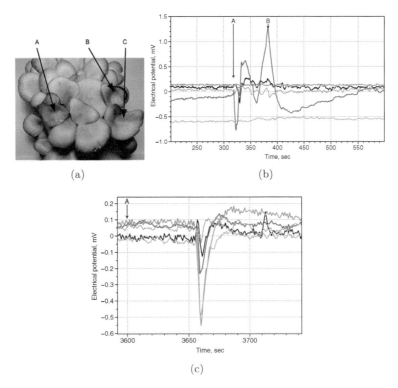

Fig. 5 Response of fruits to stimulation of neighbouring fruits. **a** Photo of the fruit cluster taken *after* experiments were completed. The stimulated fruits are indicated by arrows: 20 mg of salt ('A'), open flame of a butane lighter, temperature 600 °C–800°. for 60 s ('B'), 20 mg sugar ('C'). **b** Electrical potential of five non-stimulated fruits during stimulation of a fruit from their cluster with an open flame: start of the thermal stimulation is shown by arrow 'A', end of the stimulation by arrow 'B'. **c** Electrical potential of five fruits recorded during stimulation of a fruit from their cluster with salt, moment when salt was placed on a cap is shown by arrow labelled 'A'

($\sigma = 3.5$ mV). Duration of a spike varies from 33 to 151 s, average 71 s ($\sigma = 32$ s). Periods varies from 450 to 2870 s, average duration 953 s ($\sigma = 559$). The spikes might be caused by an osmotic function of mycelium due to saline solution intaken and transported into the caps of fruits [44] (cited by [7]).

4 Automaton Model of a Fungal Computer

To imitate propagation of depolarisation waves in the mycelium network we adopt an automaton model. The automaton models are proved to be an appropriate discrete models for spatially extended excitable media [45–47] and verified in models of calcium waves propagation [48], propagation of electrical pulses in the heart

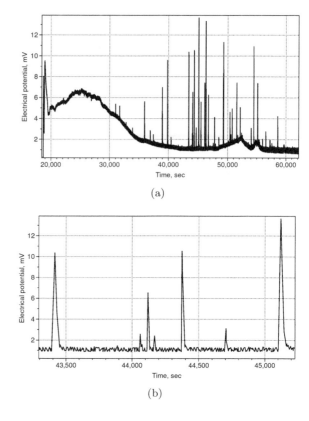

Fig. 6 Electrical response of a fruit body to the injection of saline solution in the substrate

[49–51] and simulation of action potential [52, 53]. We represent a fungal computer by an automaton $\mathcal{A} = \langle \mathbf{C}, \mathbf{Q}, r, h, \theta, \delta \rangle$, where $\mathbf{C} \subset \mathbf{R}$ is a planar set, each point $p \in \mathbf{C}$ takes states from the set $\mathbf{Q} = \{\star, \bullet, \circ\}$, excited ($\star$), refractory ($\bullet$), resting ($\circ$), and updates its state in a discrete time depending on its current state and state of its neighbourhood $u(p) = \{q \in \mathbf{C} : d(p, q) \leq r\}$; r is a neighbourhood radius, θ is an excitation threshold and δ is refractory delay. All points update their states in parallel and by the same rule:

$$p^{t+1} = \begin{cases} \star, & \text{if } (p^t = \circ) \text{ and } \sigma(p)^t > \theta \\ \bullet, & \text{if } (p^t = \circ) \text{ or } ((p^t = \bullet) \text{ and } (h_p^t > 0)) \\ \circ, & \text{otherwise} \end{cases}$$

$$h_p^{t+1} = \begin{cases} \delta, & \text{if } (p^{t+1} = \bullet) \text{ and } (p^t = \star) \\ h_p^t - 1, & \text{if } (p^{t+1} = \bullet) \text{ and } h_p^t > 0 \\ 0, & \text{otherwise} \end{cases}$$

Every resting (o) point of **C** excites ($\star$) at the moment $t + 1$ if a number of its excited neighbours at the moment t—$\sigma(p)^t = |\{q \in u(p) : q^t = \star\}|$—exceeds a threshold θ. Excited point $p^t = \star$ takes refractory state • at the next time step $t + 1$, at the same moment a counter of refractory state h_p is set to the refractory delay δ. The counter is decremented, $h_p^{t+1} = h_p^t - 1$ at each iteration until it becomes 0. When the counter h_p becomes zero the point p returns to the resting state o.

Architecture of **C** was chosen as following. We randomly distributed $2 \cdot 10^4$ points in a ring with small radius $R = 0.5$ and large radius 1 (Fig. 7a). To reflect higher density of mycelium near the propagation front and decay of mycelium inside the propagating disc we distributed points with a probability described by a quadratic function $p(R) = 3.7 \cdot R^2 - 3.6 \cdot R + 0.9$, where $R \in [0.5, 1]$ (Fig. 7b); the function reflects biomass distribution in a cross-section of a fairy ring [54, 55]. To imitate fruit bodies we distributed points in horizontal (L and R) and vertical (U and D) domains with size 0.27 by 0.023 (Fig. 7a); each domain contains 370 points distributed randomly. Distributions of a point's number of neighbours for neighbourhood radius $r = 3, \ldots, 15$ are shown in Fig. 7c. We have chosen $\theta = 4$ (Sect. 4.2.1), $\theta = 5$ (Sect. 4.2.2), $r = 10$, $\delta = 5$ in the reported experiments for the following reasons. A median radius $r = 10$ neighbourhood size is c. 600 times less than a number of points in **C** (Fig. 7d) thus a locality of the automaton state updates is assured. Excitation threshold $\theta = 4$ is a critical for $\mathcal{A}$ with $r = 10$ (Fig. 7e), i.e. it assures that excitation waves fronts propagate for at least half of the perimeter of the ring (Fig. 7a).

4.1 Automaton Action Potential

The automaton $\mathcal{A}$ supports propagation of excitation waves, fronts of which are represented by points in the state $\star$ and tails by points in state •. We assume a point in the state $\star$ has higher electrical potential than a point in the state •. To imitate a voltage difference between electrodes inserted in fruit bodies we select two domains D_1 and D_2 in each of four fruit bodies and calculate a voltage difference V between domains as follows: $V = \sum_{q \in D_1} \chi(q^t) - \sum_{q \in D_2} \chi(q^t)$, where $\chi(\circ) = 0$, $\chi(\star) = 1$, and $\chi(\bullet) = h_q^t$. This imitates an electrical potential difference between electrodes inserted in the cap and the step of a fruit, as illustrated in Fig. 2.

We excite the fungal automaton $\mathcal{A}$ by assigning points of a selected fruit body states $\star$. This is equivalent to thermal or mechanical stimulation of fruits in our laboratory experiments. We record voltage on fruit bodies at every iteration of the automation evolution. Two examples are shown in Fig. 8. For simplicity we consider $\mathcal{A}$ with only two fruit bodies: L and R. When right fruit R is stimulated (see first spike in Fig. 8c) an excitation wave propagates into the mycelium ring C and splits into two waves (Fig. 8a). Excitation waves enter fruit bodies when reach them, which is reflected in spikes of the calculated potential. If the medium was regular (as e.g. a lattice) the excitation wave-fronts would annihilate each other when colliding. However, the disorganised structure of the conductive medium leads to formation

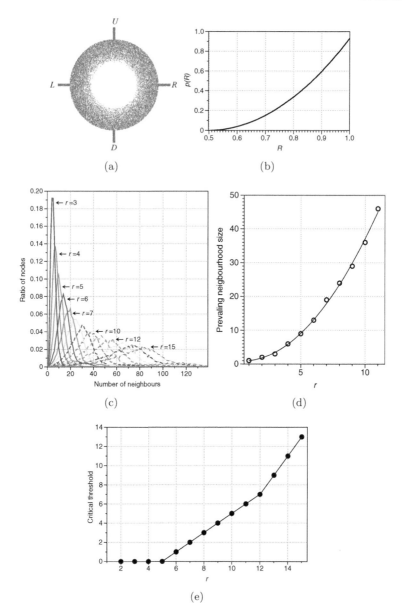

Fig. 7 Fungal computer architecture. **a** Visualisation of **C**. **b** Probability of point in **C** as a function of distance from R. **c** Distribution of a number of neighbours for $r = 3\ldots 15$. A number of neighbours depending on a neighbourhood radius r. (**b**) Prevailing number of neighbours for $r = 3\ldots 15$. (**c**) Critical values of excitation threshold θ for $r = 1, \ldots, 15$

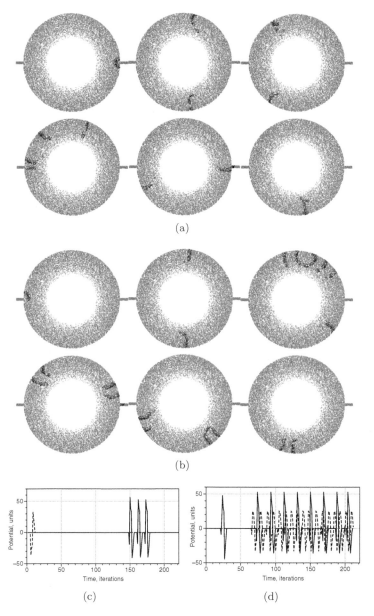

Fig. 8 Dynamics of the excitation of two fruit automaton $\mathcal{A}$: in scenarios of right R (ac) and left L (bd) fruits excited. (ab) Exemplar snapshots of the dynamics. (cd) Electrical potential measured. Dashed line is a potential measured on R and solid line is a potential measured on L

of the new excitation waves (see train of three spikes in Fig. 8c). New waves travel along the ring but eventually die out. Excitation of the left fruit L (Fig. 8b) generate two waves propagating along the ring. However, in this case, due to irregularity of the excitable medium a temporary wave generator is born in the upper part of the ring (2nd snapshot in Fig. 8b). The generator produces pairs of waves (3rd snapshots in Fig. 8b). The transition from sparse spiking to wave-packets is similar to experimental results shown in Fig. 3. In this example we witness that fungal responses to stimulation of the left and the right fruits are different. This can be employed in desings of computing schemes with fungal automata, as outlined in next section.

4.2 Logical Functions Computed by $\mathcal{A}$

Dynamics of excitation waves propagation and interaction in **C** is determined by exact configuration of the planar set. The configurations are generated at random, therefore we expect fungal automaton to implement different functions for each, or nearly configuration. This is illustrated by two following examples. Here we use four fruit bodies acting as both inputs and outputs. A logical input TRUE, or '1', is represented by excitation of a chosen fruit body. A logical output TRUE, or '1', is recognised as one or more impulses recorded at the fruit body some time interval after stimulation: we started recording 40 iterations (this parameter w introduced in Sect. 2), of automaton evolution, after stimulation and stopped recording 130 iterations. Let us consider two examples. The sets **C** are generated randomly, therefore dynamics of excitation is expected to be different in these examples.

4.2.1 First Example

In first example we consider the configuration **C** shown in Fig. 9a. Excitation dynamics for inputs $R = 0, U = 1, D = 0, L = 1$ is shown in Fig. 9 and for inputs $R = 1, U = 0, D = 1, L = 0$ in Fig. 10. When fruits U and D are stimulated (Fig. 9) the fungal automaton $\mathcal{A}$ responds with spikes on fruits L and R (Fig. 11a). Excitation dynamics is less trivial when fruits L and R are stimulated (Fig. 10): automaton $\mathcal{A}$ responds with two voltage spikes at fruit D, and a single spikes at fruits U and R (Fig. 11b). When only fruit R is stimulated the automaton $\mathcal{A}$ responds with pairs of spikes on all fruits but L (Fig. 12b). The automaton $\mathcal{A}$ responds with a spike on fruit L just before cutoff time 150. After 150th iteration two centres of spiral waves are formed and thus the fungal automaton exhibits regular trains of spikes on all fruit bodies (Fig. 12a), similar to dynamics of excitation shown in Fig. 4b.

We stimulated the fungal automaton with sixteen combinations of input variables and constructed a tabular representation of a function realised by the automaton (Table 1), where R, U, L, D are values of input variables, and $R\star, U\star, L\star, D\star$ are values of output variables. Assuming one or two impulses on the fruits represent TRUE we have the following functions implemented by the fungal automaton

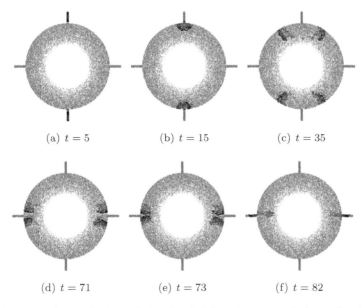

Fig. 9 Snapshots of excitation dynamics in a four-fruit fungal automaton for inputs $R = 0, U = 1, D = 0, L = 1$

$$\begin{aligned} R^\star &= \overline{R}(U + L + D) + R\overline{D} \\ U^\star &= \overline{U}(L + U + R) \\ L^\star &= \overline{L}(D + U + R) \\ D^\star &= \overline{D}(L + U + R) \end{aligned} \quad (1)$$

with equivalent circuit for fruit D shown in Fig. 13. If we assume that only two impulses represent TRUE we have $R^\star = U^\star = L^\star = 0$ and $D^\star = R\overline{D}$.

4.2.2 Second Example

In the second example, discussed below, we used a random configuration of points $\mathbf{C}$ and the automaton $\mathcal{A}$ with $\delta = 5, r = 10$ and $\theta = 5$. Dynamic of electrical potential for 15 combinations of input values is shown in Fig. 14. The response of the automaton is illustrated in the Table 2. Assuming one impulse or two impulses on the fruits

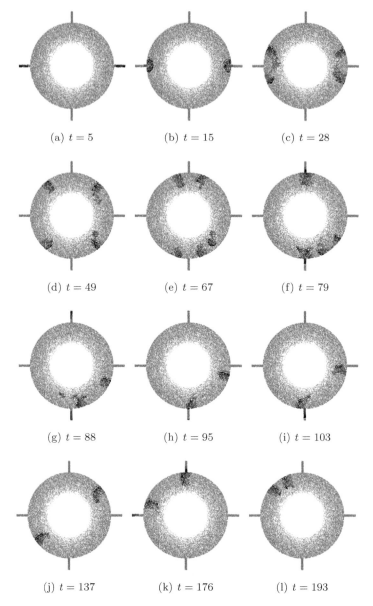

Fig. 10 Snapshots of excitation dynamics in a four-fruit fungal automaton for inputs $R = 1, U = 0, D = 0, L = 1$

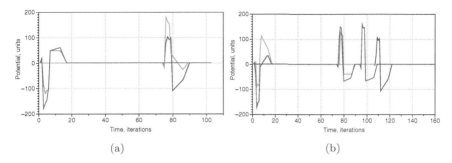

Fig. 11 Voltage measured on four fruits for inputs **a** $R = 0, U = 1, D = 0, L = 1$, see dynamics in Fig. 9, and **b** $R = 1, U = 0, D = 1, L = 0$, see dynamics in Fig. 10. Voltage recorded on fruit R is plotted with red colour, U blue, L green, D magenta

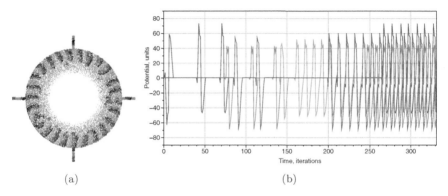

Fig. 12 Response of the fungal automaton for input values $R = 1, U = 0, D = 0, L = 0$. **a** A snapshot of the automaton taken at 350th step of evolution. **b** Voltage recorded on the fruit R is plotted with red colour, U blue, L green D magenta

symbolise '1', we have the following functions realised on each of the fruit bodies

$$\begin{aligned} R^\star &= 0 \\ U^\star &= L + D \\ L^\star &= \overline{L}D(R+U) + R\overline{U}\,\overline{L}\,\overline{D} \\ D^\star &= L(\overline{R} + D + U) \end{aligned} \tag{2}$$

Assuming only two impulses on the fruits symbolise '1', we have the following functions recorded on each of the fruit bodies.

Table 1 Table of a function realised by four-fruit automaton $\mathcal{A}$ (Fig. 9a). One impulse on a fruit is shown by '1', two impulses by '2' and no impulses by '0'

R	U	L	D	R*	U*	L*	D*
0	0	0	0	0	0	0	0
0	0	0	1	1	1	1	0
0	0	1	0	1	1	0	1
0	0	1	1	1	1	0	0
0	1	0	0	1	0	1	1
0	1	0	1	1	0	1	0
0	1	1	0	1	0	0	1
0	1	1	1	1	0	0	0
1	0	0	0	1	1	1	2
1	0	0	1	0	1	1	0
1	0	1	0	1	1	0	2
1	0	1	1	0	1	0	0
1	1	0	0	1	0	1	2
1	1	0	1	0	0	1	0
1	1	1	0	1	0	0	2
1	1	1	1	0	0	0	0

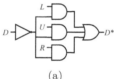

(a)

Fig. 13 Equivalent logical circuit for fruit D implemented by the fungal automaton $\mathcal{A}$, with configuration of **C** shown in Fig. 9a

$$R^\star = 0$$
$$U^\star = L + D(R + U)$$
$$L^\star = R\overline{U}\,\overline{L}D$$
$$D^\star = L\overline{D}(\overline{R} + U)$$
(3)

5 Discussion

We proposed that fungi can be used as computing devices: information is represented by spikes of electrical activity, a computation is implemented in a mycelium network and an interface is realised via fruit bodies. In laboratory experiments we

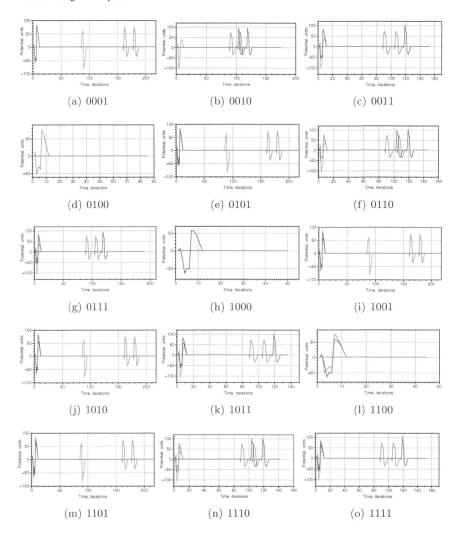

Fig. 14 Dynamics of electrical potential on fruits, in experiment with random seed 357556317, in response to stimulation of inputs. The inputs are shown in the captions in the format $(RULD)$. Spikes appearing during first 10–15 iterations are input spikes. All other spikes are outputs. Voltage recorded on fruit R is plotted with red colour, U blue, L green D magenta

demonstrated that fungi responds with spikes of electrical potential to stimulation of their fruit bodies. Thus we can input data into a fungal computer via mechanical, chemical and electrical stimulation of the fruit bodies. Electrical signalling in fungi, previously evidenced during intracellular recording of electrical potential [35, 36] is similar to the signalling in plants [7, 56]. The experimental results provided in the paper are of illustrative nature with focus on architectures of potential computing devices, a statistical analysis of spontaneous spiking behaviour of the fungi can be

Table 2 Table of a logical function realised by four-fruit automaton $\mathcal{A}$. One impulse on a fruit is shown by '1', two impulses by '2' and no impulses by '0'

R	U	L	D	R★	U★	L★	D★
0	0	0	0	0	0	0	0
0	0	0	1	0	2	1	0
0	0	1	0	0	2	0	2
0	0	1	1	0	2	0	1
0	1	0	0	0	0	0	0
0	1	0	1	0	2	1	0
0	1	1	0	0	2	0	2
0	1	1	1	0	2	0	1
1	0	0	0	0	0	0	0
1	0	0	1	0	2	2	0
1	0	1	0	0	2	1	0
1	0	1	1	0	2	0	1
1	1	0	0	0	0	0	0
1	1	0	1	0	2	1	0
1	1	1	0	0	2	0	2
1	1	1	1	0	2	0	1

found in [6]. Further extensive studies will be necessary to obtain statistical results on fungal response to a stimulation, particularly on the response' dependence on a strength of stimuli and inter-species differences in their responses.

Voltage spikes travelling along mycelium networks might be seen as analogous, but of different physical and chemical nature, to oxidation wave-fronts in a thin-layer Belousov-Zhabotinsky (BZ) medium [57, 58]. Thus, in future, we could draw some useful designs of fungal computers based on established set of experimental laboratory prototypes of of Belousov-Zhabotinsky computing devices. The prototypes produced are image processes and memory devices [59–61], logical gates implemented in geometrically constrained BZ medium [62, 63], approximation of shortest path by excitation waves [64–66], memory in BZ micro-emulsion [61], information coding with frequency of oscillations [67], on-board controllers for robots [68–70], chemical diodes [71, 72], neuromorphic architectures [39, 73–77] and associative memory [78, 79], wave-based counters [80], and other information processors [81–84]. First steps have been already made towards prototyping arithmetical circuits with BZ: simulation and experimental laboratory realisation of gates [38, 62, 63, 85–87], clocks [88] and evolving logical gates [89]. A one-bit half-adder, based on a ballistic interaction of growing patterns [90], was implemented in a geometrically-constrained light-sensitive BZ medium [91]. Models of multi-bit binary adder, decoder and comparator in BZ are proposed in [92–95]. These architectures employ crossover structures as T-shaped coincidence detectors [96] and chemical diodes [72] that heavily rely on heterogeneity of geometrically constrained space. By controlling excitability [97] in different loci of the medium we can achieve impressive results, as it is demonstrated in works related to analogs of dendritic trees [76], polymorphic logical gates [98], and experimental laboratory prototype of four-bit input, two-bit output integer square root circuits based on alternating 'conductivity' of junctions between channels [99].

Spikes of electrical potential are not the only means of implementing information processing in the fungal computers. Microfluidics could be an additional computational resource. Eukaryotic cells, including slime moulds and fungi exhibit cytoplasmic streaming [100, 101]. In experiments with slime mould *P. polycephalum* we found that when a fragment of protoplasmic tube is mechanically stimulated a cytoplasmic streaming in this fragment halts and the fragment's resistance substantially increases. Using this phenomena we designed a range of logical circuits and memory devices [102]. These designs can be adopted in prototypes of fungal computers, however, more experiments would be necessary to establish optimal ways of mechanical addressing of strands of mycelium.

5.1 Programmability

To program fungal computers we must control a geometry of mycelium network. The geometry of mycelium network can be modified by varying nutritional conditions and temperature [11, 16–18], especially a degree of branching is proportional to concentration of nutrients [20], and a wide range of chemical and physical stimuli [103]. Also, we can geometrically constrain it [23–27]. A feasibility of shaping similar networks has been demonstrated in [104]: high amplitude high frequency voltage applied between two electrodes in a network of protoplasmic tubes of *P. polycephalum* leads to abandonment of the stimulated protoplasmic without affecting the non stimulated tubes and low amplitude low frequency voltage applied between two electrodes in the network enhance the stimulated tube and encourages abandonment of other tubes [104].

5.2 Parameters of Fungal Computers

Interaction of voltage spikes, travelling along mycelium strands, at the junctions between strands is a key mechanism of the fungal computation. We can see each junction as an elementary processor of a distributed multi-processor computing network. We assume a number of junctions is proportional to a number of hyphal tips. There are estimated 10–20 tips per 1.5–3 mm [105] of a substrate. Without knowing depth of the mycelial network we go for a safest lower margin of 2D estimation: 50 tips/mm^2. Considering that the largest known fungi *Armillaria bulbosa* populates over 15 ha [2] we could assume that there could be $75 \cdot 10^{11}$ branching points, that is near a trillion of elementary processing units. With regards to a speed of computation by fungal computers, Olsson and Hansson [36] estimated that electrical activity in fungi could be used for communication with message propagation speed 0.5 mm/s (this is several orders slower than speed of a typical action potential in plants: from 0.005 m/s to 0.2 m/s [106]). Thus it would take about half-an-hour for a signal in the fungal computer to propagate one meter. The low speed of signal propagation is not a critical disadvantage of potential fungal computers, because they never meant to compete with conventional silicon devices.

5.3 Application Domains

Likely application domains of the fungal devices could be large-scale networks of mycelium which collect and analyse information about environment of soil and, possibly, air, and executes some decision making procedures. Fungi "possess almost all the senses used by humans" [103]. Fungi sense light, chemicals, gases, gravity and electric fields. Fungi show a pronounced response to changes in a substrate pH [107], demonstrate mechanosensing [108]; they sense toxic metals [109], CO_2 [110] and direction of fluid flow [111]. Fungi exhibit thigmotactic and thigmomorphogenetic responses, which might be reflects in dynamic patterns of their electrical activity [112]. Fungi are also capable for sensing chemical cues, especially stress hormones, from other species [113], thus they might be used as reporters of health and well-being of other inhabitants of the forest. Thus, fungal computers can be made an essential part of distributed large-scale environmental sensor networks in ecological research to assess not just soil quality but an over health of the ecosystems [114–116].

5.4 Further Studies

In automaton models of a fungal computer we shown that a structure of Boolean functions realised depends on a geometry of a mycelial network. In further studies we will tackle four aspects of fungal computing as follows.

First, ideas developed in the automaton model of a fungal computer should be verified in laboratory experiments with fungi. In the automaton model developed we did not take into account a full range of parameters recorded during experimental laboratory studies: origination and propagating of impulses have been imitated in dynamics of the final state machines. To keep the same physical nature of inputs and outputs we will consider stimulating fruit bodies with alternating electrical current. To cascade logical circuits implemented in clusters of fruit bodies we might need to include amplifiers in the hybrid fungi based electrical circuits.

Second, in experiments we evidenced electrical responses of fruits to thermal and chemical stimulation; in some cases we observed trains of spikes. This means we could, in principle, apply experimental findings of Physarum oscillatory logic [117], where logical values are represented by different types of stimuli, apply threshold operations to frequencies of the electrical potential oscillations, and attempt to imple-

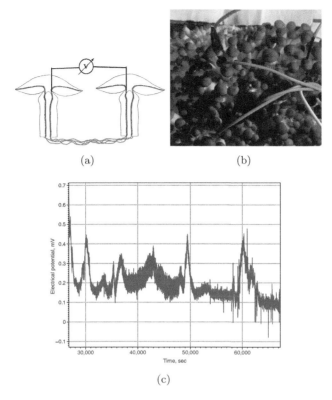

Fig. 15 Electrical potential difference between two neighbouring fruits. **a** Position of electrodes when measuring potential different between fungal bodies. **b** Part of experimental setup. **c** Exemplar plot of electrical potential

ment logical gates. Another option would be to adopt ideas of oscillatory threshold logic reported in [118]; however this might require unrealistically precise control of a geometry of mycelial networks.

Third, we might consider to measure electrical potential between fungal bodies. In a setup shown in Fig. 15ab we recorded electrical potential difference between neighbouring fruits. Example of the recorded activity is shown in Fig. 15b. Average distance between spikes is 4111 s ($\sigma = 2140$). Average duration of a spike is 287 s ($\sigma = 1515$). Average amplitude is 0.25 mV ($\sigma = 0.06$). There is a possibility that patterns of oscillation will be affected by stimulation of other fruit bodies in the cluster. This might lead to a complementary methods of computing with fungi.

Fourth, we must learn how to program a geometry of mycelial networks to be able to execute not arbitrary, as demonstrated in the automaton model, but predetermined logical circuits. Computer modelling approach may be based on formal representation of mycelial networks as a proximity graph, e.g. relative neighbourhood graph [119] (Fig. 16), and then dynamically updating the graph structure till a desired logical circuit is implemented on the graph. Connection rules in proximity

Fig. 16 Representation of a mycelium by relative neighbourhood graph with 2000 nodes. Black discs are fruit bodies

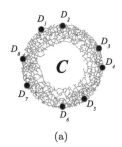

(a)

graphs are fixed, therefore the graph structure can be updated only by adding or removing nodes. New set of nodes can be added to a living mycelial network by placing sources of nutrients. However, due to a very slow growth rate of mycelium, this could be unfeasible. Thus, best way would be to focus only on removing parts of the mycelial network. When parts of a network are removed the network will re-route locally, and a set of logical functions implemented by the network will change.

References

1. Carlile, M.J., Watkinson, S.C., Gooday, G.W.: The fungi. Gulf Professional Publishing (2001)
2. Smith, M.L., Bruhn, J,N., Anderson, J.B.: The fungus Armillaria bulbosa is among the largest and oldest living organisms. Nature **356**(6368), 428 (1992)
3. Dai, Y.-C., Cui, B.-K.: Fomitiporia ellipsoidea has the largest fruiting body among the fungi. Fungal Biol. **115**(9), 813–814 (2011)
4. Adamatzky, A.: Physarum Machines: Computers from Slime Mould, vol. 74. World Scientific (2010)
5. Adamatzky, A. (eds.): Advances in Physarum Machines: Sensing and Computing with Slime Mould. Springer (2016)
6. Adamatzky, A.: On spiking behaviour of oyster fungi Pleurotus djamor. Sci. Rep. 7873 (2018)
7. Gallé, A., Lautner, S., Flexas, J., Fromm, J.: Environmental stimuli and physiological responses: the current view on electrical signalling. Environ. Exp. Botany **114**, 15–21 (2015)
8. Molliard, M.M.: De l'action du *Marasmius Oreades Fr.* sur la végétation. Bulletin De La Societe Botanique De France **57**(1), 62–69 (1910)
9. Shantz, H.L., Piemeisel, R.L.: Fungus fairy rings in eastern Colorado and their effects on vegetation. J. Agric. Res. **XI**(5), 191–245 (1917)
10. Dowson, C.G., Rayner, A.D.M., Boddy, L.: Inoculation of mycelial cord-forming basidiomycetes into woodland soil and litter II. Resource capture and persistence. New Phytol. **109**(3):343–349 (1988)
11. Boddy, L., Wells, J.M., Culshaw, C., Donnelly, D.P.: Fractal analysis in studies of mycelium in soil. Geoderma **88**(3):301–328 (1999)
12. Watkinson, S.C.: The relation between nitrogen nutrition and formation of mycelial strands in *Serpula lacrimans*. Trans. British Mycol. Soc. **64**(2), 195–200 (1975)
13. Dowson, C.G., Rayner, A.D.M., Boddy, L.: Outgrowth patterns of mycelial cord-forming basidiomycetes from and between woody resource units in soil. Microbiology **132**(1), 203–211, (1986)
14. Jennings, D.H.: The Physiology of Fungal Nutrition. Cambridge University Press (1995)
15. Boddy, L., Hynes, J., Bebber, D.P., Fricker, M.D.: Saprotrophic cord systems: dispersal mechanisms in space and time. Mycoscience **50**(1):9–19 (2009)

16. Hoa, H.T., Wang, C.-L.: The effects of temperature and nutritional conditions on mycelium growth of two oyster mushrooms (Pleurotus ostreatus and Pleurotus cystidiosus). Mycobiology **43**(1), 14–23 (2015)
17. Rayner, A.D.M.: The challenge of the individualistic mycelium. Mycologia 48–71 (1991)
18. Regalado, C.M., Crawford, J.W., Ritz, K., Sleeman, B.D.: The origins of spatial heterogeneity in vegetative mycelia: a reaction-diffusion model. Mycol. Res. **100**(12), 1473–1480 (1996)
19. Bolton, R.G., Boddy, L.: Characterization of the spatial aspects of foraging mycelial cord systems using fractal geometry. Mycol. Res. **97**(6), 762–768 (1993)
20. Ritz, K.: Growth responses of some soil fungi to spatially heterogeneous nutrients. FEMS Microbiol. Ecol. **16**(4), 269–279 (1995)
21. Adamatzky, A.: Developing proximity graphs by Physarum polycephalum: does the plasmodium follow the toussaint hierarchy? Parallel Proc. Lett. **19**(01), 105–127 (2009)
22. Adamatzky, A. (ed.).: Bioevaluation of World Transport Networks. World Scientific (2012)
23. Hanson, K.L., Nicolau, D.V. Jr., Filipponi, L., Wang, L., Lee, A.P., Nicolau, D.V.: Fungi use efficient algorithms for the exploration of microfluidic networks. Small **2**(10), 1212–1220 (2006)
24. Held, M., Edwards, C., Nicolau, D.V.: Examining the behaviour of fungal cells in microconfined mazelike structures. In: Imaging, Manipulation, and Analysis of Biomolecules, Cells, and Tissues VI, vol. 6859, p. 68590U. International Society for Optics and Photonics (2008)
25. Held, M., Edwards, C., Nicolau, D.V.: Fungal intelligence; or on the behaviour of microorganisms in confined micro-environments. J. Phys.: Conf. Ser. **178**, 012005. IOP Publishing (2009)
26. Held, M., Lee, A.P., Edwards, C., Nicolau, D.V.: Microfluidics structures for probing the dynamic behaviour of filamentous fungi. Microelectron. Eng. **87**(5–8), 786–789 (2010)
27. Held, M., Edwards, C., Nicolau, D.V.: Probing the growth dynamics of neurospora crassa with microfluidic structures. Fungal Biol. **115**(6), 493–505 (2011)
28. Nakagaki, T., Yamada, H., Tóth, Á.: Intelligence: maze-solving by an amoeboid organism. Nature **407**(6803), 470 (2000)
29. Nakagaki, T.: Smart behavior of true slime mold in a labyrinth. Res. Microbiol. **152**(9), 767–770 (2001)
30. Nakagaki, T., Yamada, H., Toth, A.: Path finding by tube morphogenesis in an amoeboid organism. Biophys. Chem. **92**(1–2), 47–52 (2001)
31. Nakagaki, T., Iima, M., Ueda, T., Nishiura, Y., Saigusa, T., Tero, A., Kobayashi, R., Showalter, K.: Minimum-risk path finding by an adaptive amoebal network. Phys. Rev. Lett. **99**(6), 068104 (2007)
32. Tero, A., Takagi, S., Saigusa, T., Ito, K., Bebber, D.P., Fricker, M.D., Yumiki, K., Kobayashi, R., Nakagaki, T.: Rules for biologically inspired adaptive network design. Science **327**(5964), 439–442 (2010)
33. Shirakawa, T., Adamatzky, A., Gunji, Y.-P., Miyake, Y.: On simultaneous construction of voronoi diagram and delaunay triangulation by physarum polycephalum. Int. J. Bifurc. Chaos **19**(09), 3109–3117 (2009)
34. Jones, J., Adamatzky, A.: Computation of the travelling salesman problem by a shrinking blob. Nat. Comput. **13**(1), 1–16 (2014)
35. Slayman, C.L., Long, W.S., Gradmann, D.: Action potentials. In: *Neurospora crassa*, a mycelial fungus. Biochimica et Biophysica Acta (BBA)—Biomembranes **426**(4), 732–744 (1976)
36. Olsson, S., Hansson, B.S.: Action potential-like activity found in fungal mycelia is sensitive to stimulation. Naturwissenschaften **82**(1), 30–31 (1995)
37. Jones, J., Adamatzky, A.: Slime mould inspired generalised Voronoi diagrams with repulsive fields (2015). arXiv:1503.06973
38. Adamatzky, A.: Collision-based computing in Belousov-Zhabotinsky medium. Chaos Sol. Fract. **21**(5), 1259–1264 (2004)
39. Gorecki, J., Gorecka, J.N., Igarashi, Y.: Information processing with structured excitable medium. Nat. Comput. **8**(3), 473–492 (2009)

40. Schütte, K.H.: Translocation in the fungi. New Phytologist **55**(2), 164–182 (1956)
41. Shah, Z.A., Ashraf, M., Ishtiaq, M.: Comparative study on cultivation and yield performance of oyster mushroom (pleurotus ostreatus) on different substrates (wheat straw, leaves, saw dust). Pakistan J. Nutr. **3**(3), 158–160 (2004)
42. Sánchez, C.: Cultivation of pleurotus ostreatus and other edible mushrooms. Appl. Microbiol. Biotechnol. **85**(5), 1321–1337 (2010)
43. Oei, P. et al.: Mushroom Cultivation: Appropriate Technology for Mushroom Growers. Number Ed. 3. Backhuys Publishers (2003)
44. Mummert, H., Gradmann, D.: Voltage dependent potassium fluxes and the significance of action potentials in acetabularia. Biochimica et Biophysica Acta (BBA)-Biomembranes **443**(3), 443–450 (1976)
45. Markus, M., Hess, B.: Isotropic cellular automaton for modelling excitable media. Nature **347**(6288), 56 (1990)
46. Gerhardt, M., Schuster, H., Tyson, J.J.: A cellular automation model of excitable media including curvature and dispersion. Science **247**(4950), 1563–1566 (1990)
47. Weimar, J.R., Tyson, J.J., Watson, L.T.: Diffusion and wave propagation in cellular automaton models of excitable media. Physica D: Nonlinear Phenomena **55**(3–4), 309–327 (1992)
48. Lechleiter, J., Girard, S., Peralta, E., Clapham, D.: Spiral calcium wave propagation and annihilation in xenopus laevis oocytes. Science **252**(5002), 123–126 (1991)
49. Saxberg, B.E.H., Cohen, R.J.: Cellular automata models of cardiac conduction. In: Theory of Heart, pp. 437–476. Springer (1991)
50. Dowle, M., Mantel, R.M., Barkley, D.: Fast simulations of waves in three-dimensional excitable media. Int. J. Bifurc. Chaos **7**(11), 2529–2545 (1997)
51. Siregar, P., Sinteff, J.P., Julen, N., Le Beux, P.: An interactive 3d anisotropic cellular automata model of the heart. Comput. Biomed. Res. **31**(5), 323–347 (1998)
52. Ye, P., Entcheva, E., Grosu, R., Smolka, S.A.: Efficient modeling of excitable cells using hybrid automata. In: Proceedings of the CMSB, vol. 5, pp. 216–227 (2005)
53. Atienza, F.A., Carrión, J.R., Alberola, A.G., Álvarez, J.L.R., Muñoz, J.J.S., Sánchez, J.M., Chávarri, M.V.: A probabilistic model of cardiac electrical activity based on a cellular automata system. Revista Española de Cardiología (English Edition) **58**(1), 41–47 (2005)
54. Karst, N., Dralle, D., Thompson, S.: Spiral and rotor patterns produced by fairy ring fungi. PloS One **11**(3), e0149254 (2016)
55. Dahlberg, A., Stenlid, J.: Spatiotemporal patterns in ectomycorrhizal populations. Canadian J. Botany **73**(S1), 1222–1230 (1995)
56. Pickard, B.G.: Action potentials in higher plants. Bot. Rev. **39**(2), 172–201 (1973)
57. Belousov, B.P.: A periodic reaction and its mechanism. Compil. Abstr. Radiat. Med. **147**(145), 1 (1959)
58. Zhabotinsky, A.M.: Periodic processes of malonic acid oxidation in a liquid phase. Biofizika **9**(306–311), 11 (1964)
59. Kuhnert, L.: A new optical photochemical memory device in a light-sensitive chemical active medium. Nature (1986)
60. Kuhnert, L., Agladze, K.I., Krinsky, V.I.: Image processing using light-sensitive chemical waves (1989)
61. Kaminaga, A., Vanag, V.K., Epstein, I.R.: A reaction–diffusion memory device. Angewandte Chemie Int. Ed. **45**(19), 3087–3089 (2006)
62. Steinbock, O., Kettunen, P., Showalter, K.: Chemical wave logic gates. J. Phys. Chem. **100**(49), 18970–18975 (1996)
63. Sielewiesiuk, J., Górecki, J.: Logical functions of a cross junction of excitable chemical media. J. Phys. Chem. A **105**(35), 8189–8195 (2001)
64. Steinbock, O., Tóth, Á., Showalter, K.: Navigating complex labyrinths: optimal paths from chemical waves. Science 868–868 (1995)
65. Rambidi, N.G., Yakovenchuk, D.: Chemical reaction-diffusion implementation of finding the shortest paths in a labyrinth. Phys. Rev. E **63**(2), 026607 (2001)

66. Andrew Adamatzky and Benjamin de Lacy Costello: Collision-free path planning in the Belousov-Zhabotinsky medium assisted by a cellular automaton. Naturwissenschaften **89**(10), 474–478 (2002)
67. Gorecki, J., Gorecka, J.N., Adamatzky, A.: Information coding with frequency of oscillations in Belousov-Zhabotinsky encapsulated disks. Phys. Rev. E **89**(4), 042910 (2014)
68. Adamatzky, A., de Lacy Costello, B., Melhuish, C., Ratcliffe, N.: Experimental implementation of mobile robot taxis with onboard Belousov–Zhabotinsky chemical medium. Mater. Sci. Eng.: C **24**(4), 541–548 (2004)
69. Yokoi, H., Adamatzky, A., de Lacy Costello, B., Melhuish, C.: Excitable chemical medium controller for a robotic hand: Closed-loop experiments. Int. J. Bifurc. Chaos **14**(09), 3347–3354 (2004)
70. Vazquez-Otero, A., Faigl, J., Duro, N., Dormido, R.: Reaction-diffusion based computational model for autonomous mobile robot exploration of unknown environments. IJUC **10**(4), 295–316 (2014)
71. Agladze, K., Aliev, R.R., Yamaguchi, T., Yoshikawa, K.: Chemical diode. J. Phys. Chem. **100**(33), 13895–13897 (1996)
72. Igarashi, Y., Gorecki, J.: Chemical diodes built with controlled excitable media. IJUC **7**(3), 141–158 (2011)
73. Gorecki, J., Gorecka, J.N.: Information processing with chemical excitations–from instant machines to an artificial chemical brain. Int. J. Unconv. Comput. **2**(4) (2006)
74. Stovold, J., O'Keefe, S.: Simulating neurons in reaction-diffusion chemistry. In: International Conference on Information Processing in Cells and Tissues, pp. 143–149. Springer (2012)
75. Gentili, P.L., Horvath, V., Vanag, V.K., Epstein, I.R.: Belousov-Zhabotinsky "chemical neuron" as a binary and fuzzy logic processor. IJUC **8**(2), 177–192 (2012)
76. Takigawa-Imamura, H., Motoike, I.N.: Dendritic gates for signal integration with excitability-dependent responsiveness. Neural Netw. **24**(10), 1143–1152 (2011)
77. Gruenert, G., Gizynski, K., Escuela, G., Ibrahim, B., Gorecki, J., Dittrich, P.: Understanding networks of computing chemical droplet neurons based on information flow. Int. J. Neural Syst. **25**(07), 1450032 (2015)
78. Stovold, J., O'Keefe, S.: Reaction–diffusion chemistry implementation of associative memory neural network. Int. J. Parallel Emerg. Distrib. Syst. 1–21 (2016)
79. Stovold, J., O'Keefe, S.: Associative memory in reaction-diffusion chemistry. Adv. Unconven. Comput. 141–166. Springer (2017)
80. Gorecki, J., Yoshikawa, K., Igarashi, Y.: On chemical reactors that can count. J. Phys. Chem. A **107**(10), 1664–1669 (2003)
81. Yoshikawa, K., Motoike, I., Ichino, T., Yamaguchi, T., Igarashi, Y., Gorecki, J., Gorecka, J.N.: Basic information processing operations with pulses of excitation in a reaction-diffusion system. IJUC **5**(1), 3–37 (2009)
82. Escuela, G., Gruenert, G., Dittrich, P.: Symbol representations and signal dynamics in evolving droplet computers. Nat. Comput. **13**(2), 247–256 (2014)
83. Gruenert, G., Gizynski, K., Escuela, G., Ibrahim, B., Gorecki, J., Dittrich, P.: Understanding networks of computing chemical droplet neurons based on information flow. Int. J. Neural Syst. 1450032 (2014)
84. Gorecki, J., Gizynski, K., Guzowski, J., Gorecka, J.N., Garstecki, P., Gruenert, G., Dittrich, P.: Chemical computing with reaction-diffusion processes. Phil. Trans. R. Soc. A **373**(2046), 20140219 (2015)
85. Andrew Adamatzky and Benjamin de Lacy Costello: Binary collisions between wave-fragments in a sub-excitable Belousov-Zhabotinsky medium. Chaos Sol. Fract. **34**(2), 307–315 (2007)
86. Toth, R., Stone, C., de Lacy Costello, B., Adamatzky, A., Bull, L.: Simple collision-based chemical logic gates with adaptive computing. Theoretical and Technological Advancements in Nanotechnology and Molecular Computation: Interdisciplinary Gains: Interdisciplinary Gains, p. 162 (2010)

87. Adamatzky, A., De Lacy, B., Costello, L.B., Holley, J.: Towards arithmetic circuits in subexcitable chemical media. Israel J. Chem. **51**(1), 56–66 (2011)
88. de Lacy Costello, B., Toth, R., Stone, C., Adamatzky, A., Bull, L.: Implementation of glider guns in the light-sensitive Belousov-Zhabotinsky medium. Physical Rev. E **79**(2), 026114 (2009)
89. Toth, R., Stone, C., Adamatzky, A., de Lacy Costello, B., Bull, L.: Experimental validation of binary collisions between wave fragments in the photosensitive Belousov–Zhabotinsky reaction. Chaos Sol. Fract. **41**(4), 1605–1615 (2009)
90. Adamatzky, A.: Slime mould logical gates: exploring ballistic approach (2010). arXiv:1005.2301
91. De Lacy, B., Costello, A.A., Jahan, I., Zhang, L.: Towards constructing one-bit binary adder in excitable chemical medium. Chem. Phys. **381**(1), 88–99 (2011)
92. Sun, M.-Z., Zhao, X.: Multi-bit binary decoder based on Belousov-Zhabotinsky reaction. J. Chem. Phys. **138**(11), 114106 (2013)
93. Zhang, G.-M., Wong, I., Chou, M.-T., Zhao, X.: Towards constructing multi-bit binary adder based on Belousov-Zhabotinsky reaction. J. Chem. Phys. **136**(16), 164108 (2012)
94. Sun, M.-Z., Zhao, X.: Crossover structures for logical computations in excitable chemical medium. Int. J. Unconv. Comput. (2015)
95. Guo, S., Sun, M.-Z., Han, X.: Digital comparator in excitable chemical media. Int. J. Unconv. Comput. (2015)
96. Gorecka, J., Gorecki, J.: T-shaped coincidence detector as a band filter of chemical signal frequency. Phys. Rev. E **67**(6), 067203 (2003)
97. Igarashi, Y., Gorecki, J., Gorecka, J.N.: Chemical information processing devices constructed using a nonlinear medium with controlled excitability. Unconv. Comput. 130–138. Springer (2006)
98. Adamatzky, A., de Lacy Costello, B., Bull, L.: On polymorphic logical gates in subexcitable chemical medium. Int. J. Bifur. Chaos **21**(07), 1977–1986 (2011)
99. Stevens, W.M., Adamatzky, A., Jahan, I., de Lacy Costello, B.: Time-dependent wave selection for information processing in excitable media. Phys. Rev. E 85(6), 066129 (2012)
100. Cole, L., Orlovich, D.A., Ashford, A.E.: Structure, function, and motility of vacuoles in filamentous fungi. Fungal Gen. Biol. **24**(1–2), 86–100 (1998)
101. Goldstein, R.E., Tuval, I., van de Meent, J.-W.: Microfluidics of cytoplasmic streaming and its implications for intracellular transport. Proc. Nat. Acad. Sci. **105**(10), 3663–3667 (2008)
102. Adamatzky, A., Schubert, T.: Slime mold microfluidic logical gates. Mater. Today **17**(2), 86–91 (2014)
103. Bahn, Y.-S., Xue, C., Idnurm, A., Rutherford, J.C., Heitman, J., Cardenas, M.E.: Sensing the environment: lessons from fungi. Nat. Rev. Microbiol. **5**(1), 57 (2007)
104. Whiting, J.G.H., Jones, J., Bull, L., Levin, M., Adamatzky, A.: Towards a Physarum learning chip. Sci. Rep. **6**:19948 (2016)
105. Trinci, A.P.J.: A study of the kinetics of hyphal extension and branch initiation of fungal mycelia. Microbiology **81**(1), 225–236 (1974)
106. Fromm, J., Lautner, S.: Electrical signals and their physiological significance in plants. Plant Cell Environ. **30**(3), 249–257 (2007)
107. Van Aarle, I.M., Olsson, P.A., Söderström, B.: Arbuscular mycorrhizal fungi respond to the substrate ph of their extraradical mycelium by altered growth and root colonization. New Phytologist **155**(1), 173–182 (2002)
108. Kung, C.: A possible unifying principle for mechanosensation. Nature **436**(7051), 647 (2005)
109. Fomina, M., Ritz, K., Gadd, G.M.: Negative fungal chemotropism to toxic metals. FEMS Microbiol. Lett. **193**(2), 207–211 (2000)
110. Bahn, Y.-S., Mühlschlegel, F.A.: Co2 sensing in fungi and beyond. Curr. opin. Microbiol. **9**(6), 572–578 (2006)
111. Ki-Bong, O., Nishiyama, T., Sakai, E., Matsuoka, H., Kurata, H.: Flow sensing in mycelial fungi. J. Biotechnol. **58**(3), 197–204 (1997)

112. Jaffe, M.J., Leopold, A.C., Staples, R.C.: Thigmo responses in plants and fungi. Am. J. Botany **89**(3), 375–382 (2002)
113. Howitz, K.T., Sinclair, D.A.: Xenohormesis: sensing the chemical cues of other species. Cell **133**(3), 387–391 (2008)
114. Rundel, P.W., Graham, E.A., Allen, M.F., Fisher, J.C., Harmon, T.C.: Environmental sensor networks in ecological research. New Phytologist **182**(3), 589–607 (2009)
115. Schloter, M., Nannipieri, P., Sørensen, S.J., van Elsas, J.D.: Microbial indicators for soil quality. Biol. Fertil. Soils **54**(1), 1–10 (2018)
116. Vogt, K.A., Publicover, D.A., Bloomfield, J., Perez, J.M., Vogt, D.J., Silver, W.L.: Belowground responses as indicators of environmental change. Environ. Exp. Botany **33**(1), 189–205 (1993)
117. Whiting, J.G.H., de Lacy Costello, B.P.J., Adamatzky, A.: Slime mould logic gates based on frequency changes of electrical potential oscillation. Biosystems **124**, 21–25 (2014)
118. Borresen, J., Lynch, S.: Oscillatory threshold logic. PloS One **7**(11), e48498 (2012)
119. Toussaint, G.T.: The relative neighbourhood graph of a finite planar set. Pattern Recogn. **12**(4), 261–268 (1980)

On Boolean Gates in Fungal Colony

Andrew Adamatzky, Martin Tegelaar, Han A. B. Wosten, Alexander E. Beasley, and Richard Mayne

Abstract A fungal colony maintains its integrity via flow of cytoplasm along mycelium network. This flow, together with possible coordination of mycelium tips propagation, is controlled by calcium waves and associated waves of electrical potential changes. We propose that these excitation waves can be employed to implement a computation in the mycelium networks. We use FitzHugh-Nagumo model to imitate propagation of excitation in a single colony of *Aspergillus niger*. Boolean values are encoded by spikes of extracellular potential. We represent binary inputs by electrical impulses on a pair of selected electrodes and we record responses of the colony from sixteen electrodes. We derive sets of two-inputs-on-output logical gates implementable the fungal colony and analyse distributions of the gates.

1 Introduction

A vibrant field of unconventional computing aims to employ space-time dynamics of physical, chemical and biological media to design novel computational techniques, architectures and working prototypes of embedded computing substrates and devices. Interaction-based computing devices, is one of the most diverse and promising families of the unconventional computing structures. They are based on interactions of fluid streams, signals propagating along conductors or excitation wave-fronts, see e.g. [1–3, 3–9, 9–12]. Typically, logical gates and their cascade implemented in an excitable medium are 'handcrafted' to address exact timing and type of interactions between colliding wave-fronts [1, 2, 13–23]. The artificial design of logical circuits might be suitable when chemical media or functional materials are used. However, the approach might be not feasible when embedding computation in living systems, where the architecture of conductive pathways may be difficult to alter or control. In such situations an opportunistic approach to outsourcing computation can

---

A. Adamatzky (✉) · A. E. Beasley · R. Mayne
Unconventional Computing Laboratory, UWE, Bristol, UK
e-mail: andrew.adamatzky@uwe.ac.uk

M. Tegelaar · H. A. B. Wosten
Microbiology Department, University of Utrecht, Utrecht, The Netherlands

© The Author(s), under exclusive license to Springer Nature Switzerland AG 2023
A. Adamatzky (ed.), *Fungal Machines*, Emergence, Complexity and Computation 47,
https://doi.org/10.1007/978-3-031-38336-6_18

Fig. 1 Exemplar spikes of extracellular electrical potential propagating in fungal mycelium

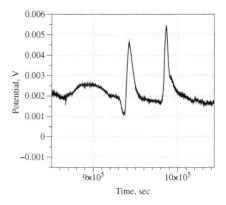

be adopted. The system is perturbed via two or more input loci and its dynamics if recorded at one or more output loci. A wave-front appearing at one of the output loci is interpreted as logical truth or '1'. Thus the system with relatively unknown structure implements a mapping $\{0, 1\}^n \rightarrow \{0, 1\}^m$, where n is a number of input loci and m is a number of output loci, $n, m > 0$ [24, 25]. The approach belong to same family of computation outsourcing techniques as *in materio* computing [26–30] and reservoir computing [31–35].

Fungal colonies are characterised by rich typology of mycelium networks [36–40] in some cases affine to fractal structures [41–46]. Rich morphological features might imply rich computational abilities and thus worse to analyse from realising Boolean functions point of view. In numerical experiments we study implementation of logical gates via interaction of numerous travelling excitation waves, seen as as action potentials, on an image of a real fungal colony. Action potential-like spikes of electrical potential have been discovered using intra-cellular recording of mycelium of *Neurospora crassa* [47] and further confirmed in intra-cellular recordings of action potential in hypha of *Pleurotus ostreatus* and *Armillaria bulbosa* [48] and in extra-cellular recordings of fruit bodies of and substrates colonized by mycelium of *Pleurotus ostreatus* [49] (Fig. 1). While the exact nature of the travelling spikes remains uncertain we can speculate, by drawing analogies with oscillations of electrical potential of slime mould *Physarum polycephalum* [50–53], that the spikes in fungi are triggered by calcium waves, reversing of cytoplasmic flow, translocation of nutrients and metabolites. Studies of electrical activity of higher plants can brings us even more clues. Thus, the plants use the electrical spikes for a long-distance communication aimed to coordinate an activity of their bodies [54–56]. The spikes of electrical potential in plants relate to a motor activity [57–60], responses to changes in temperature [61], osmotic environment [62] and mechanical stimulation [63, 64].

2 Methods

2.1 Colony Imaging

Aspergillus niger strain AR9#2 [65], expressing Green Fluorescent Protein (GFP) from the glucoamylase (*glaA*) promoter, was grown at 30°C on minimal medium (MM) [66] with 25 mM xylose and 1.5% agarose (MMXA). MMXA cultures were grown for three days, after which conidia were harvested using saline-Tween (0.8% NaCl and 0.005% Tween-80). 250 ml liquid cultures were inoculated with $1.25 \cdot 10^9$ freshly harvested conidia and grown at 200 rpm and 30°C in 1 L Erlenmeyer flasks in complete medium (CM) (MM containing 0.5% yeast extract and 0.2% enzymatically hydrolyzed casein) supplemented with 25 mM xylose (repressing glaA expression). Mycelium was harvested after 16 h and washed twice with PBS. Ten g of biomass (wet weight) was transferred to MM supplemented with 25 mM maltose (inducing glaA expression).

Fluorescence of GFP was localised in micro-colonies using a DMI 6000 CS AFC confocal microscope (Leica, Mannheim, Germany). Micro-colonies were fixed overnight at 4°C in 4% paraformaldehyde in PBS, washed twice with PBS and taken up in 50 ml PBS supplemented with 150 mM glycine to quench autofluorescence. Micro-colonies were then transferred to a glass bottom dish (Cellview™, Greiner Bio-One, Frickenhausen, Germany, PS, 35/10 MM) and embedded in 1% low melting point agarose at 45°C. Micro-colonies were imaged at 20× magnification (HC PL FLUOTAR L 20 × 0.40 DRY). GFP was excited by white light laser at 472 nm using 50% laser intensity (0.1 kW/cm2) and a pixel dwell time of 72 ns. Fluorescent light emission was detected with hybrid detectors in the range of 490–525 nm. Pinhole size was 1 Airy unit. Z-stacks of imaged micro-colonies were made using 100 slices with a slice thickness of 8.35 μm. 3D projections were made with Fiji [67].

2.2 Numerical Modelling

We used still image of the colony as a conductive template. The image of the fungal colony (Fig. 2a) was projected onto a 1000 × 960 nodes grid. The original image $M = (m_{ij})_{1 \leq j \leq n_i, 1 \leq j \leq n_j}$, $m_{ij} \in \{r_{ij}, g_{ij}, b_{ij}\}$, where $n_i = 1000$ and $n_j = 960$, and $1 \leq r, g, b \leq 255$ (Fig. 2a), was converted to a conductive matrix $C = (m_{ij})_{1 \leq i, j \leq n}$ (Fig. 2b) derived from the image as follows: $m_{ij} = 1$ if $r_{ij} < 20$, $g_{ij} > 40$ and $b_{ij} < 20$; a dilution operation was applied to C.

FitzHugh-Nagumo (FHN) equations [68–70] is a qualitative approximation of the Hodgkin-Huxley model [71] of electrical activity of living cells:

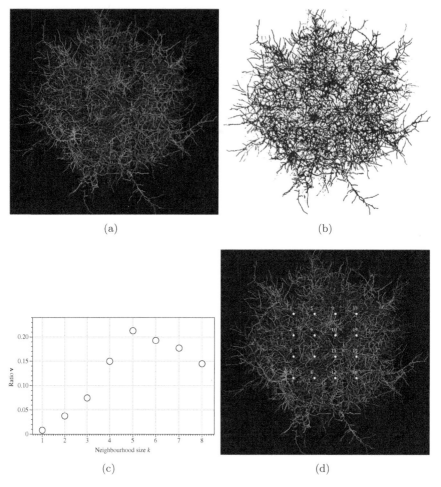

Fig. 2 Image of the fungal colony, 1000×960 pixels used as a template conductive for FHN. **a** Original image, mycelium is seen as green pixels. **b** Conductive matrix C, conductive pixels are black. **c** Distribution of neigbourhood sizes. **d** Configuration of electrodes

$$\frac{\partial v}{\partial t} = c_1 u(u-a)(1-u) - c_2 uv + I + D_u \nabla^2 \quad (1)$$

$$\frac{\partial v}{\partial t} = b(u-v), \quad (2)$$

where u is a value of a trans-membrane potential, v a variable accountable for a total slow ionic current, or a recovery variable responsible for a slow negative feedback, I is a value of an external stimulation current. The current through intra-cellular spaces is approximated by $D_u \nabla^2$, where D_u is a conductance. Detailed explanations of the 'mechanics' of the model are provided in [72], here we shortly repeat some insights.

The term $D_u \nabla^2 u$ governs a passive spread of the current. The terms $c_2 u(u-a)(1-u)$ and $b(u-v)$ describe the ionic currents. The term $u(u-a)(1-u)$ has two stable fixed points $u = 0$ and $u = 1$ and one unstable point $u = a$, where a is a threshold of an excitation.

We integrated the system using the Euler method with the five-node Laplace operator, a time step $\Delta t = 0.015$ and a grid point spacing $\Delta x = 2$, while other parameters were $D_u = 1, a = 0.13, b = 0.013, c_1 = 0.26$. We controlled excitability of the medium by varying c_2 from 0.05 (fully excitable) to 0.015 (non excitable). Boundaries are considered to be impermeable: $\partial u/\partial \mathbf{n} = 0$, where $\mathbf{n}$ is a vector normal to the boundary.

The waves of excitation propagated on conductive nodes of the grid of C, in addition to the parameter c_2, excitability of each conductive node was dependent on a number k of its immediate conductive neighbours. Distribution of neighbourhood sizes are shown in Fig. 2c.

To show dynamics of excitation in the network we simulated electrodes by calculating a potential p_x^t at an electrode location x as $p_x = \sum_{y:|x-y|<2}(u_x - v_x)$. Configuration of electrodes $1, \cdots, 16$ is shown in Fig. 2d. The configuration selected imitates a micro-electrode array. The numerical integration code written in Processing was inspired by previous methods of numerical integration of FHN and our own computational studies of the impulse propagation in biological networks [25, 70, 72–74]. Time-lapse snapshots provided were recorded at every 100th time step, and we display sites with $u > 0.04$; videos and figures were produced by saving a frame of the simulation every 100th step of the numerical integration and assembling the saved frames into the video with a play rate of 30 fps. Videos are available at [75].

3 Results

While implementing numerical experiments we selected a range of the network excitability (Sect. 3.1) and then realised sets of logical gates for excitability values selected (Sect. 3.2).

3.1 Effect of Excitability on Overall Activity

For $c_2 < 0.0945$ any source of excitation triggers excitation dynamics which occupies all parts of the network accessible, via mycelial strands, from the source. Due to the high level of excitability the network remains in the excitable state (Fig. 3a). For values c_2 from 0.094 to 0.0965 we observe propagation of 'classical' excitation wave-fronts resembling circular, target and spiral waves in a continuous medium. Examples of wave-fronts propagating in networks with excitability levels $c_2 = 0.095$ and $c_2 = 0.096$, excited at the same loci shown in Fig. 4a, are shown in Figs. 4 and 5. In the network with c_2 there are many pathways for propagation of the exci-

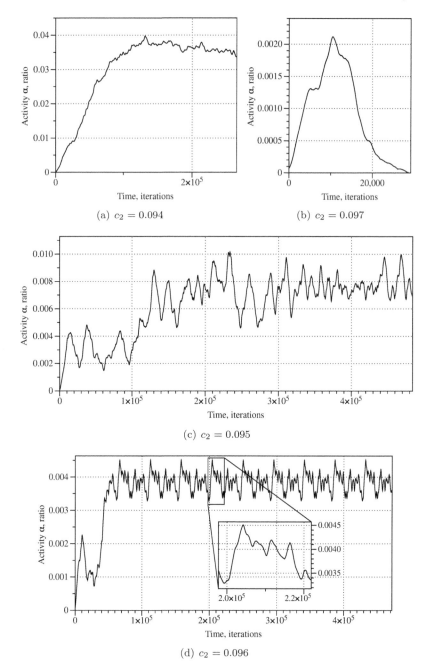

Fig. 3 Dynamics of the activity α for various values of excitability c_2, the values are shown in sub-captions. For every iteration t we measured the activity of the network as a number of conductive nodes x with $u_x^t > 0.1$

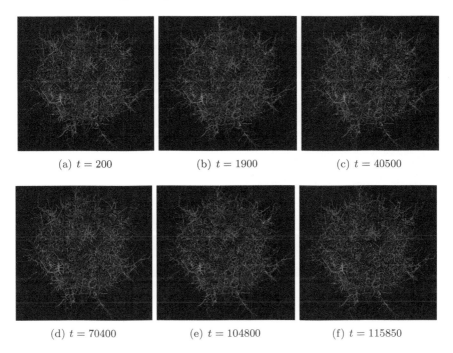

Fig. 4 Snapshots of excitation dynamics for $c_2 = 0.095$

tation wave-fronts, therefore, despite being fully deterministic, the network exhibit disordered oscillations of its activity (Fig. 3c). A number of conductive pathways decreases when c_2 increases from 0.095 to 0.096. Thus many propagating wave-fronts become, relatively, quickly confined to a limited domains of the network, where they continue 'circling' indefinitely. A set of regular oscillations of activity becomes evidence after a number of iterations (Fig. 3d). The coverage of the network by excitation wave-fronts reduced with increase of c_2 from 0.095 to 0.096 (Fig. 6a, b) and becomes localised when c_2 reaches 0.097 (Figs. 6c and 3b). Thus we used networks with $c_2 = 0.095$ or 0.096 for implementation of Boolean functions.

3.2 Distribution of Boolean Gates

Input Boolean values are encoded as follows. We earmark two sites of the network as dedicated inputs, x and y, and represent logical TRUE, or '1', as an excitation, or an impulse injected in the network via electrodes. If $x = 1$ then the site corresponding to x is excited, if $x = 0$ the site is not excited (Fig. 8).

Assume that each spike represents logical TRUE and that spikes occurring within less than $2 \cdot 10^2$ iterations happen simultaneously. By selecting specific intervals of

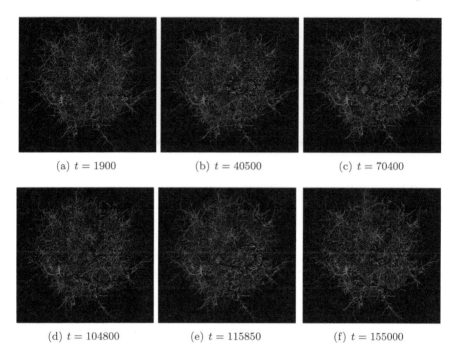

Fig. 5 Snapshots of excitation dynamics for $c_2 = 0.096$. Compare **c** and **e**: the pattern of excitation returns to the exact point of the cycle

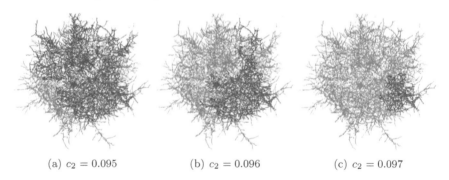

Fig. 6 Coverage of the network for excitability c_2 **a** 0.095, **b** 0.096, **c** 0.097. If the a pixel p of the image was excited, $u_p^t > 0.1$, it is assumed to be covered and coloured red in the pictures (abc); the pixels which never were excited are coloured gray

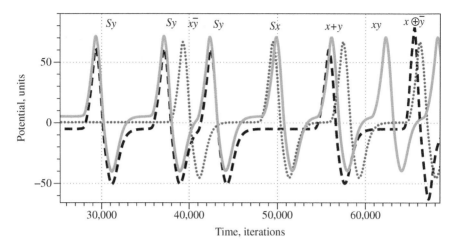

Fig. 7 Fragment of electrical potential record on electrode 7 in response to inputs (01), black dashed line, (10), red dotted line, (11), solid green line, entered as impulses via electrodes $E_x = 5$ and $E_y = 15$. See locations of electrodes in Fig. 2d. To make the individual plots visible in places of exact overlapping, we added potential -5 to recording in response to input (01) and and potential 5 to recording in response to input (11)

recordings we can realise several gates in a single site of recording. In this particular case we assumed that spikes are separated if their occurrences lie more than 10^3 iterations apart. An example is shown in Fig. 7.

Numbers of Boolean gates detected on the electrodes for selected pairs of input electrodes are shown in Table 1. We see that select x and select y gates, Sx and Sy are most frequent. They usually are detected with the same frequency (Table 1d–f), however there are examples of input electrode pairs, where one of the select gates is found much more often than another. This is most visible for the pair $(E_x, E_y) = (3, 13)$ where Sx dominates (Table 1a), and the pair $(7, 14)$ and (Table 1bc) where Sy dominates. Next common gates in the hierarchy are $\overline{x}y$ and $x\overline{y}$. The gates xy and $x + y$ are detected with nearly the same frequency with gate $x + y$ being slightly more common. The $x \oplus y$ is the most rare gate.

The sub-tables Table 1d–f show how excitability of the network affects numbers of gates detected. The networks with high excitability, $c_2 = 0.094$, and low excitability, $c_2 = 0.0096$ realise smaller number of gates then that realised by sub-excitable network, $c_2 = 0.095$.

Overall distribution (average of outputs of input electrode pairs (3,13), (5,15), (7,14), (4,13), (13,7)) of a ratio of gates discovered is shown in Fig. 9. This is accompanied by distributions of gates discovered in experimental laboratory reservoir computing with slime mould *Physarum polycephalum* [76], succulent plant [77] and numerical modelling experiments on computing with protein verotoxin [78], actin bundles network [24], and actin monomer [79]. All the listed distributions have very

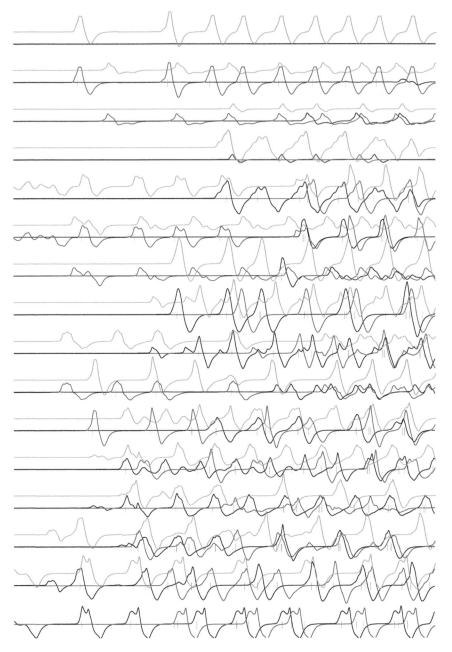

Fig. 8 Recording of electrical potential from all electrodes in responses to inputs in response to inputs (01), black line, (10), red line, (11), green line, injected as spikes via electrodes $E_x = 5$ and $E_y = 15$

Table 1 Numbers of Boolean gates detected for selected pairs of input electrodes E_x and E_y. The gates indicated are also known in other notations as follows. $x+y$ is OR gate, Sy is SELECT input y gate, $x \oplus y$ is XOR gate, Sx is SELECT input s gate, $\bar{x}y$ is NOT AND gate, $x\bar{y}$ is AND NOT gate, xy is AND gate

(a) $E_x = 3$, $E_y = 13$, $c_2 = 0.095$

E	$x+y$	Sy	$x \oplus y$	Sx	$\bar{x}y$	$x\bar{y}$	xy	Total
0	0	0	0	0	0	0	0	0
1	0	0	0	2	0	0	0	2
2	0	0	0	0	0	0	0	0
3	0	0	0	0	0	0	0	0
4	1	0	0	7	1	0	0	9
5	0	0	0	2	2	0	0	4
6	0	0	0	2	0	0	0	2
7	1	0	0	8	2	0	0	11
8	1	0	0	6	1	0	0	8
9	0	0	0	0	1	0	0	1
10	0	1	1	2	0	1	2	7
11	0	0	0	4	2	0	0	6
12	0	0	0	3	2	0	0	5
13	1	5	0	0	0	1	0	7
14	2	5	0	1	0	1	0	9
15	0	1	0	5	2	0	0	8
Total	6	12	1	42	13	3	2	79

(b) $E_x = 7$, $E_y = 14$, $c_2 = 0.095$

E	$x+y$	Sy	$x \oplus y$	Sx	$\bar{x}y$	$x\bar{y}$	xy	Total
0	0	0	0	0	0	0	0	0
1	0	0	0	5	0	0	0	5
2	0	0	0	0	0	0	0	0
3	0	0	0	0	0	0	0	0
4	0	0	0	6	1	0	0	7
5	0	0	0	6	0	0	0	6
6	0	0	0	2	0	0	0	2
7	1	0	0	4	2	0	0	7
8	0	0	0	7	0	0	0	7
9	0	0	0	0	0	0	0	0
10	1	0	1	5	0	0	0	7
11	1	0	0	4	3	0	0	8
12	2	0	0	3	0	0	0	5
13	1	5	0	1	0	2	0	9
14	0	4	0	2	1	2	1	10
15	0	0	0	8	2	0	0	10
Total	6	9	1	53	9	4	1	83

(continued)

Table 1 (continued)

(c) $E_x = 7, E_y = 14, c_2 = 0.094$

E	$x+y$	Sy	$x \oplus y$	Sx	$\bar{x}y$	$x\bar{y}$	xy	Total
0	0	0	0	0	0	0	0	0
1	0	0	0	2	1	0	0	3
2	0	0	0	0	0	0	0	0
3	0	0	0	0	0	0	0	0
4	0	0	0	1	3	0	0	4
5	0	0	0	1	1	0	0	2
6	0	0	0	1	1	0	0	2
7	0	0	0	0	2	0	0	2
8	0	0	0	3	4	0	0	7
9	0	0	0	0	1	2	0	3
10	0	0	0	1	1	0	0	2
11	0	0	0	0	6	0	0	6
12	1	0	1	2	3	1	0	8
13	0	2	0	2	0	1	0	5
14	1	3	0	0	3	0	2	9
15	0	2	0	0	0	1	0	3
Total	2	7	1	13	26	5	2	56

(d) $E_x = 5, E_y = 15, c_2 = 0.094$

E	$x+y$	Sy	$x \oplus y$	Sx	$\bar{x}y$	$x\bar{y}$	xy	Total
0	0	0	0	0	0	0	0	0
1	0	0	0	1	1	0	0	2
2	0	0	0	0	0	0	0	0
3	0	0	0	0	0	0	0	0
4	0	1	0	0	2	1	1	5
5	0	0	0	0	1	0	0	1
6	0	0	0	0	1	0	0	1
7	1	2	0	0	0	4	0	7
8	0	1	0	1	2	1	0	5
9	0	0	0	1	2	0	0	3
10	0	0	0	2	0	0	1	3
11	1	5	0	0	1	1	1	9
12	0	6	0	0	1	2	0	9
13	2	0	2	1	1	1	1	8
14	0	1	0	1	0	5	0	7
15	0	0	0	0	0	1	0	1
Total	4	16	2	7	12	16	4	61

(continued)

Table 1 (continued)

(e) $E_x = 5, E_y = 15, c_2 = 0.095$

E	$x+y$	Sy	$x \oplus y$	Sx	$\bar{x}y$	$x\bar{y}$	xy	Total
0	0	0	0	0	0	0	0	0
1	0	0	0	8	0	0	0	8
2	0	0	0	0	0	0	0	0
3	0	0	0	0	0	0	0	0
4	1	4	0	0	0	2	0	7
5	3	0	0	4	0	0	0	7
6	0	0	0	0	1	0	0	1
7	1	3	1	1	0	1	1	8
8	0	5	0	1	0	2	0	8
9	0	0	0	3	0	0	0	3
10	1	0	2	4	2	0	2	11
11	2	4	0	2	2	0	1	11
12	1	7	0	0	0	3	0	11
13	3	1	0	2	0	1	0	7
14	1	5	0	0	0	6	0	12
15	1	3	1	2	2	1	1	11
Total	14	32	4	27	7	16	5	105

(f) $E_x = 5, E_y = 15, c_2 = 0.096$

E	$x+y$	Sy	$x \oplus y$	Sx	$\bar{x}y$	$x\bar{y}$	xy	Total
0	0	0	0	0	0	0	0	0
1	0	0	0	2	1	0	0	3
2	0	0	0	0	0	0	0	0
3	0	0	0	0	0	0	0	0
4	1	3	0	0	0	1	0	5
5	0	0	0	4	2	0	0	6
6	0	0	0	2	1	0	0	3
7	1	4	0	1	0	1	0	7
8	1	5	0	0	0	1	0	7
9	0	0	0	7	0	0	0	7
10	0	1	0	5	1	1	0	8
11	2	3	0	1	1	1	1	9
12	0	4	0	0	0	0	0	4
13	0	2	0	2	0	2	0	6
14	1	3	0	2	0	3	0	9
15	0	3	0	2	0	2	0	7
Total	6	28	0	28	6	12	1	81

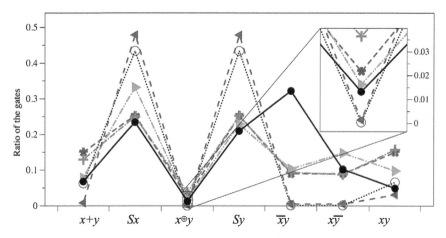

Fig. 9 Comparative ratios of Boolean gates discovered in mycelium network, black disc and solid line; slime mould *Physarum polycephalum* [76], black circle and dotted line; succulent plant [77], red snowflake and dashed line; single molecule of protein verotoxin [78], light blue '+' and dash-dot line; actin bundles network [24], green triangle pointing right and dash-dot-dot line; actin monomer [79], magenta triangle pointing left and dashed line. Area of XOR gate is magnified in the insert. Lines are to guide eye only

similar structure with gates selecting one of the inputs in majority, followed by OR gate, NOT- AND an AND- NOT gates. The gate AND is usually underrepresented in experimental and modelling experiments. The gate XOR is a rare find.

4 Discussion

We have demonstrated how sets of logical gates can be implemented in single colony mycelium networks via initiation of electrical impulses. The impulses travel in the network, interact with each other (annihilate, reflect, change their phase). Thus for different combinations of input impulses and record different combinations of output impulses, which in some cases can be interpreted as representing two-inputs-one-output functions.

To estimate a speed of computation we refer to Olsson and Hansson's [48] original study, in which they proposed that electrical activity in fungi could be used for communication with message propagation speed 0.5 mm/s. Diameter of the colony (Fig. 2a), which experimental laboratory images has been used to run FHN model, is c. 1.7 mm. Thus, it takes the excitation waves initiated at a boundary of the colony up to 3–4 s to span the whole mycelium network (this time is equivalent to c. 70K iterations of the numerical integration model). In 3–4 sec the mycelium network can compute up to a hundred logical gates. This gives us the rate of a gate per 0.03 s, or, in terms of frequency this will be c. 30 Hz. The mycelium network computing can not compete with existing silicon architecture however its application domain can be a unique of living biosensors (a distribution of gates realised might be affected by environmental conditions) [80] and computation embedded into structural elements where fungal materials are used [81–83].

References

1. Steinbock, O., Kettunen, P., Showalter, K.: Chemical wave logic gates. J. Phys. Chem. **100**(49), 18970–18975 (1996)
2. Sielewiesiuk, J., Górecki, J.: Logical functions of a cross junction of excitable chemical media. J. Phys. Chem. A **105**(35), 8189–8195 (2001)
3. Adamatzky, A., de Lacy Costello, B., Melhuish, C., Ratcliffe, N.: Experimental implementation of mobile robot taxis with onboard Belousov–Zhabotinsky chemical medium. Mater. Sci. Eng.: C **24**(4), 541–548 (2004)
4. Hiroshi Yokoi, Andy Adamatzky, Ben de Lacy Costello, and Chris Melhuish. Excitable chemical medium controller for a robotic hand: Closed-loop experiments. *International Journal of Bifurcation and Chaos*, 14(09):3347–3354, 2004
5. Vazquez-Otero, A., Faigl, J., Duro, N., Dormido, R.: Reaction-diffusion based computational model for autonomous mobile robot exploration of unknown environments. IJUC **10**(4), 295–316 (2014)
6. Igarashi, Y., Gorecki, J.: Chemical diodes built with controlled excitable media. IJUC **7**(3), 141–158 (2011)
7. Gorecki, J., Gorecka, J.N.: Information processing with chemical excitations–from instant machines to an artificial chemical brain. Int. J. Unconvent. Comput. **2**(4) (2006)
8. Gorecki, J., Gorecka, J.N., Igarashi, Y.: Information processing with structured excitable medium. Nat. Comput. **8**(3), 473–492 (2009)
9. Stovold, J., O'Keefe, S.: Simulating neurons in reaction-diffusion chemistry. In: International Conference on Information Processing in Cells and Tissues, pp. 143–149. Springer
10. Gentili, P.L., Horvath, V., Vanag, V.K., Epstein, I.R.: Belousov-Zhabotinsky "chemical neuron" as a binary and fuzzy logic processor. IJUC **8**(2), 177–192 (2012)
11. Takigawa-Imamura, H., Motoike, I.N.: Dendritic gates for signal integration with excitability-dependent responsiveness. Neural Netw. **24**(10), 1143–1152 (2011)
12. Gruenert, G., Gizynski, K., Escuela, G., Ibrahim, B., Gorecki, J., Dittrich, P.: Understanding networks of computing chemical droplet neurons based on information flow. Int. J. Neural Syst. **25**(07), 1450032 (2015)
13. Adamatzky, A.: Collision-based computing in Belousov-Zhabotinsky medium. Chaos, Solitons & Fractals **21**(5), 1259–1264 (2004)
14. Andrew Adamatzky and Benjamin de Lacy Costello: Binary collisions between wave-fragments in a sub-excitable Belousov-Zhabotinsky medium. Chaos, Solitons & Fractals **34**(2), 307–315 (2007)
15. Toth, R., Stone, C., de Lacy Costello, B., Adamatzky, A., Bull, L.: Simple collision-based chemical logic gates with adaptive computing. Theoretical and Technological Advancements in Nanotechnology and Molecular Computation: Interdisciplinary Gains: Interdisciplinary Gains, p. 162 (2010)
16. Adamatzky, A., De Lacy, B., Costello, L.B., Holley, J.: Towards arithmetic circuits in sub-excitable chemical media. Isr. J. Chem. **51**(1), 56–66 (2011)
17. de Lacy Costello, B., Toth, R., Stone, C., Adamatzky, A., Bull, L.: Implementation of glider guns in the light-sensitive Belousov-Zhabotinsky medium. Phys. Rev. E **79**(2), 026114 (2009)
18. Sun, M.-Z., Zhao, X.: Multi-bit binary decoder based on Belousov-Zhabotinsky reaction. J. Chem. Phys. **138**(11), 114106 (2013)

19. Zhang, G.-M., Wong, I., Chou, M.-T., Zhao, X.: Towards constructing multi-bit binary adder based on Belousov-Zhabotinsky reaction. J. Chem. Phys. **136**(16), 164108 (2012)
20. Sun, M-Z., Zhao, X.: Crossover structures for logical computations in excitable chemical medium. Int. J. Unconvent. Comput. (2015)
21. Guo, S., Sun, M-Z., Han, X.: Digital comparator in excitable chemical media. Int. J. Unconvent. Comput. (2015)
22. Adamatzky, A., de Lacy Costello, B., Bull, L.: On polymorphic logical gates in subexcitable chemical medium. Int. J. Bifurcat. Chaos **21**(07), 1977–1986 (2011)
23. Stevens, W.M., Adamatzky, A., Jahan, I., de Lacy Costello, B.: Time-dependent wave selection for information processing in excitable media. Phys. Rev. E **85**(6), 066129 (2012)
24. Adamatzky, A., Huber, F., Schnauß, J.: Computing on actin bundles network. Sci. Rep. **9**(1), 1–10 (2019)
25. Adamatzky, A.: Plant leaf computing. Biosystems (2019)
26. Miller, J.F., Downing, K.: Evolution in materio: Looking beyond the silicon box. In: Proceedings 2002 NASA/DoD Conference on Evolvable Hardware, pp. 167–176. IEEE (2002)
27. Miller, J.F., Harding, S.L., Tufte, G.: Evolution-in-materio: evolving computation in materials. Evolut. Intell. **7**(1), 49–67 (2014)
28. Stepney, S.: Co-designing the computational model and the computing substrate. In: International Conference on Unconventional Computation and Natural Computation, pp. 5–14. Springer (2019)
29. Miller, J.F., Hickinbotham, S.J., Amos, M.: In materio computation using carbon nanotubes. In: Computational Matter, pp. 33–43. Springer (2018)
30. Julian Francis Miller: The alchemy of computation: designing with the unknown. Nat. Comput. **18**(3), 515–526 (2019)
31. Verstraeten, D., Schrauwen, B., d'Haene, M., Stroobandt, D.: An experimental unification of reservoir computing methods. Neural Netw. **20**(3), 391–403 (2007)
32. Lukoševičius, M., Jaeger, H.: Reservoir computing approaches to recurrent neural network training. Comput. Sci. Rev. **3**(3), 127–149 (2009)
33. Dale, M., Miller, J.F., Stepney, S.: Reservoir computing as a model for in-materio computing. In: Advances in Unconventional Computing, pp. 533–571. Springer (2017)
34. Konkoli, Z., Nichele, S., Dale, M., Stepney, S.: Reservoir computing with computational matter. In: Computational Matter, pp. 269–293. Springer (2018)
35. Dale, M., Miller, J.F., Stepney, S., Trefzer, M.A.: A substrate-independent framework to characterize reservoir computers. Proc. Roy. Soc. A **475**(2226), 20180723 (2019)
36. Hitchcock, D., Glasbey, C.A., Ritz, K.: Image analysis of space-filling by networks: Application to a fungal mycelium. Biotechnol. Tech. **10**(3), 205–210 (1996)
37. Giovannetti, M., Sbrana, C., Avio, L., Strani, P.: Patterns of below-ground plant interconnections established by means of arbuscular mycorrhizal networks. New Phytol. **164**(1), 175–181 (2004)
38. Fricker, M., Boddy, L., Bebber, D.: Network organisation of mycelial fungi. In: Biology of the Fungal Cell, pp. 309–330. Springer (2007)
39. Fricker, M.D., Heaton, L.L., Jones, N.S., Boddy, L.: The mycelium as a network. The Fungal Kingdom, pp. 335–367 (2017)
40. Islam, M.R., Tudryn, G., Bucinell, R., Schadler, L., Picu, R.C.: Morphology and mechanics of fungal mycelium. Sci. Rep. **7**(1), 1–12 (2017)
41. Obert, M., Pfeifer, P., Sernetz, M.: Microbial growth patterns described by fractal geometry. J. Bacteriol. **172**(3), 1180–1185 (1990)
42. Patankar, D.B., Liu, T.C., Oolman, T.: A fractal model for the characterization of mycelial morphology. Biotechnol. Bioeng. **42**(5), 571–578 (1993)
43. Bolton, R.G., Boddy, L.: Characterization of the spatial aspects of foraging mycelial cord systems using fractal geometry. Mycol. Res. **97**(6), 762–768 (1993)
44. Mihail, J.D., Obert, M., Bruhn, J.N., Taylor, S.J.: Fractal geometry of diffuse mycelia and rhizomorphs of armillaria species. Mycol. Res. **99**(1), 81–88 (1995)

45. Boddy, L., Wells, J.M., Culshaw, C., Donnelly, D.P.: Fractal analysis in studies of mycelium in soil. Geoderma **88**(3), 301–328 (1999)
46. Papagianni, M.: Quantification of the fractal nature of mycelial aggregation in aspergillus niger submerged cultures. Microb. Cell Fact. **5**(1), 5 (2006)
47. Slayman, C.L., Long, W.S., Gradmann, D.: Action potentials. Neurospora crassa, a mycelial fungus. Biochimica et Biophysica Acta (BBA) — Biomembranes **426**(4), 732–744 (1976)
48. Olsson, S., Hansson, B.S.: Action potential-like activity found in fungal mycelia is sensitive to stimulation. Naturwissenschaften **82**(1), 30–31 (1995)
49. Adamatzky, A.: On spiking behaviour of oyster fungi pleurotus djamor. Sci. Rep. **8**(1), 1–7 (2018)
50. Iwamura, T.: Correlations between protoplasmic streaming and bioelectric potential of a slime mold. Physarum polycephalum. Shokubutsugaku Zasshi **62**(735–736), 126–131 (1949)
51. Kamiya, N., Abe, S.: Bioelectric phenomena in the myxomycete plasmodium and their relation to protoplasmic flow. J. Colloid Sci. **5**(2), 149–163 (1950)
52. Kishimoto, U.: Rhythmicity in the protoplasmic streaming of a slime mold, *Physarum polycephalum*. I. a statistical analysis of the electric potential rhythm. J. General Physiol. **41**(6), 1205–1222 (1958)
53. Meyer, R., Stockem, W.: Studies on microplasmodia of Physarum polycephalum V: electrical activity of different types of microplasmodia and macroplasmodia. Cell Biol. Int. Rep. **3**(4), 321–330 (1979)
54. Trebacz, K., Dziubinska, H., Krol, E.: Electrical signals in long-distance communication in plants. In: Communication in Plants, pp. 277–290. Springer (2006)
55. Fromm, J., Lautner, S.: Electrical signals and their physiological significance in plants. Plant, cell Environ. **30**(3), 249–257 (2007)
56. Zimmermann, M.R., Mithöfer, A.: Electrical long-distance signaling in plants. In: Long-Distance Systemic Signaling and Communication in Plants, pp. 291–308. Springer
57. Simons, P.J.: The role of electricity in plant movements. New Phytol. **87**(1), 11–37 (1981)
58. Fromm, J.: Control of phloem unloading by action potentials in mimosa. Physiol. Plant. **83**(3), 529–533 (1991)
59. Sibaoka, T.: Rapid plant movements triggered by action potentials. Bot. magazine= Shokubutsu-gaku-zasshi **104**(1), 73–95 (1991)
60. Volkov, A.G., Foster, J.C., Ashby, T.A., Walker, R.K., Johnson, J.A., Markin, V.S.: Mimosa pudica: electrical and mechanical stimulation of plant movements. Plant, cell Environ. **33**(2), 163–173 (2010)
61. Minorsky, P.V.: Temperature sensing by plants: a review and hypothesis. Plant, Cell Environ. **12**(2), 119–135 (1989)
62. Volkov, A.G.: Green plants: electrochemical interfaces. J. Electroanal. Chem. **483**(1–2), 150–156 (2000)
63. Roblin, G.: Analysis of the variation potential induced by wounding in plants. Plant Cell Physiol. **26**(3), 455–461 (1985)
64. Pickard, B.G.: Action potentials in higher plants. Bot. Rev. **39**(2), 172–201 (1973)
65. Vinck, A., de Bekker, C., Ossin, A., Ohm, R.A., de Vries, R.P., Wösten, H.A.: Heterogenic expression of genes encoding secreted proteins at the periphery of Aspergillus niger colonies. Environ. Microbiol. **13**(1), 216–225 (2011)
66. De Vries, R.P., Burgers, K., van de Vondervoort, P.J., Frisvad, J.C., Samson, R.A., Visser, J.: A new black aspergillus species, a. vadensis, is a promising host for homologous and heterologous protein production. Appl. Environ. Microbiol. **70**(7), 3954–3959 (2004)
67. Schindelin, J., Arganda-Carreras, I., Frise, E., Kaynig, V., Longair, M., Pietzsch, T., Preibisch, S., Rueden, C., Saalfeld, S., Schmid, B., et al.: Fiji: an open-source platform for biological-image analysis. Nat. Methods **9**(7), 676–682 (2012)
68. FitzHugh, R.: Impulses and physiological states in theoretical models of nerve membrane. Biophys. J. **1**(6), 445–466 (1961)
69. Nagumo, J., Arimoto, S., Yoshizawa, S.: An active pulse transmission line simulating nerve axon. Proc. IRE **50**(10), 2061–2070 (1962)

70. Pertsov, A.M., Davidenko, J.M., Salomonsz, R., Baxter, W.T. Jalife, J.: Spiral waves of excitation underlie reentrant activity in isolated cardiac muscle. Circ. Res. **72**(3), 631–650 (1993)
71. Beeler, G.W., Reuter, H.: Reconstruction of the action potential of ventricular myocardial fibres. J. Physiol. **268**(1), 177–210 (1977)
72. Rogers, J.M., McCulloch, A.D.: A collocation-Galerkin finite element model of cardiac action potential propagation. IEEE Trans. Biomed. Eng. **41**(8), 743–757 (1994)
73. Hammer, P.: Spiral waves in monodomain reaction-diffusion model (2009)
74. Adamatzky, A.: On interplay between excitability and geometry (2019). arXiv:1904.06526
75. Adamatzky, A., Tegelaar, M., Wosten, H., Powell, A., Beasley, A.: Supplementary materials. on Boolean gates in fungal colony (2020). http://doi.org/10.5281/zenodo.3678131
76. Harding, S., Koutník, J., Schmidhuber, J., Adamatzky, A.: Discovering boolean gates in slime mould. In: Inspired by Nature, pp. 323–337. Springer (2018)
77. Adamatzky, A., Harding, S., Erokhin, V., Mayne, R., Gizzie, N., BaluškaF., Mancuso, S., Sirakoulis, G.C.: Computers from plants we never made: speculations. In: Inspired by Nature, pp. 357–387. Springer (2018)
78. Adamatzky, A.: Computing in verotoxin. Chem. Phys. Chem. **18**(13), 1822–1830 (2017)
79. Adamatzky, A.: Logical gates in actin monomer. Sci. Rep. **7**(1), 1–14 (2017)
80. Manzella, V., Gaz, C., Vitaletti, A., Masi, E., Santopolo, L., Mancuso, S., Salazar, D., De Las Heras, J.J.: Plants as sensing devices: the PLEASED experience. In: Proceedings of the 11th ACM conference on embedded networked sensor systems, pp. 1–2 (2013)
81. Ross, P.: Your rotten future will be great. The Routledge Companion to Biology in Art and Architecture, p. 252 (2016)
82. Heisel, F., Lee, J., Schlesier, K., Rippmann, M., Saeidi, N., Javadian, A., Nugroho, A.R., Van Mele, T., Block, P., Hebel, D.E.: Design, cultivation and application of load-bearing mycelium components. Int. J. Sustain. Energy Dev. **6**(2) (2018)
83. Dahmen, J.: Soft matter: responsive architectural operations. Technoetic Arts **14**(1–2), 113–125 (2016)

Electrical Frequency Discrimination by Fungi *Pleurotus Ostreatus*

Dawid Przyczyna, Konrad Szacilowski, Alessandro Chiolerio, and Andrew Adamatzky

Abstract We stimulate mycelian networks of oyster fungi *Pleurotus ostreatus* with low frequency sinusoidal electrical signals. We demonstrate that the fungal networks can discriminate between frequencies in a fuzzy-like or threshold based manner. Details about the mixing of frequencies by the mycelium networks are provided. The results advance the novel field of fungal electronics and pave ground for the design of living, fully recyclable, electron devices.

1 Introduction

Fungal electronics aims to design bio-electronic devices with living networks of fungal mycelium [1] and proposes novel and original designs of information and signal processing systems. The living fungal electronic devices offer fault-tolerance and self-repairability featured in living systems, non-linear electrical properties (memfractance, capacitance, photoreactance, electrical oscillations) necessary for implementing analog electronic, neuromorphic and even digital (spike based) circuits, and the fungal circuits are capable of electrical responding to mechanical, optical, chemical and electrical stimulation. Mycelium bound composites (grain or hemp substrates colonised by fungi) are environmentally sustainable growing bio-materials [2–4]. They have been already used in insulation panels [5–9], packaging materials [10, 11], building materials and architectures [12] and wearables [2, 13–16]. To make the fungal materials functional we need to embed flexible electronic devices into the materials. Hyphae of fungal mycelium spanning the mycelium bound composites

---

D. Przyczyna · K. Szacilowski
Academic Centre for Materials and Nanotechnology, AGH University of Science and Technology, Krakow, Poland

A. Chiolerio
Istituto Italiano di Tecnologia, Center for Converging Technologies, Soft Bioinspired Robotics, Via Morego 30, Genova 16165, Italy

A. Chiolerio · A. Adamatzky (✉)
Unconventional Computing Lab, UWE, Bristol, UK
e-mail: andrew.adamatzky@uwe.ac.uk

© The Author(s), under exclusive license to Springer Nature Switzerland AG 2023
A. Adamatzky (ed.), *Fungal Machines*, Emergence, Complexity and Computation 47, https://doi.org/10.1007/978-3-031-38336-6_19

can play a role of unconventional electronic devices. interestingly, their topology is very similar to conducting polymer dendrites [17, 18]. These properties originate not only from common topology [19] but also from complex charge carrier transport phenomena. Therefore, it is not surprising that electrical properties of mycelial hyphae and conducting polymer filaments have similar electrical properties: proton hopping and ionic transport in hyphae *verses* ionic and electronic transport in polymers. Such transport duality must result in highly nonlinear voltage/current characteristics, which in turn, upon AC stimulation must result in generation of complex Fourier patterns in resulting current, as well as other phenomena relevant from the point of view of unconventional computing, e.g. stochastic resonance [20].

We have already demonstrated that we achieved in implementing memristors [21], oscillators [22], photo-sensors [23], pressure sensors [24], and Boolean logical circuits [25] with living mycelium networks. Due to nonlinead electric response of fungal tissues, they are ideally suited for transformation of low-frequency AC signals. The study presented is devoted to frequency discriminators and transformers, which are a significant contribution to the field of fungal electronics.

Electrical communication in mycelium networks is an almost unexplored topic. Fungi exhibit oscillations of extracellular electrical potential, which can be recorded via differential electrodes inserted into a substrate colonised by mycelium or directly into sporocarps [26–28]. We used iridium-coated stainless steel sub-dermal needle electrodes (Spes Medica S.r.l., Italy), with twisted cables. In experiments with recording of electrical potential of oyster fungi *Pleurotus djamor* we discovered two types of spiking activity: high-frequency 6 mHz and low-freq 1 mHz [28] ones. While studying other species of fungi, *Ganoderma resinaceum*, we found that the most common signature of an electrical potential spike is 2–3 mHz [22]. In both species of fungi we observed bursts of spikes within trains of impulses similar to that observed in animal central nervous system [29, 30]. In [31] we demonstrated that information-theoretical complexity of fungal electrical activity exceeds the complexity of European languages. In [32] we analysed the electrical activity of *Omphalotus nidiformis*, *Flammulina velutipes*, *Schizophyllum commune* and *Cordyceps militaris*. We assumed that the spikes of electrical activity could be used by fungi to communicate and process information in mycelium networks and demonstrated that distributions of fungal word lengths match that of human languages. Taking all the above into account it would be valuable to analyse the electrical reactions of fungi to strings of electrical oscillations, featuring frequencies matching those of the supposed fungal language.

2 Methods

A slab of substrate, 200 g, grains and hemp colonised by *Pleurotus ostreatus* (Ann Miller's Speciality Mushrooms, UK, https://www.annforfungi.co.uk/shop/oyster-grain-spawn/) was placed at the bottom of a 5 l plastic container. Measurements were performed in a classic two electrode setup. Electric contacts to the fungi sample were made using iridium-coated stainless steel sub-dermal needle electrodes (purchased by Spes Medica S.r.l., Italy), with twisted cables. Signal was applied with 4050B Series Dual Channel Function/Arbitrary Waveform Generators (B&K Precision Corporation) with a 16-bit vertical resolution. Signals featuring a series of frequencies—1–10 mHz with a 1 mHz step and 10–100 mHz with a 10 mHz step—have been applied between two points of the fungi and measured with two differential channels on ADC-24 (purchased by Pico Technology, UK) high-resolution data logger with a 24-bit analog-to-digital converter. We have chosen these particular intervals of frequencies because they well cover frequencies of action-potential spiking behaviour of a range of fungi species [22, 28, 32].

For these frequencies, the sinusoidal signal was applied along two paths separately. Finally, mixing of signals was performed for 1 mHz base frequency applied on Path 1 and a series of frequencies on the Path 2. Frequencies used on Path 2 are 2, 5 and 7 mHz). Fast Fourier transform (FFT) was calculated with Origin Pro software. Blackman window function was used as it is best suitable for the representation of amplitudes [33]. Fuzzy sets for inference of new input data were constructed using "fuzzylogic 1.2.0" Python package.

3 Results

A response of the fungi sample to electrical stimulation is shown in Fig. 1a. In all measurements, electrical activity with frequency 50–200 mHz was observed even when substrates were not stimulated. This activity is attributed to endogenous oscillations of electrical potential of fungi [22, 28, 32].

Exemplary generations of higher harmonics are shown in Fig. 2b. In some cases presented on Fig. 3, 2nd harmonic is more damped than the 3rd harmonic. Generally, for frequencies below 10 mHz, higher amplitudes were observed for 3rd harmonic versus the 2nd.

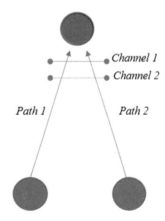

Fig. 1 Diagram showing connecting points to the fungi sample. The blue circles represent two places for the input signal injection whereas the red circle represents the ground

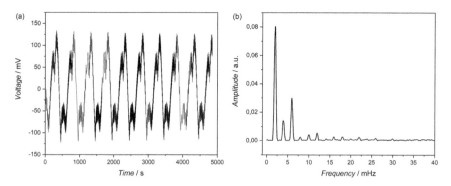

Fig. 2 Exemplary response of the fungal sample to 2 mHz, 10 Vpp sinusoidal electrical stimulation (**a**) and FFT for the same response

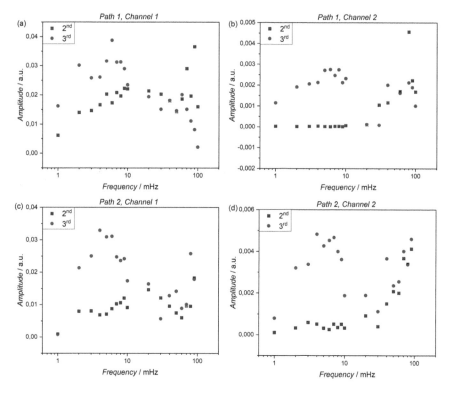

Fig. 3 Collection of 2nd and 3rd harmonic amplitudes obtained for the measured fungi response, for two signal paths and two differential channels

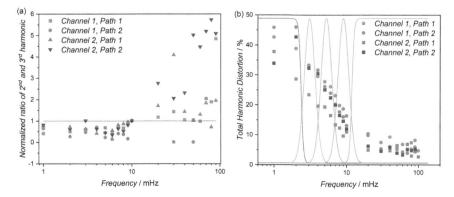

Fig. 4 Harmonic distributions. **a** Normalised ratios of 2nd verses 3rd harmonics for analysed signals. Straight line marks threshold frequency. **b** Total Harmonic Distortion calculated for fungi sample. Proposition of the fuzzy sets is included in the background

The ratio of the 2nd to 3rd harmonic amplitudes was calculated to better illustrate the changes between them (Fig. 4a). The calculated ratios were then normalised to the ratio of harmonics at 10mHz. Points at 30 and 50 mHz in 1 path, and 2 channel were treated as outliers because the ratios at these frequencies were disproportionally larger than those at other frequencies, which disturbed data visualisation. Besides, the omitted data points in the presented graph still support the observation that in general, below 10 mHz, the ratio of the 2nd and 3rd harmonics are smaller than for higher frequencies.

In the next step, Total Harmonic Distortion (THD) of the measured signal was calculated (Fig. 4b). THD is the ratio between the fundamental frequency amplitude V_0 and the amplitude of higher harmonics V_n:

$$\text{THD}_F = \frac{\sqrt{V_2^2 + V_3^2 + V_4^2 + \ldots}}{V_1} \quad (1)$$

where V_n is the nth amplitude of the frequency of successive higher harmonic peaks observed in the Fourier spectra. Furthermore, normalisation to 100% of the THD parameter can be applied as follows:

$$\text{THD}_R = \frac{\text{THD}_F}{\sqrt{1 + \text{THD}_F^2}}, \quad (2)$$

where R in THD_R stands for "root mean square".

For the frequencies below 10 mHz, higher values of THD (up to 45.9%) can be observed in relation to higher frequencies, which tend to exhibit lower THD values (below 10%). The THD of a pure signal ranges between different values, for example a square wave features a THD of 48.3% and a triangular wave features a

THD of 12.1%. This result may suggest changes in the dominant conductivity type: slower signals are more distorted and faster signals are much less distorted. Lower THD values are obtained, when the generation of higher harmonics of the modulated signal is low, hence the fungi sample has lower effect on its transformation. This effect is a consequence of a dual electric charge transport mechanism in mycelium. Furthermore, the changes occurring at low frequencies indicate, that slow physical phenomena (as diffusion) are critically responsible for the distortion of electric signals. This effect is similar to those observed in the case of solid-state memristor, however in the latter case the dependence is opposite [34]. It can be concluded that in the studied case at high frequencies only one, faster conductivity mode plays a significant role. Therefore, the nonlinear character of electric transport is much less pronounced and signal can apparently "fly through" the sample and can be transmitted across a macroscopic distance with low distortion.

As the changes of THD parameter below 10mHz occurs in a rather continuous manner, arbitrary linguistic (very low, low, medium, high, very high, etc.) could be defined for ranges of obtained values. To cope with uncertainty of the classification, we can employ fuzzy sets theory [35]. Following, membership function could be specified for the allocation of data into sets so that fuzzification of data could be implemented and allow for inference of given new input data into proper category [36]. Proposition for such sets is depicted in the background of Fig. 4b. Two sigmoidal sets were selected for the boundary and three Gaussian sets for the center of the data.

The results demonstrate that, based on increase of the THD parameter or on the amplitude values of 2nd and 3rd harmonic components, signal discrimination based on its frequency could be realised.

After analysis of single signal paths, signals were applied to the two signal paths at once. Results show that with increasing frequency, further damping of the 2nd harmonic is achieved. Furthermore, satellite frequencies appear around base frequencies as well as around higher harmonics. For example, on the Fig. 5b, for the mixing of 1 and 5 mHz signal, higher frequencies—9 mHz and 11 mHz—around damped 10 mHz 2nd harmonic are present. This effect is present as well for the 1 mHz

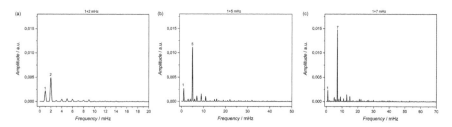

Fig. 5 Result of frequency mixing in the fungi samples. For each measurement, base 1 mHz driving signal was used on Path 1 (Fig. 1). For each successive measurement, higher frequency signal—2, 5 and 7 mHz—was applied to the Path 2

and 7 mHz mixed frequencies. The results indicate a nontrivial frequency mixing scheme, which may results in transport phenomena within highly branched network of mycelial hyphae. Such a transport can be a topic of future experiments [37, 38].

4 Conclusion

We demonstrated that fungal mycelium networks modify frequencies of external electrical inputs. Damping of 2nd harmonic and amplification of the 3nd harmonic amplitudes below 10mHz allow for frequency discrimination in a threshold manner. The frequency discrimination could occur in a continuous manner, with the help of the concepts of fuzzy logic based on THD parameter.

References

1. Adamatzky, A., Gandia, A., Chiolerio, A.: Fungal sensing skin. Fungal Biol. Biotechnol. **8**(1), 1–6 (2021)
2. Karana, E., Blauwhoff, D., Hultink, E-J., Camere, S.: When the material grows: A case study on designing (with) mycelium-based materials. Int. J. Des. **12**(2) (2018)
3. Jones, M., Mautner, A., Luenco, S., Bismarck, A., John, S.: Engineered mycelium composite construction materials from fungal biorefineries: a critical review. Mater. Des. **187**, 108397 (2020)
4. Cerimi, K., Akkaya, C.A., Pohl, C., Schmidt, B., Neubauer, P.: Fungi as source for new bio-based materials: a patent review. Fungal Biol. Biotechnol. **6**(1), 1–10 (2019)
5. Pelletier, M.G., Holt, G.A., Wanjura, J.D., Bayer, E., McIntyre, G.: An evaluation study of mycelium based acoustic absorbers grown on agricultural by-product substrates. Ind. Crops Prod. **51**, 480–485 (2013)
6. Elsacker, E., Vandelook, S., Van Wylick, A., Ruytinx, J., De Laet, L., Peeters, E.: A comprehensive framework for the production of mycelium-based lignocellulosic composites. Sci. Total Environ. **725**, 138431 (2020)
7. Dias, P.P., Jayasinghe, L.B., Waldmann, D.: Investigation of mycelium-miscanthus composites as building insulation material. Results in Mater. **10**, 100189 (2021)
8. Wang, F., Li, H.Q., Kang, S.S., Bai, Y.F., Cheng, G.Z., Zhang, G.Q.: The experimental study of mycelium/expanded perlite thermal insulation composite material for buildings. Sci. Technol. Eng. **2016**, 20 (2016)
9. Juan Pablo Cárdenas-R. Thermal insulation biomaterial based on hydrangea macrophylla. In *Bio-Based Materials and Biotechnologies for Eco-Efficient Construction*, pages 187–201. Elsevier, 2020
10. Holt, G.A., Mcintyre, G., Flagg, D., Bayer, E., Wanjura, J.D., Pelletier, M.G.: Fungal mycelium and cotton plant materials in the manufacture of biodegradable molded packaging material: Evaluation study of select blends of cotton byproducts. J. Biobased Mater. Bioenergy **6**(4), 431–439 (2012)
11. Mojumdar, A., Behera, H.T., Ray, L.: Mushroom mycelia-based material: an environmental friendly alternative to synthetic packaging. Microb. Polym. 131–141 (2021)
12. Andrew Adamatzky, Phil Ayres, Gianluca Belotti, and Han Wösten. Fungal architecture position paper. *International Journal of Unconventional Computing*, 14, 2019
13. Adamatzky, A., Nikolaidou, A., Gandia, A., Chiolerio, A., Dehshibi, M.M.: Reactive fungal wearable. Biosystems **199**, 104304 (2021)

14. Silverman, J., Cao, H., Cobb, K.: Development of mushroom mycelium composites for footwear products. Cloth. Text. Res. J. **38**(2), 119–133 (2020)
15. Appels, F.V.W.: The use of fungal mycelium for the production of bio-based materials. Ph.D. thesis, Universiteit Utrecht (2020)
16. Jones, M., Gandia, A., John, S., Bismarck, A.: Leather-like material biofabrication using fungi. Nat. Sustain. 1–8 (2020)
17. Cucchi, M., Kleemann, H., Tseng, H., Ciccone, G., Lee, A., Pohl, D., Leo, K.: Directed growth of dendritic polymer networks for organic electrochemical transistors and artificial synapses. Adv. Electron. Mat. **7**, 2100586 (2021)
18. Janzakova, K., Kumar, A., Ghazal, M., Susloparova, A., Coffinier, Y., Alibart, F., Pecqueur, S.: Analog programing of conducting-polymer dendritic interconnections and control of their morphology. Nat. Commun. **12**, 6898 (2021)
19. Pismen, L.: Morphogenesis Deconstructed. Springer, An integrated view of the generation of forms (2020)
20. Kasai, S., Inoue, S., Okamoto, S., Sasaki, K., Yin, X., Kuroda, R., Sato, M., Wakamiya, R., Saito, K.: Detection and Control of Charge State in Single Molecules Toward Informatics in Molecule Networks. Springer (2017)
21. Beasley, A.E., Abdelouahab, M-S., Lozi, R., Tsompanas, M.A., Powell, A.L., Adamatzky, A.: Mem-fractive properties of mushrooms. Bioinspiration & Biomimetics **16**(6), 066026 (2022)
22. Adamatzky, A., Gandia, A.: On electrical spiking of ganoderma resinaceum. Biophys. Rev. Lett. 1–9 (2021)
23. Beasley, A.E., Powell, A.L., Adamatzky, A.: Fungal photosensors (2020). arXiv:2003.07825
24. Adamatzky, A., Gandia, A.: Living mycelium composites discern weights via patterns of electrical activity. J. Bioresources Bioprod. **7**(1), 26–32 (2022)
25. Roberts, N., Adamatzky, A.: Mining logical circuits in fungi. Sci. Rep. **12**(1), 15930 (2022)
26. Slayman, C.L., Long, W.S., Gradmann, D.: Action potentials. In: Neurospora crassa, a mycelial fungus. Biochimica et Biophysica Acta (BBA) — Biomembranes **426**(4), 732–744 (1976)
27. Olsson, S., Hansson, B.S.: Action potential-like activity found in fungal mycelia is sensitive to stimulation. Naturwissenschaften **82**(1), 30–31 (1995)
28. Adamatzky, A.: On spiking behaviour of oyster fungi pleurotus djamor. Sci. Rep. **8**(1), 1–7 (2018)
29. Cocatre-Zilgien, J.H., Delcomyn, F.: Identification of bursts in spike trains. J. Neurosci. Methods **41**(1), 19–30 (1992)
30. Legendy, C.R., Salcman, M.: Bursts and recurrences of bursts in the spike trains of spontaneously active striate cortex neurons. J. Neurophysiol. **53**(4), 926–939 (1985)
31. Mohammad Mahdi Dehshibi and Andrew Adamatzky: Electrical activity of fungi: spikes detection and complexity analysis. Biosystems **203**, 104373 (2021)
32. Adamatzky, A.: Language of fungi derived from their electrical spiking activity. Roy. Soc. Open Sci. **9**(4), 211926 (2022)
33. Dactron, L.: Understanding FFT windows. Application Note. LDS group (2003)
34. Przyczyna, D., Hess, G., Szaciłowski, K.: KNOWM memristors in a bridge synapse delay-based reservoir computing system for detection of epileptic seizures. Int. J. Parallel Emergent Distrib. Syst. **37**(5), 512–527 (2022)
35. Zadeh, L.A.: Fuzzy sets. Inf. Control **8**(3), 338–353 (1965)
36. Mendel, J.M.: Fuzzy logic systems for engineering: a tutorial. Proc. IEEE **83**(3), 345–377 (1995)
37. Zanin, M., Sun, X., Wandelt, S.: Studying the topology of transportation systems through complex networks: handle with care. J. Adv. Transp. **2018** (2018)
38. Xiong, K., Liu, Z., Zeng, C., Li, B.: Thermal-siphon phenomenon and thermal/electric conduction in complex networks. Natl. Sci. Rev. **7**(2), 270–277 (2020)

On Electrical Gates on Fungal Colony

Alexander E. Beasley, Phil Ayres, Martin Tegelaar,
Michail-Antisthenis Tsompanas, and Andrew Adamatzky

Abstract Mycelium networks are promising substrates for designing unconventional computing devices providing rich topologies and geometries where signals propagate and interact. Fulfilling our long-term objectives of prototyping electrical analog computers from living mycelium networks, including networks hybridised with nanoparticles, we explore the possibility of implementing Boolean logical gates based on electrical properties of fungal colonies. We converted a 3D image-data stack of *Aspergillus niger* fungal colony to an Euclidean graph and modelled the colony as resistive and capacitive (RC) networks, where electrical parameters of edges were functions of the edges' lengths. We found that AND, OR and AND- NOT gates are implementable in RC networks derived from the geometrical structure of the real fungal colony.

1 Introduction

Fungi are demonstrated to be at the forefront of environmentally sustainable biomaterials [1–3] used in manufacturing of acoustic [4–6] and thermal [7–12] insulation panels, packaging materials [13–15] and adaptive wearables [1, 16–19]. In our project 'Fungal architectures' [20] we proposed to grow mycelium bound composites into monolithic building elements [21]. The composite would combine living mycelium, capable of sensing light, chemicals, gases, gravity and electric fields [22–28], with dead mycelium functionalised using nanoparticles and polymers. These living build-

---

A. E. Beasley
Centre for Engineering Research, University of Hertfordshire, Hatfield, UK

A. E. Beasley · M.-A. Tsompanas
Unconventional Computing Laboratory, UWE, Bristol, UK

P. Ayres
The Centre for Information Technology and Architecture, Royal Danish Academy, Copenhagen, Denmark

M. Tegelaar · A. Adamatzky (✉)
Microbiology Department, University of Utrecht, Utrecht, The Netherlands
e-mail: andrew.adamatzky@uwe.ac.uk

© The Author(s), under exclusive license to Springer Nature Switzerland AG 2023
A. Adamatzky (ed.), *Fungal Machines*, Emergence, Complexity and Computation 47,
https://doi.org/10.1007/978-3-031-38336-6_20

ing structures would have embedded bioelectronics electronics [29–31], implement sensorial fusion and decision making in the mycelium networks [32] and be able to grow monolithic buildings from the functionalised fungal substrate [21].

A decision making feature requires inference logical circuits to be embedded directly into mycelium bonded composites. To check what range and frequencies of logical gates could be implemented in the mycelium bound composites we adopted an approach developed originally in [32, 33]. The technique is based on selecting a pair of input sites, applying all possible combinations of inputs to the sites and recording outputs on a set of the selected output sites. The approach belongs to same family of computation outsourcing techniques as *in materio* computing [34–38] and reservoir computing [39–43]. In our previous studies [32] we demonstrated that logical circuits can be derived from electrical spiking activity of the fungal colony. The approach, whilst elegant theoretically, might lack practical applications because the spiking activity of living fungi is of very low frequency, e.g. a spike per 20 min [44, 45]. Thus, we decided to explore electrical properties of the fungal colony, because the electrical analog implementation of logical gates is notoriously fast. In the numerical experiments described here, '0' and '1' signals are represented by low and high voltage applied to the input sites.

2 Methods

Aspergillus niger strain AR9#2 [46], expressing Green Fluorescent Protein (GFP) from the glucoamylase (*glaA*) promoter, was grown at 30 °C on minimal medium (MM) [47] with 25 mM xylose and 1.5% agarose (MMXA). MMXA cultures were grown for three days, after which conidia were harvested using saline-Tween (0.8% NaCl and 0.005% Tween-80). 250 ml liquid cultures were inoculated with $1.25 \cdot 10^9$ freshly harvested conidia and grown at 200 rpm and 30 °C in 1 L Erlenmeyer flasks in complete medium (CM) (MM containing 0.5% yeast extract and 0.2% enzymatically hydrolyzed casein) supplemented with 25 mM xylose (repressing glaA expression). Mycelium was harvested after 16 h and washed twice with PBS. Ten g of biomass (wet weight) was transferred to MM supplemented with 25 mM maltose (inducing glaA expression).

Fluorescence of GFP was localised in micro-colonies using a DMI 6000 CS AFC confocal microscope (Leica, Mannheim, Germany). Micro-colonies were fixed overnight at 4 °C in 4% paraformaldehyde in PBS, washed twice with PBS and taken up in 50 ml PBS supplemented with 150 mM glycine to quench autofluorescence. Micro-colonies were then transferred to a glass bottom dish (Cellview™, Greiner Bio-One, Frickenhausen, Germany, PS, 35/10 MM) and embedded in 1% low melting point agarose at 45 °C. Micro-colonies were imaged at 20× magnification (HC PL FLUOTAR L 20 × 0.40 DRY). GFP was excited by white light laser at 472 nm using 50% laser intensity (0.1 kW/cm2) and a pixel dwell time of 72 ns. Fluorescent light emission was detected with hybrid detectors in the range of 490–525 nm. Pinhole size was 1 Airy unit.

On Electrical Gates on Fungal Colony 303

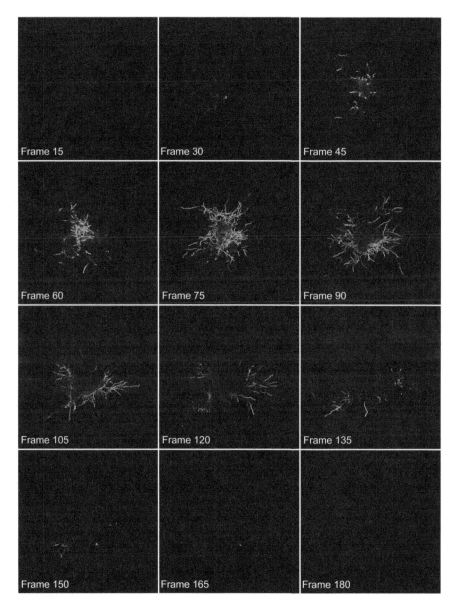

Fig. 1 Z-slices of the fungal colony of *Aspergillus niger* imaged by fluorescence microscopy

3D projections were made with Fiji [48] (Fig. 1). Conversion of the imaged microcolonies to graph data was accomplished using a publicly available ImageJ macro.[1] The macro was run on the Fiji (ImageJ) platform, version 1.52, with the supplementary 3D ImageJ Suite installed to provide enhanced 3D capabilities [49]. To run the macro, initialisation parameters for expected hypha radius were given as 3 μm [50]; detection sensitivity threshold was set to 8; minimal vessel volume was set to 100 pixels. Key processing tasks performed by the macro were: morphological closing of tubular structures, pre-filtering to enhance filamentous voxels, segmentation of tubular structures, skeletonisation and analysis of the network. Results of the segmentation included a Z-stack 3D visualisation of the network with identified branching and end-points, and tabular data including the number of disjoint networks together with constituent branch segments defined by start and end points given in voxel coordinates.

The tabular data was then processed using a custom Python script to convert voxel coordinates to real-world coordinates. The set of all vertices was determined and all branch start and end points were indexed. A weighted graph was generated using the NetworkX library for python [51], with graph nodes defined by the vertex index, edges defined between vertex pairs and edge weights given as the euclidean distance of the branch segment (Fig. 2). Graph topology could then be determined using the NetworkX degree function. Shortest weighted paths between source and sink vertices could also be found, allowing a direct correlation to resistive networks.

The 3D graph was converted to a resistive and capacitive (RC) network, whose magnitudes are a function of the length of the connections. Resistances were in the order of kOhms and capacitance were in the order of pF. Separate models were created with the RC connections modelled either in series or in parallel modules. The networks were parsed for the order one nodes, which are considered to the extent of the sample. The positive voltage and ground nodes were randomly assigned from the sample and 1000 networks are created in each arrangement for analysis. SPICE analysis consisted of transient analysis using a two voltage pulses of 60 mV on the randomly assigned positive nodes with the following parameters: T_{delay} = 10 s for V_1 and 20 s for V_2, T_{rise} = 0.001 s, T_{fall} = 0.001 s, T_{on} = 10 s for V_1 and 20 s for V_2, T_{off} = 20 s for V_1 and 20 s for V_2, N_{cycles} = 2 for V_1 and 1 for V_2. Circuit analysis was transient analysis for 40 s in steps of 1 ms. The voltage at each node and current through each link were measured every 1 ms of the simulation. We modelled the fungal colony in serial RC networks and parallel RC networks. The output voltages have been binarised with the threshold θ: $V > theta$ symbolises logical TRUE otherwise FALSE.

---

[1] The macro was developed by the Advanced Digital Microscopy Core Facility, IRB Barcelona, to process Z-stack data for blood vessel segmentation and network analysis, see details in adm.irbbarcelona.org/bioimage-analysis/image-j-fiji and biii.eu/blood-vessel-segmentation-and-network-analysis.

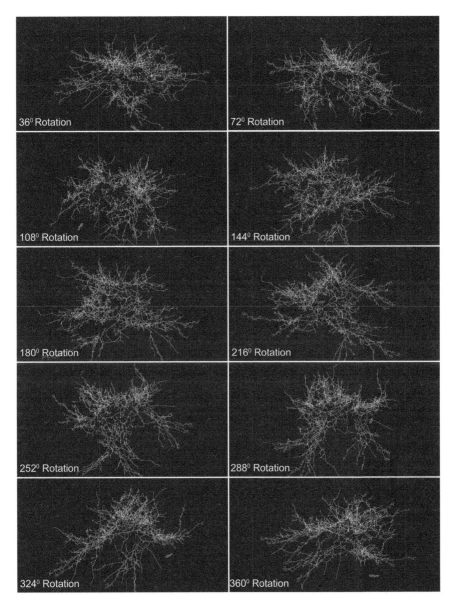

Fig. 2 Perspective views of the 3D Graph. Each frame shows the graph after a 36° rotation around the z-axis with origin located approximately in the centre of the colony, on the x-y plane indicated with registration marks

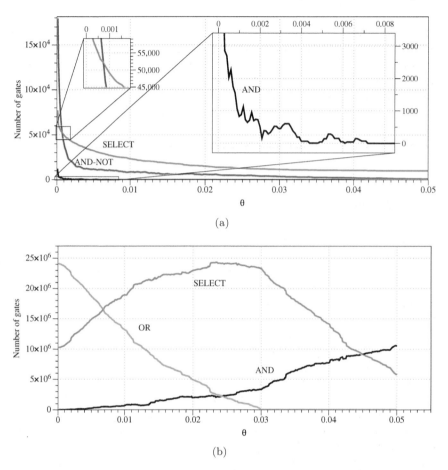

Fig. 3 Occurrences of the gates from the groups AND, black, OR, green, AND- NOT, red, and SELECT, blue, for $\theta \in [0.0001, 0.05]$, with θ increment 0.0001, in **a** fungal colony modelled with serial RC networks, **b** fungal colony modelled with parallel RC networks

3 Results

In general, there are 16 possible logical gates realisable for two inputs and one output. The gates implying input 0 and evoking a response 1, i.e. $f(0, 0) = 1$, are not realisable because the fungal circuit simulated is passive. The remaining 8 gates are AND, OR, AND–NOT (x AND NOT y and NOT x AND y), SELECT (SELECT x and SELECT y) and XOR.

No XOR gates have been found in neither of the RC models of the fungal colony.

In the model of serial RC networks we found gates AND, SELECT and AND-NOT; no OR gates have been found. The number n of the gates discovered decreases by a power low with increase of θ: $n_{\text{AND- NOT}} = 72 \cdot x^{-0.98}$, $n_{\text{SELECT}} = 2203 \cdot x^{-0.48}$,

$n_{\text{AND}} = 0.02 \cdot x^{-1.6}$. Frequency of AND gate oscillates, as shown in zoom insert in Fig. 3a, more likely due to its insignificant presence in the samples. The oscillations reach near zero base when θ exceeds 0.001.

In the model of parallel RC networks we found only gates AND, SELECT and OR. The number of OR gates decreases quadratically and becomes nil when $\theta > 0.03$. The number of AND gates increases near linearly, $n_{\text{AND}} = -1.72 \cdot 10^6 + 2.25 \cdot 10^8 \cdot x$, with increase of θ. The number of SELECT gates reaches its maximum at $\theta = 0.023$, and then starts to decreases with the further increase of θ: $n_{\text{SELECT}} = 9.61 \cdot 10^6 + 1.21 \cdot 10^9 \cdot x - 2.7 \cdot x^2$.

4 Discussion

By simulating a fungal colony as an electrical network we discovered families of Boolean gates realisable in the network. Voltage values have been binarised via threshold θ. All non-active, i.e. $f(0, 0) \neq 1$, gates but XOR have been discovered and their dynamics in relation to θ. The systems of gates discovered are functionally complete and therefore we can speculate that an arbitrary logical circuit can be realised in living fungal networks by encoding Boolean values in differences of electrical potential. The XOR gates have not been observed in our models. This is unsurprising as the XOR gate is the most rare gate to be discovered in non-linear systems [52–54]. A disadvantage of the electrical analog logical circuits in living fungal colonies would be that the colony requires maintenance and have a relatively short life span. A way forward would be to 'imprint' the colony in other long-living materials. This can be done, for example, by means of biolithography as previously tested on slime mould *Physarum polycephalum* [55].

References

1. Karana, E., Blauwhoff, D., Hultink, E.-J., Camere, S.: When the material grows: a case study on designing (with) mycelium-based materials. Int. J. Des. **12**(2) (2018)
2. Jones, M., Mautner, A., Luenco, S., Bismarck, A., John, S.: Engineered mycelium composite construction materials from fungal biorefineries: a critical review. Mater. Des. **187**, 108397 (2020)
3. Cerimi, K., Akkaya, K.C., Pohl, C., Schmidt, B., Neubauer, P.: Fungi as source for new bio-based materials: a patent review. Fungal Biol. Biotechnol. **6**(1), 1–10 (2019)
4. Pelletier, M.G., Holt, G.A., Wanjura, J.D. Bayer, E., McIntyre, G.: An evaluation study of mycelium based acoustic absorbers grown on agricultural by-product substrates. Ind. Crops Prod. **51**, 480–485 (2013)
5. Elsacker, E., Vandelook, S., Van Wylick, A., Ruytinx, J., De Laet, L., Peeters, E.: A comprehensive framework for the production of mycelium-based lignocellulosic composites. Sci. Total Environ. **725**, 138431 (2020)
6. Robertson, O., et al.: Fungal future: a review of mycelium biocomposites as an ecological alternative insulation material. In: DS 101: Proceedings of NordDesign 2020, Lyngby, Denmark, 12th-14th August 2020, pp. 1–13 (2020)

7. Yang, Z., Zhang, F., Still, B., White, M., Amstislavski, P.: Physical and mechanical properties of fungal mycelium-based biofoam. J. Mater. Civ. Eng. **29**(7), 04017030 (2017)
8. Xing, Y., Brewer, M., El-Gharabawy, H., Griffith, G., Jones, P.: Growing and testing mycelium bricks as building insulation materials. In: IOP Conference Series: Earth and Environmental Science, vol. 121, pp. 022032. IOP Publishing (2018)
9. Girometta, C., Picco, A.M., Baiguera, R.M., Dondi, D., Babbini, S., Cartabia, M., Pellegrini, M., Savino, E.: Physico-mechanical and thermodynamic properties of mycelium-based biocomposites: a review. Sustainability **11**(1), 281 (2019)
10. Dias, P.P., Jayasinghe, L.B., Waldmann, D.: Investigation of mycelium-miscanthus composites as building insulation material. Results Mater. **10**, 100189 (2021)
11. Wang, F., Li, H.-G., Kang, S.-S., Bai, Y.-F., Cheng, G.-Z., Zhang, G.-G.: The experimental study of mycelium/expanded perlite thermal insulation composite material for buildings. Sci. Technol. Eng. **2016**, 20 (2016)
12. Cárdenas-R, J.P.: Thermal insulation biomaterial based on hydrangea macrophylla. In: Bio-Based Materials and Biotechnologies for Eco-Efficient Construction, pp. 187–201. Elsevier (2020)
13. Holt, G.A., Mcintyre, G., Flagg, D., Bayer, E., Wanjura, J.D., Pelletier, M.G.: Fungal mycelium and cotton plant materials in the manufacture of biodegradable molded packaging material: evaluation study of select blends of cotton byproducts. J. Biobased Mater. Bioenergy **6**(4), 431–439 (2012)
14. Sivaprasad, S., Byju, S.K., Prajith, C., Shaju, J., Rejeesh, C.R.: Development of a novel mycelium bio-composite material to substitute for polystyrene in packaging applications. Mater. Today Proc. (2021)
15. Mojumdar, A., Behera, H.T., Ray, L.: Mushroom mycelia-based material: an environmental friendly alternative to synthetic packaging. Microb. Polym. 131–141 (2021)
16. Adamatzky, A., Nikolaidou, A., Gandia, A., Chiolerio, A., Dehshibi, M.M.: Reactive fungal wearable. Biosystems **199**, 104304 (2021)
17. Silverman, J., Cao, H., Cobb, K.: Development of mushroom mycelium composites for footwear products. Cloth. Text. Res. J. **38**(2), 119–133 (2020)
18. Appels, F.V.W.: The use of fungal mycelium for the production of bio-based materials. PhD thesis, Universiteit Utrecht (2020)
19. Jones, M., Gandia, A., John, S., Bismarck, A.: Leather-like material biofabrication using fungi. Nat. Sustain. 1–8 (2020)
20. Adamatzky, A., Ayres, P., Belotti, G., Wösten, H.: Fungal architecture position paper. Int. J. Unconv. Comput. **14** (2019)
21. Adamatzky, A., Gandia, A., Ayres, P., Wösten, H., Tegelaar, M.: Adaptive fungal architectures. LINKs-series **5**, 66–77 (2021)
22. Bahn, Y.-S., Xue, C., Idnurm, A., Rutherford, J.C., Heitman, J., Cardenas, M.E.: Sensing the environment: lessons from fungi. Nat. Rev. Microbiol. **5**(1), 57 (2007)
23. Van Aarle, I.M., Olsson, P.A., Söderström, B.: Arbuscular mycorrhizal fungi respond to the substrate ph of their extraradical mycelium by altered growth and root colonization. New Phytol. **155**(1), 173–182 (2002)
24. Kung, C.: A possible unifying principle for mechanosensation. Nature **436**(7051), 647 (2005)
25. Fomina, M., Ritz, K., Gadd, G.M.: Negative fungal chemotropism to toxic metals. FEMS Microbiol. Lett. **193**(2), 207–211 (2000)
26. Bahn, Y.-S., Mühlschlegel, F.A.: Co2 sensing in fungi and beyond. Current Opin. Microbiol. **9**(6), 572–578 (2006)
27. Jaffe, M.J., Carl Leopold, A., Staples, R.C.: Thigmo responses in plants and fungi. Am. J. Bot. **89**(3), 375–382 (2002)
28. Howitz, K.T., Sinclair, D.A.: Xenohormesis: sensing the chemical cues of other species. Cell **133**(3), 387–391 (2008)
29. Beasley, A.E., Powell, A.L., Adamatzky, A.: Capacitive storage in mycelium substrate (2020). arXiv:2003.07816

30. Beasley, A.E., Abdelouahab, M.-S., Lozi, R., Powell, A.L., Adamatzky, A.: Mem-fractive properties of mushrooms (2020). arXiv:2002.06413
31. Beasley, A.E., Powell, A.L., Adamatzky, A.: Fungal photosensors (2020). arXiv:2003.07825
32. Adamatzky, A., Tegelaar, M., Wosten, H.A.B., Powell, A.L., Beasley, A.E., Mayne, R.: On Boolean gates in fungal colony. Biosystems **193**, 104138 (2020)
33. Siccardi, S., Adamatzky, A., Tuszyński, J., Huber, F., Schnauß, J.: Actin networks voltage circuits. Phys. Rev. E **101**(5), 052314 (2020)
34. Miller, J.F., Downing, K.: Evolution in materio: looking beyond the silicon box. In: Proceedings 2002 NASA/DoD Conference on Evolvable Hardware, pp. 167–176. IEEE (2002)
35. Miller, J.F., Harding, S.L., Tufte, G.: Evolution-in-materio: evolving computation in materials. Evol. Intell. **7**(1), 49–67 (2014)
36. Stepney, S.: Co-designing the computational model and the computing substrate. In: International Conference on Unconventional Computation and Natural Computation, pp. 5–14. Springer (2019)
37. Miller, J.F., Hickinbotham, S.J., Amos, M.: In materio computation using carbon nanotubes. In: Computational Matter, pp. 33–43. Springer (2018)
38. Julian Francis Miller: The alchemy of computation: designing with the unknown. Nat. Comput. **18**(3), 515–526 (2019)
39. Verstraeten, D., Schrauwen, B., d'Haene, M., Stroobandt, D.: An experimental unification of reservoir computing methods. Neural Netw. **20**(3), 391–403 (2007)
40. Lukoševičius, M., Jaeger, H.: Reservoir computing approaches to recurrent neural network training. Comput. Sci. Rev. **3**(3), 127–149 (2009)
41. Dale, M., Miller, J.F., Stepney, S.: Reservoir computing as a model for in-materio computing. In: Advances in Unconventional Computing, pp. 533–571. Springer (2017)
42. Konkoli, Z., Nichele, S., Dale, M., Stepney, S.: Reservoir computing with computational matter. In: Computational Matter, pp. 269–293. Springer (2018)
43. Dale, M., Miller, J.F., Stepney, S., Trefzer, M.A.: A substrate-independent framework to characterize reservoir computers. Proc. R. Soc. A **475**(2226), 20180723 (2019)
44. Adamatzky, A.: On spiking behaviour of oyster fungi pleurotus djamor. Sci. Rep. **8**(1), 1–7 (2018)
45. Adamatzky, A., Gandia, A.: On electrical spiking of ganoderma resinaceum. Biophys. Rev. Lett. 1–9 (2021)
46. Vinck, A., de Bekker, C., Ossin, A., Ohm, R.A., de Vries, R.P., Wösten, H.A.B.: Heterogenic expression of genes encoding secreted proteins at the periphery of Aspergillus niger colonies. Environ. Microbiol. **13**(1), 216–225 (2011)
47. De Vries, R.P., Burgers, K., van de Vondervoort, P.J.I., Frisvad, J.C., Samson, R.A., Visser, J.: A new black aspergillus species, a. vadensis, is a promising host for homologous and heterologous protein production. Appl. Environ. Microbiol. **70**(7), 3954–3959 (2004)
48. Schindelin, J., Arganda-Carreras, I., Frise, E., Kaynig, V., Longair, M., Pietzsch, T., Preibisch, S., Rueden, C., Saalfeld, S., Schmid, B., et al.: Fiji: an open-source platform for biological-image analysis. Nat. Methods **9**(7), 676–682 (2012)
49. Ollion, J., Cochennec, J., Loll, F., Escudé, C., Boudier, T.: Tango: a generic tool for high-throughput 3d image analysis for studying nuclear organization. Bioinformatics **29**(14), 1840–1841 (2013)
50. Tegelaar, M., Wösten, H.A.B.: Functional distinction of hyphal compartments. Sci. Rep. **7**(1), 1–6 (2017)
51. Hagberg, A., Swart, P., Chult, D.S.: Exploring network structure, dynamics, and function using NetworkX. Technical report, Los Alamos National Lab.(LANL), Los Alamos, NM (United States) (2008)
52. Adamatzky, A., Bull, L.: Are complex systems hard to evolve? Complexity **14**(6), 15–20 (2009)

53. Siccardi, S., Tuszynski, J.A., Adamatzky, A.: Boolean gates on actin filaments. Phys. Lett. A **380**(1), 88–97 (2016)
54. Harding, S., Koutník, J., Schmidhuber, J., Adamatzky, A.: Discovering Boolean gates in slime mould. In: Inspired by Nature, pp. 323–337. Springer (2018)
55. Berzina, T., Dimonte, A., Adamatzky, A., Erokhin, V., Iannotta, S.: Biolithography: Slime mould patterning of polyaniline. Appl. Surf. Sci. **435**, 1344–1350 (2018)

Mining Logical Circuits in Fungi

Nic Roberts and Andrew Adamatzky

Abstract Living substrates are capable for nontrivial mappings of electrical signals due to the substrate nonlinear electrical characteristics. This property can be used to realise Boolean functions. Input logical values are represented by amplitude or frequency of electrical stimuli. Output logical values are decoded from electrical responses of living substrates. We demonstrate how logical circuits can be implemented in mycelium bound composites. The mycelium bound composites (fungal materials) are getting growing recognition as building, packaging, decoration and clothing materials. Presently the fungal materials are passive. To make the fungal materials adaptive, i.e. sensing and computing, we should embed logical circuits into them. We demonstrate experimental laboratory prototypes of many-input Boolean functions implemented in fungal materials from oyster fungi *P. ostreatus*. We characterise complexity of the functions discovered via complexity of the space-time configurations of one-dimensional cellular automata governed by the functions. We show that the mycelium bound composites can implement representative functions from all classes of cellular automata complexity including the computationally universal. The results presented will make an impact in the field of unconventional computing, experimental demonstration of purposeful computing with fungi, and in the field of intelligent materials, as the prototypes of computing mycelium bound composites.

1 Introduction

The fungi are one of the largest, the oldest, most adaptive and widely distributed group of organisms [1]. Smallest fungi are single cells. The largest mycelium spreads in hectares [2]. When growing in a bulk medium of wood or plant shavings fungi bind the medium in a solid monolith with outstanding mechanical properties. The

N. Roberts
Department of Engineering and Technology, University of Huddersfield, Huddersfield, UK

N. Roberts · A. Adamatzky (✉)
Unconventional Computing Laboratory, UWE, Bristol, UK
e-mail: andrew.adamatzky@uwe.ac.uk

mycelium bound composites are seen as future environmentally sustainable growing biomaterials [3–6]. They are already used in acoustic [7–9] and thermal [10–15] insulation panels and cladding, materials for packaging [16–18] and wearables [3, 19–22]. The currently used fungal materials are passive and inert because the fungi in the composites are dead and treated to prevent decay. To make the fungal materials adaptive and intelligent we must either (1) leave part of the fungal materials alive, or (2) dope the materials with functional nanoparticles and polymers.

Why do we need to compute with fungi? The research is undertaken in the frame for the FUNGAR (www.fungar.eu), acronym for Fungal Architectures, a EU Horizon 2020 research project that seeks to develop a fully integrated structural and computational living monolith by using fungal mycelium. The goal, to advance towards the realisation of full-scale intelligent bio-buildings and other functional bio-structures. Distributions of Boolean gates depends on environmental and physiological conditions of the mycelium bound composites and therefore will provide a computational characterisation of the fungal material states. This distribution of logical functions will be somewhat analogous to Kolmogorov complexity of the living building materials.

Fungal colonies are characterised by rich typology of mycelium networks [23–27] in some cases similar to fractal structures [28–33]. Rich morphological features might imply rich computational abilities and thus worth to analyse from a realising Boolean functions point of view. To implement logical functions we adopted a theoretical approach developed in [34, 35]. The technique is based on selecting a pair of input sites, applying all possible combinations of inputs, where logical values are represented by electrical characteristics of input signals, to the sites and recording outputs, represented by electrical responses of the substrate, on a set of the selected output sites. The approach belong to the family of reservoir computing [36–40] and *in materio* computing [41–45] techniques of analysing computational properties of physical and biological substrates.

2 Methods

A hemp shavings substrate was colonised by the mycelium of the grey oyster fungi, *P. ostreatus* (Ann Miller's Speciality Mushrooms Ltd, UK). Recordings were carried out in a stable indoor environment with the temperature remaining stable at $22 \pm 0.5°$ and relative humidity of air $40 \pm 5\%$. The humidity of the substrate colonised by fungi was kept at c. 70–80%.

Hardware was developed that was capable of sending sequences of 4 bit strings to a mycelium substrate. The strings were encoded as step voltage inputs where –5 V denoted a logical 0 and 5 V a logical 1. The hardware was based around an Arduino Mega 2560 (Elegoo, China) and a series of programmable signal generators, AD9833 (Analog, USA). The 4 input electrodes were 1 mm diameter platinum rods inserted to a depth of 50 mm in the substrate in a straight line with a separation of 20 mm. Data acquisition (DAQ) probes were placed in a parallel line 50 mm away separated

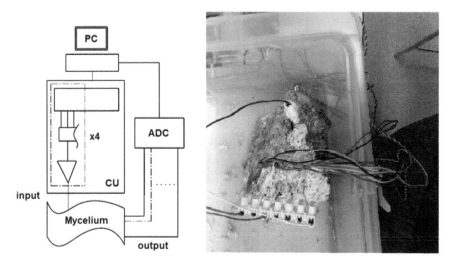

Fig. 1 Left: Schematic of the mycelium communications system; PC–laptop for generating sequences; CU–control unit, dashed section is a breakdown of a single channel; ADC–analogue to digital converter. Right: experimental set up

by 10 mm. The electron sink and source was placed 50 mm on from DAQ probes. There were 7 DAQ differential inputs from the mycelium substrate to a Pico 24 (Pico Technology, UK) analogue-to-digital converter (ADC), the 8th channel was used to pass a pulse to the ADC on every input state change, see Fig. 1 for a schematic of the apparatus. The substrate and probes were placed in a semi-sealed container. After each experimental repeat the substrate was sprayed with water, left for an hour and then the next repeat was conducted. There were a total of 14 repeats.

A sequence of 4 bit strings counting up from binary *0000* to *1111*, with a state change every hour, were passed into the substrate, see Fig. 2 for timing details. In all 14 repeats of the experiment were done on the same substrate to capture changes in structure of the growing mycelium. Samples from 7 channels were taken at 1 Hz over the whole duration of a given experimental run. Peaks for each channel were located for a set of 32 thresholds, from 20 to 175 mV with step 5 mV, for each input state, *0000 to 1111*.

The voltage spiking events occur at the scale of seconds usually during state transitions which happen every hour which is in line with the decay time after a spike. Boolean strings were extracted from the data, where a logic '1' was noted for a channel if it had a peak outside the threshold band for a particular state else, a value of '0' was recorded, the polarity of the peak was not considered.

The strings for each experimental repeat were stored in their respective Boolean table. To extract state graphs, a state/node was defined as the string of output values from each channel at each input state, transitions/edges were defined as a change in input state. This led to a total of 448 state graphs. The sum of products (SOP)

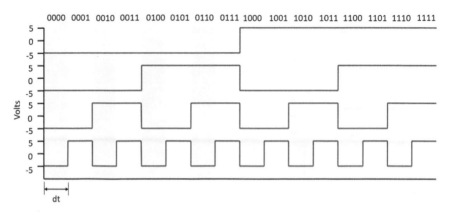

Fig. 2 Timing diagram and associated Boolean strings for four inputs into the mycelium substrate, time step is one hour

Boolean functions were calculated for each output channel. For each repeat there were 7 channels and 32 thresholds giving total of 3136 individual truth tables.

See Fig. 3 for SOP extraction. If a peak is found in Fig. 3a during an input state then this is considered a logical 1, highlighted in yellow in table Fig. 3b are the thresholded values for channel 5, the resulting truth table is then reduced to a sum products shown below the table.

3 Results

We have discovered total of 3136 4-inputs-1-output Boolean functions. Figure 4 shows the Boolean function distribution. The two peak values were logical FALSE, $n = 238$, and logical TRUE, $n = 237$. The highest occurring non-trivial gate was $\overline{A} + \overline{B} + \overline{C} + \overline{D}$, $n = 145$. The top 16 occurring non-trivial Boolean functions are listed in Table 1. The only single gate functions found were for NAND ($\overline{A} + \overline{B} + \overline{C} + \overline{D}$), $n = 145$, OR ($A + B + C + D$), $n = 46$, and AND ($ABCD$), $n = 8$.

Let us discuss complexity of the functions discovered (Table 1) via complexity of the space-time configurations of one-dimensional cellular automata governed by the functions. We consider an array Z of finite state machines, called cells, where every cell takes states '0' or '1' and updates its state depending on the states of its four immediate neighbours. All cells update their states by the same rule and in discrete time. For example, a cell with index i, $x_i \in Z$, updates its state at time t as a function of states of its four neighbours: $x^{t+1} = f(x_{i-2}^t, x_{i-1}^t, x_{i+1}^t, x_{i+2}^t)$. To map functions from Table 1 to the rules governing the cellular automata we assume that A corresponds to x_{i-2}^t, B to x_{i-1}^t, C to x_{i+1}^t and D to x_{i+1}^t. For example, a cell x_i of cellular automaton governed by the function F_5 (Table 1) updates its state as $x^{t+1} = \overline{x_{i-2}}x_{i-1} + x_{i+1}\overline{x_{i+2}} + \overline{x_{i-2}}x_{i+2}$.

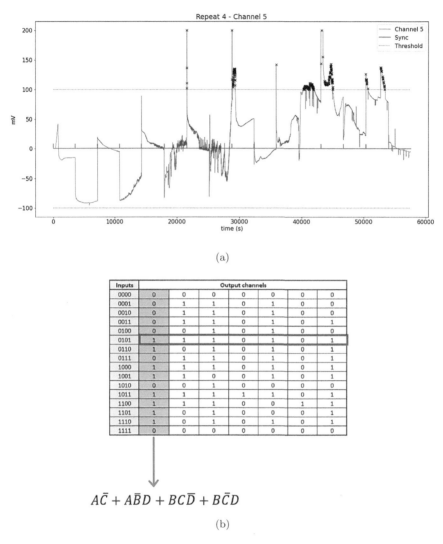

Fig. 3 Workflow example. **a** The measurements taken by channel 5 of the DAQ in blue, the synchronisation signal is shown red which marks the state change, threshold band shown in green, peaks outside this band are highlighted with 'x' marker. **b** The truth and the function extracted

Automaton governed by F_1, F_6, F_8 fall into absorbing state where all cells are in state '0'. The automaton governed by rule F_9 falls into the state where all cells are in state '1'. Space-time configurations, random initial conditions and absorbing boundaries, of automata governed by other rules are shown in Fig. 5. We characterise a complexity of the space-time patterns via Lempel-Ziv complexity Lempel-Ziv complexity (compressibility) LZ. The LZ complexity is evaluated by a size of concentration profiles saved as PNG files of the configurations. This is sufficient because

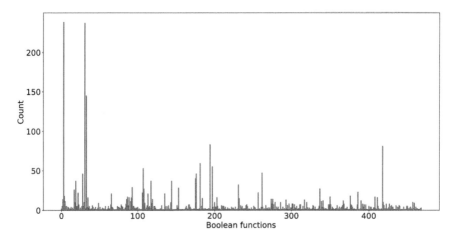

Fig. 4 Counts of realised Boolean functions discovered in laboratory experiments. Horizontal axis is a decimal representation of functions. Vertical axis is a number of functions discovered in experiments

Table 1 Top 16 highest occurring Boolean functions

Count		Boolean function
145	F_1	$\overline{A} + \overline{B} + \overline{C} + \overline{D}$ (NAND)
83	F_2	$A\overline{B} + A\overline{C} + A\overline{D} + \overline{A}B + B\overline{C} + B\overline{D} + \overline{A}C + \overline{B}C + C\overline{D} + \overline{A}D + \overline{B}D + \overline{C}D$
81	F_3	$AC\overline{D} + \overline{A}B\overline{C} + \overline{A}\overline{B}C + \overline{A}BD$
59	F_4	$A\overline{C} + A\overline{D} + \overline{A}C + C\overline{D} + \overline{A}D + \overline{B}D + \overline{C}D$
55	F_5	$\overline{A}B + C\overline{D} + \overline{A}D$
53	F_6	$A\overline{B}CD$
47	F_7	$B\overline{D} + C\overline{D} + \overline{A}D + \overline{B}CD$
46	F_8	$AB\overline{C}\overline{D}$
46	F_9	$A + B + C + D$ (OR)
40	F_{10}	$A\overline{B} + A\overline{D} + \overline{A}B + B\overline{D} + \overline{A}D + \overline{B}D + \overline{C}D$
37	F_{11}	$A\overline{B}\overline{C}D$
37	F_{12}	$A\overline{D} + \overline{A}B + B\overline{C} + \overline{A}D + \overline{B}CD$
37	F_{13}	$A\overline{B} + A\overline{C} + A\overline{D} + \overline{A}D + \overline{B}D + \overline{C}D\overline{A}BC + BC\overline{D}$
32	F_{14}	$A\overline{D} + \overline{A}B + B\overline{D} + \overline{A}C + C\overline{D} + \overline{A}D + A\overline{B}\overline{C} + \overline{B}CD$
29	F_{15}	$\overline{C} + A\overline{B} + A\overline{D} + \overline{A}B + B\overline{D}AD + \overline{B}D$
28	F_{16}	$\overline{A}B + \overline{A}C + \overline{B}D + BC\overline{D} + A\overline{B}\overline{C}$

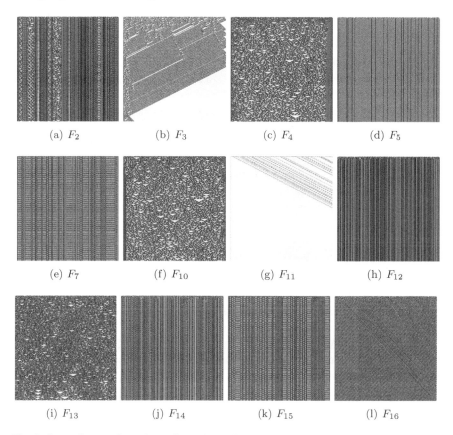

Fig. 5 Space-time configurations of one-dimensional cellular automata governed by functions from Table 1. An automaton has 500 cells and evolves for 500 iterations. Initial configurations has a random uniform distribution of cells in state '1' where each cell takes the state '1' with a probability $\frac{1}{2}$

the 'deflation' algorithm used in PNG lossless compression [46–48] is a variation of the classical Lempel–Ziv 1977 algorithm [49]. The frequency of the functions occurrence in the experimental circuit mining versus LZ complexity of the functions is shown in Fig. 6. We can see that there is no correlation between how often a function can be found and how complexity the function is. Thus, e.g. the function F_{13} (Table 1) generates most complex space-time configuration (Fig. 5i) yet it is in the mid-range of the frequency of experimental occurrence. The less complex functions F_5, F_7, F_{12}, F_{15} span the interval [29, 55] counts of occurrences in experimental laboratory mining.

Let us consider positions of the functions Table 1 in the Wolfram classification [50] of cellular automaton behaviour. Functions F_1, F_6, F_8, F_9 and F_11 belong to the class I, the class of automata exhibiting a dull dynamics and evolving to a stable state where all cells are in the same state. Functions F_2, F_7, F_{12}, F_{14}, F_{15} belong to the class

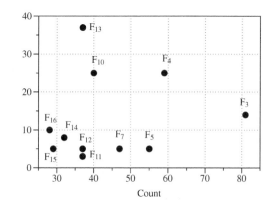

Fig. 6 Frequency of functions from Table 1 versus LZ complexity, measured via compressibility of the space-time configurations of cellular automata governed by the functions. Functions F_1, F_6, F_8 and F_9 are not displayed because their LZ is near zero

II: the automata fall into global cells do not update their state or update them cyclically from '0' to '1'. Functions F_4, F_{10} and F_{13} belong to class III: the space-time dynamics is characterised by quasi-random behaviour and difficult predictability of the successions of the global states. These functions generate the most complex, as evaluated by LZ measure, space-time configurations. Function F_2 shows an interesting example of the function belonging to classes II and III. Two functions F_3 and F_{16} belong to class IV: the space-time dynamics of automata show gliders (compact patterns translating in space) with non-trivial interactions between the gliders. The automata governed by rules F_3 and F_{16} are computationally universal, because it is possible to implement an arbitrary logical circuit via collisions between the gliders, see e.g. [51, 52].

4 Discussion

Mycelium bound composites transform electrical signals in a non-linear manner due to mem-fractive and capacitive properties of the fungal tissue [53]. Whilst exact biophysical mechanisms of the signal transformation by the mycelium remain unknown we can explore the non-linear properties of this living substrate to implement logical circuits. In experimental laboratory studies we demonstrated that mycelium bound composites implement a wide range of Boolean circuits. Analyses of the functions extracted in terms of space-time dynamics of cellular automata helped us to order the functions in several classes of complexity and pinpoint the functions supporting a universal computation. It would be possible to concatenate outputs from the different channels to create another layer of logic gate outputs. The current study looked at single output systems via SOP but the potential of using multiple outputs in parallel is there.

The first ever prototype of the fungal reservoir computer, presented in the chapter, demonstrates that a computation can be embedded into living materials. The research presented also pinpointed a high degree of variability in the logical circuits implemented by the fungi. This is because the live mycelium remain in the continuous process of growth and reconfiguration. To decrease the variability of the results we could consider to functionalise the mycelium networks with semi-conductive particles and polymers and allow the mycelium to dry. The resulting networks will have a permanent structure which will guarantee repeatability of the experimental circuits discovered. This will be a topic of our future studies.

References

1. Carlile, M.J., Watkinson, S.C., Gooday, G.W.: The Fungi. Gulf Professional Publishing (2001)
2. Smith, M.L., Bruhn, J.N., Anderson, J.B.: The fungus Armillaria bulbosa is among the largest and oldest living organisms. Nature, **356**(6368), 428 (1992)
3. Karana, E., Blauwhoff, D., Hultink, E.-J., Camere, S.: When the material grows: a case study on designing (with) mycelium-based materials. Int. J. Des. **12**(2) (2018)
4. Jones, M., Mautner, A., Luenco, S., Bismarck, A., John, S.: Engineered mycelium composite construction materials from fungal biorefineries: a critical review. Mater. Des. **187**, 108397 (2020)
5. Cerimi, K., Akkaya, K.C., Pohl, C., Schmidt, B., Neubauer, P.: Fungi as source for new bio-based materials: a patent review. Fungal Biol. Biotech. **6**(1), 1–10 (2019)
6. Adamatzky, A., Gandia, A., Ayres, P., Wösten, H., Tegelaar, M.: Adaptive fungal architectures. LINKs-series **5**, 66–77 (2021)
7. Pelletier, M.G., Holt, G.A., Wanjura, J.D., Bayer, E., McIntyre, G.: An evaluation study of mycelium based acoustic absorbers grown on agricultural by-product substrates. Ind. Crops Prod. **51**, 480–485 (2013)
8. Elsacker, E., Vandelook, S., Van Wylick, A., Ruytinx, J., De Laet, L., Peeters, E.: A comprehensive framework for the production of mycelium-based lignocellulosic composites. Sci. Total Environ. **725**, 138431 (2020)
9. Robertson, O., et al.: Fungal future: a review of mycelium biocomposites as an ecological alternative insulation material. In: DS 101: Proceedings of NordDesign 2020, Lyngby, Denmark, 12th-14th August 2020, pp. 1–13 (2020)
10. Yang, Z., Zhang, F., Still, B., White, M., Amstislavski, P.: Physical and mechanical properties of fungal mycelium-based biofoam. J. Mater. Civ. Eng. **29**(7), 04017030 (2017)
11. Xing, Y., Brewer, M., El-Gharabawy, H., Griffith, G., Jones, P.: Growing and testing mycelium bricks as building insulation materials. In: IOP Conference Series: Earth and Environmental Science, vol. 121, pp. 022032. IOP Publishing (2018)
12. Girometta, C., Picco, A.M., Baiguera, R.M., Dondi, D., Babbini, S., Cartabia, M., Pellegrini, M., Savino, E.: Physico-mechanical and thermodynamic properties of mycelium-based biocomposites: a review. Sustainability **11**(1), 281 (2019)
13. Dias, P.P., Jayasinghe, L.B., Waldmann, D.: Investigation of mycelium-miscanthus composites as building insulation material. Results Mater. **10**, 100189 (2021)
14. Wand, F., Li, H.-G., Kang, S.-S., Bai, Y.-F., Cheng, G.-Z., Zhang, G.-Q.: The experimental study of mycelium/expanded perlite thermal insulation composite material for buildings. Sci. Technol. Eng. **2016**, 20 (2016)
15. Cárdenas-R, J.P.: Thermal insulation biomaterial based on hydrangea macrophylla. In: Bio-Based Materials and Biotechnologies for Eco-Efficient Construction, pp. 187–201. Elsevier (2020)
16. Holt, G.A., Mcintyre, G., Flagg, D., Bayer, E., Wanjura, J.D., Pelletier, M.G.: Fungal mycelium and cotton plant materials in the manufacture of biodegradable molded packaging material:

evaluation study of select blends of cotton byproducts. J. Biobased Mater. Bioenergy **6**(4), 431–439 (2012)
17. Sivaprasad, S., Byju, S.K., Prajith, C., Shaju, J., Rejeesh, C.R.: Development of a novel mycelium bio-composite material to substitute for polystyrene in packaging applications. Mater. Today Proc. (2021)
18. Mojumdar, A., Behera, H.T., Ray, L.: Mushroom mycelia-based material: an environmental friendly alternative to synthetic packaging. Microb. Poly. 131–141 (2021)
19. Adamatzky, A., Nikolaidou, A., Gandia, A., Chiolerio, A., Dehshibi, M.M.: Reactive fungal wearable. Biosystems **199**, 104304 (2021)
20. Silverman, J., Cao, H., Cobb, K.: Development of mushroom mycelium composites for footwear products. Cloth. Text. Res. J. **38**(2), 119–133 (2020)
21. Appels, F.V.W.: The use of Fungal Mycelium for the Production of Bio-Based Materials. PhD thesis, Universiteit Utrecht (2020)
22. Jones, M., Gandia, A., John, S., Bismarck, A.: Leather-like material biofabrication using fungi. Nat. Sustain. 1–8 (2020)
23. Hitchcock, D., Glasbey, C.A., Ritz, K.: Image analysis of space-filling by networks: application to a fungal mycelium. Biotechnol. Tech. **10**(3), 205–210 (1996)
24. Giovannetti, M., Sbrana, C., Avio, L., Strani, P.: Patterns of below-ground plant interconnections established by means of arbuscular mycorrhizal networks. New Phytol. **164**(1), 175–181 (2004)
25. Fricker, M., Boddy, L., Bebber, D.: Network organisation of mycelial fungi. In: Biology of the Fungal Cell, pp. 309–330. Springer (2007)
26. Fricker, M.D., Heaton, L.L.M., Jones, N.S., Boddy, L.: The mycelium as a network. In: The Fungal Kingdom, pp. 335–367 (2017)
27. Islam, M.R., Tudryn, G., Bucinell, R., Schadler, L., Picu, R.C.: Morphology and mechanics of fungal mycelium. Sci. Rep. **7**(1), 1–12 (2017)
28. Obert, M., Pfeifer, P., Sernetz, M.: Microbial growth patterns described by fractal geometry. J. Bacteriol. **172**(3), 1180–1185 (1990)
29. Patankar, D.B., Liu, T.-C., Oolman, T.: A fractal model for the characterization of mycelial morphology. Biotechnol. Bioeng. **42**(5), 571–578 (1993)
30. Bolton, R.G., Boddy, L.: Characterization of the spatial aspects of foraging mycelial cord systems using fractal geometry. Mycol. Res. **97**(6), 762–768 (1993)
31. Mihail, J.D., Obert, M., Bruhn, J.N., Taylor, S.J.: Fractal geometry of diffuse mycelia and rhizomorphs of armillaria species. Mycol. Res. **99**(1), 81–88 (1995)
32. Boddy, L., Wells, J.M., Culshaw, C., Donnelly, D.P.: Fractal analysis in studies of mycelium in soil. Geoderma **88**(3), 301–328 (1999)
33. Papagianni, M.: Quantification of the fractal nature of mycelial aggregation in aspergillus niger submerged cultures. Microb. Cell Fact. **5**(1), 5 (2006)
34. Adamatzky, A., Tegelaar, M., Wosten, H.A.B., Powell, A.L., Beasley, A.E., Mayne, R.: On Boolean gates in fungal colony. Biosystems **193**, 104138 (2020)
35. Siccardi, S., Adamatzky, A.: Actin quantum automata: communication and computation in molecular networks. Nano Commun. Netw. **6**(1), 15–27 (2015)
36. Verstraeten, D., Schrauwen, B., d'Haene, M., Stroobandt, D.: An experimental unification of reservoir computing methods. Neural Netw. **20**(3), 391–403 (2007)
37. Lukoševičius, M., Jaeger, H.: Reservoir computing approaches to recurrent neural network training. Comput. Sci. Rev. **3**(3), 127–149 (2009)
38. Dale, M., Miller, J.F., Stepney, S.: Reservoir computing as a model for in-materio computing. In: Advances in Unconventional Computing, pp. 533–571. Springer (2017)
39. Konkoli, Z., Nichele, S., Dale, M., Stepney, S.: Reservoir computing with computational matter. In: Computational Matter, pp. 269–293. Springer (2018)
40. Dale, M., Miller, J.F., Stepney, S., Trefzer, M.A.: A substrate-independent framework to characterize reservoir computers. Proc. R. Soc. A **475**(2226), 20180723 (2019)
41. Miller, J.F., Downing, K.: Evolution in materio: looking beyond the silicon box. In: Proceedings 2002 NASA/DoD Conference on Evolvable Hardware, pp. 167–176. IEEE (2002)

42. Miller, J.F., Harding, S.L., Tufte, G.: Evolution-in-materio: evolving computation in materials. Evol. Intell. **7**(1), 49–67 (2014)
43. Stepney, S.: Co-designing the computational model and the computing substrate. In: International Conference on Unconventional Computation and Natural Computation, pp. 5–14. Springer (2019)
44. Miller, J.F., Hickinbotham, J.F., Amos, M.: In materio computation using carbon nanotubes. In: Computational Matter, pp. 33–43. Springer (2018)
45. Julian Francis Miller: The alchemy of computation: designing with the unknown. Nat. Comput. **18**(3), 515–526 (2019)
46. Roelofs, G., Koman, R.: PNG: The Definitive Guide. O'Reilly & Associates, Inc. (1999)
47. Howard, P.G.: The Design and Analysis of Efficient Lossless Data Compression Systems. PhD thesis, Citeseer (1993)
48. Deutsch, P., Gailly, J.: Zlib compressed data format specification version 3.3. Technical report, RFC 1950, May, 1996
49. Ziv, J., Lempel, A.: A universal algorithm for sequential data compression. IEEE Trans. Inf. Theory **23**(3), 337–343 (1977)
50. Wolfram, S.: Statistical mechanics of cellular automata. Rev. Mod. Phys. **55**(3), 601 (1983)
51. Martínez, G.J., Adamatzky, A., McIntosh, H.V.: Phenomenology of glider collisions in cellular automaton rule 54 and associated logical gates. Chaos, Solitons & Fractals **28**(1), 100–111 (2006)
52. Martínez, G.J., Adamatzky, A., Stephens, C.R., Hoeflich, A.F.: Cellular automaton supercolliders. Int. J. Modern Phys. C **22**(04), 419–439 (2011)
53. Beasley, A.E., Abdelouahab, M.-S., Lozi, R., Powell, A.L., Adamatzky, A.: Mem-fractive properties of mushrooms (2020). arXiv:2002.06413

Fungal Automata

Andrew Adamatzky, Eric Goles, Genaro J. Martínez, Michail-Antisthenis Tsompanas, Martin Tegelaar, and Han A. B. Wosten

Abstract We study a cellular automaton (CA) model of information dynamics on a single hypha of a fungal mycelium. Such a filament is divided in compartments (here also called cells) by septa. These septa are invaginations of the cell wall and their pores allow for flow of cytoplasm between compartments and hyphae. The septal pores of the fungal phylum of the Ascomycota can be closed by organelles called Woronin bodies. Septal closure is increased when the septa become older and when exposed to stress conditions. Thus, Woronin bodies act as informational flow valves. The one dimensional fungal automata is a binary state ternary neighbourhood CA, where every compartment follows one of the elementary cellular automata (ECA) rules if its pores are open and either remains in state '0' (first species of fungal automata) or its previous state (second species of fungal automata) if its pores are closed. The Woronin bodies closing the pores are also governed by ECA rules. We analyse a structure of the composition space of cell-state transition and pore-state transitions rules, complexity of fungal automata with just few Woronin bodies, and exemplify several important local events in the automaton dynamics.

1 Introduction

The fungal kingdom represents organisms colonising all ecological niches [1] where they play a key role [2–5]. Fungi can consist of a single cell, can form enormous underground networks [6] and can form microscopic fruit bodies or fruit bodies

A. Adamatzky (✉) · G. J. Martínez · M.-A. Tsompanas
Unconventional Computing Laboratory, UWE, Bristol, UK
e-mail: andrew.adamatzky@uwe.ac.uk

E. Goles
Faculty of Engineering and Science, University of Adolfo Ibáñez, Santiago, Chile

G. J. Martínez
High School of Computer Science, National Polytechnic Institute, Mexico City, Mexico

M. Tegelaar · H. A. B. Wosten
Microbiology Department, University of Utrecht, Utrecht, The Netherlands

© The Author(s), under exclusive license to Springer Nature Switzerland AG 2023
A. Adamatzky (ed.), *Fungal Machines*, Emergence, Complexity and Computation 47,
https://doi.org/10.1007/978-3-031-38336-6_22

weighting up to half a ton [7]. The underground mycelium network can be seen as a distributed communication and information processing system linking together trees, fungi and bacteria [8]. Mechanisms and dynamics of information processing in mycelium networks form an unexplored field with just a handful of papers published related to space exploration by mycelium [9, 10], patterns of electrical activity of fungi [11–13] and potential use of fungi as living electronic and computing devices [14–16].

Filamentous fungi grow by means of hyphae that grow at their tip and that branch sub-apically. Hyphae may be coenocytic or divided in compartments by septa. Filamentous fungi in the phylum *Ascomycota* have porous septa that allow for cytoplasmic streaming [17, 18]. Woronin bodies plug the pores of these septa after hyphal wounding to prevent excessive bleeding of cytoplasm [19–24]. In addition, they plug septa of intact growing hyphae to maintain intra-and inter-hyphal heterogeneity [25–28].

Woronin bodies can be located in different hyphal positions (Fig. 1a). When first formed, Woronin bodies are generally localised to the apex [29–31]. Subsequently, Woronin bodies are either transported to the cell cortex (*Neurospora crassa, Sordaria fimicola*) or to the septum (*Aspergillus oryzae, Aspergillus nidulans, Aspergillus fumigatus, Magnaporthe grisea, Fusarium oxysporum, Zymoseptoria tritici*) index-Aspergillus fumigatus) where they are anchored with a leashin tether and largely immobile until they are translocated to the septal pore due to cytoplasmic flow or ATP depletion [23, 24, 27, 29, 30, 32–35]. Woronin bodies that are not anchored at the cellular cortex or the septum, are located in the cytoplasm and are highly mobile (*Aspergillus fumigatus, Aspergillus nidulans, Zymoseptoria tritici*) [27, 29, 31]. Septal pore occlusion can be induced by bulk cytoplasmic flow [27] or developmental [36] and environmental cues, like puncturing of the cell wall, high temperature, carbon and nitrogen starvation, high osmolarity and low pH. Interestingly, high environmental pH reduces the proportion of occluded apical septal pores [28].

Aiming to lay a foundation of an emerging paradigm of fungal intelligence—distributed sensing and information processing in living mycelium networks—we decided to develop a formal model of mycelium and investigate a role of Woronin bodies in potential information dynamics in the mycelium.

2 Fungal Automata $\mathcal{M}$

A fungal automaton is a one-dimensional cellular automaton with binary cell states and ternary, including central cell, cell neighbourhood, governed by two elementary cellular automata (ECA) rules, namely the cell state transition rule f and the Woronin bodies adjustment rule g: $\mathcal{M} = \langle \mathbf{N}, u, \mathbf{Q}, f, g \rangle$. Each cell x_i has a unique index $i \in \mathbf{N}$. Its state is updated from $\mathbf{Q} = \{0, 1\}$ in discrete time depending of its current state x_i^t, the states of its left x_{i-1}^t and right neighbours x_{i+1}^t and the state of cell x's Woronin body w. Woronin bodies take states from $\mathbf{Q}$: $w^t = 1$ means Woronin bodies (Fig. 1) in cell x blocks the pores and the cell has no communication with its neighbours, and

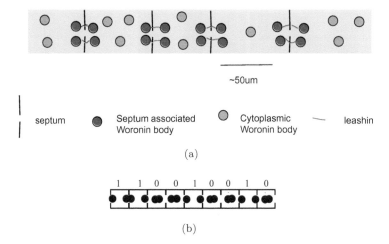

Fig. 1 a A biological scheme of a fragment of a fungal hypha of an ascomycete, where we can see septa and associated Woronin bodies. **b** A scheme representing states of Woronin bodies: '0' open, '1' closed.eps

$w^t = 0$ means that Woronin bodies in cell x do not block the pores. Woronin bodies update their states $g(\cdot)$, $w^{t+1} = g(u(x)^t)$, depending on the state of neighbourhood $u(x)^t$. Cells x update their states by function $f(\cdot)$ if their Woronin bodies do not block the pores (Fig. 2).

Two species of fungal automata are considered $\mathcal{M}_1$, where each cell updates its state as following:

$$x^{t+1} = \begin{cases} 0 & \text{if } w^t = 1 \\ f(u(x)^t) & \text{otherwise} \end{cases}$$

and $\mathcal{M}_2$ where each cell updates its state as following:

$$x^{t+1} = \begin{cases} x^t & \text{if } w^t = 1 \\ f(u(x)^t) & \text{otherwise} \end{cases}$$

where $w^t = g(u(x)^t)$.

State '1' in the cells of array x symbolises metabolites, signals exchanged between cells. Where pores in a cell are open the cell updates its state by ECA rule $f : \{0, 1\}^3 \to \{0, 1\}$.

In automaton $\mathcal{M}_1$, when Woronin bodies block the pores in a cell, the cell does not update its state and remains in the state '0' and left and right neighbours of the cells can not detect any 'cargo' in this cell. In automaton, $\mathcal{M}_2$, where Woronin bodies block the pores in a cell, the cell does not update its state and remains in its current state. In real living mycelium glucose and possibly other metabolites [26] can still

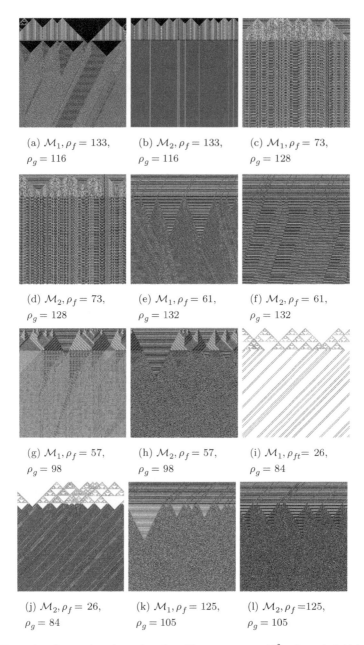

Fig. 2 Examples of space-time dynamics of $\mathcal{M}$. The automata are 10^3 cells each. Initial configuration is random with probability of a cell x to be in state '1', $x^0 = 1$, equals 0.01. Each automaton evolved for 10^3 iterations. Binary values of ECA rules f and g are shown in sub-captions. Rule g is applied to every iteration starting from 200th. Cells in state '0' are white, in state '1' are black, cells with Woronin bodies blocking pores are red. Indexes of cells increase from the left to the right, iterations are increasing from the to the bottom

cross the septum even when septa are closed by Woronin bodies, but we can ignore this fact in present abstract model.

Both species are biologically plausible and, thus, will be studied in parallel. The rules for closing and opening Woronin bodies are also ECA rules $g : \{0, 1\}^3 \to \{0, 1\}$. If $g(u(x)^t) = 0$ this means that pores are open, if $g(u(x)^t) = 1$ Woronin bodies block the pores. Examples of space-time configurations of both species of $\mathcal{M}$ are shown in Fig. 1.

3 Properties of Composition $f \circ g$

Predecessor sets

Let $\mathbf{F} = \{h : \{0, 1\}^3 \to \{0, 1\}\}$ be a set of all ECA functions. Then for any composition $f \circ g$, where $f, g \in \mathbf{F}$, can be converted to a single function $h \in \mathbf{F}$. For each $h \in \mathbf{F}$ we can construct a set $\mathbf{P}(h) = \{f \circ g \in \mathbf{F} \times \mathbf{F} \mid f \circ g \to h\}$. The sets $\mathbf{P}(h)$ for each $h \in \mathbf{F}$ are available online.[1]

A size of $\mathbf{P}(h)$ for each h is shown in Fig. 3c. The functions with largest size of $\mathbf{P}(h)$ are rule 0 in automaton $\mathcal{M}_1$ and rule 51 (only neighbourhood configurations (010, 011, 110, 111 are mapped into 1) in $\mathcal{M}_2$.

Size σ of $\mathbf{P}(h)$ vs a number γ of functions h having set $\mathbf{P}(h)$ of size σ is shown for automata $\mathcal{M}_1$ and $\mathcal{M}_2$ in Table 1a.

With regards to Wolfram classification [37], sizes of $\mathbf{P}(h)$ for rules from Class III vary from 9 to 729 in $\mathcal{M}_1$ (Table 1b). Rule 126 would be the most difficult to obtain in $\mathcal{M}_1$ by composition two ECA rules chosen at random, it has only 9 'predecessor' $f \circ g$ pairs. Rule 18 would be the easiest, for Class III rules, to be obtained, it has 729 predecessors, in both $\mathcal{M}_1$ (Table 1b) and $\mathcal{M}_2$ (Table 1d). In $\mathcal{M}_1$, one rule, rule 41, from the class IV has 243 $f \circ g$ predecessors, and all other rules in that class have 81 (Table 1c). From Class IV rule 54 has the largest number of predecessors in $\mathcal{M}_2$, and thus can be considered as most common (Table 1d).

Diagonals

In automaton $\mathcal{M}_1$ for any $f \in \mathbf{F}$ $f \circ f = 0$. Assume $f : \{0, 1\}^3 \to 1$ then Woronin bodies close the pores and, thus, second application of f produces state '0'. If $f : \{0, 1\}^3 \to 0$ then Woronin bodes does not close pores but yet second application of the f produce state '0'.

For automaton $\mathcal{M}_2$ a structure of diagonal mapping $f \circ f \to h$, where $f, h \in \mathbf{F}$ is shown in Table 2. The set of the diagonal outputs $f \circ f$ consists of 16 rules: (0, 1, 2, 3), (16, 17, 18, 19), (32, 33, 34, 35), (48, 49, 40, 51). These set of rules can be reduced to the following rule. Let $C(x^t) = [u(x)^t = (111)] \vee [u(x)^t = (111)]$ and $B(x^t) = [u(x)^t = (011)] \vee [u(x)^t = (010)]$. Then $x^t = 1$ if $C(x)^t \vee C(x)^t \wedge B(x^t)$.

---

[1] https://figshare.com/s/b7750ee3fe6df7cbe228.

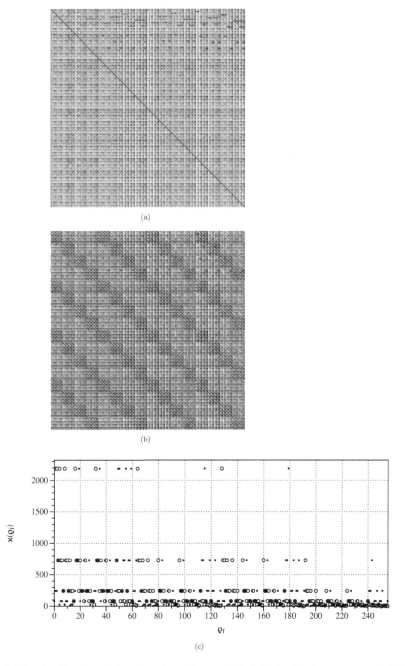

Fig. 3 Mapping $\mathbf{F} \times \mathbf{F} \to \mathbf{F}$ for automaton $\mathcal{M}_1$ **a** and $\mathcal{M}_2$ **b** is visualised as an array of pixels, $\mathbf{P} = (p)_{0 \leq \rho_f \leq 255, 0 \leq \rho_f \leq 255}$. An entry at the intersection of any ρ_f and ρ_g is a coloured as follows: red if $p_{\rho_f \rho_g} = p_{\rho_g \rho_f}$, blue if $\rho_g = p_{\rho_g \rho_f}$, green if $\rho_f = p_{\rho_g \rho_f}$. **c** Sizes of $\mathbf{P}(h)$ sets for $\mathcal{M}_1$, circle, and $\mathcal{M}_2$, solid discs, are shown for every function h apart of rule 0 ($\mathcal{M}_1$) and rule 51 ($\mathcal{M}_2$)

Fungal Automata 329

Table 1 Characterisations of automaton mapping $\mathbf{F} \times \mathbf{F} \to \mathbf{F}$. (a) Size σ of $\mathbf{P}(h)$ vs a number γ of functions h having set $\mathbf{P}(h)$ of size σ. T (b) Sizes of sets $\mathbf{P}(h)$ for rules from Wolfram class III. (b) Sizes of sets $\mathbf{P}(h)$ for rules from Wolfram class IV.

(a) Rules per $\|\mathbf{P}(h)\|$		(b) $\mathcal{M}_1$: Class III rules		(c) $\mathcal{M}_1$: Class IV rules	
σ	γ	Rule	σ	Rule	σ
1	1	18	729	41	243
3	8	22, 146	243	54, 106, 110	81
9	28	30, 45, 60, 90, 105, 150	81		
27	56	122	27		
81	70	126	9		
243	56				
729	28				
2187	8				
6561	1				
(d) $\mathcal{M}_2$: Class III rules		(e) $\mathcal{M}_2$: Class IV rules			
Rule	σ	Rule	σ		
18	729	41	243		
22, 146	243	54	729		
30, 45, 60 90, 105, 150	81	106	81		
122	243	110	27		
126	81				

Commutativity

In automaton $\mathcal{M}_1$, for any $f, g \in \mathbf{F}$ $f \circ g \neq g \circ f$ only if $f \neq g$. In automaton $\mathcal{M}_2$ there are 32768 pairs of function which $\circ$ is commutative, their distribution visualised in red in Fig. 3b.

Identities and zeros

In automaton $\mathcal{M}_1$ there are no left or right identities, neither right zeros in $\langle \mathbf{F}, \mathbf{F}, \circ \rangle$. The only left zero is the rule 0. In automaton $\mathcal{M}_2$ there are no identities or zeros at all.

Associativity

In automaton $\mathcal{M}_1$ there 456976 triples $\langle f, g, h \rangle$ on which operation $\circ$ is associative: $(f \circ g) \circ h = f \circ (g \circ h)$. The ratio of associative triples to the total number of triples is 0.027237892. There are 104976 associative triples in $\mathcal{M}_2$, a ratio of 0.006257057.

Table 2 Diagonals of automaton $\mathcal{M}_2$

$f \circ f$	f
0	0, 1, 2, 3, 16, 17, 18, 19, 32, 33, 34, 35, 48, 49, 50, 51
1	128, 129, 130, 131, 144, 145, 146, 147, 160, 161, 162, 163, 176, 177, 178, 179
2	64, 65, 66, 67, 80, 81, 82, 83, 96, 97, 98, 99, 112, 113, 114, 115
3	192, 193, 194, 195, 208, 209, 210, 211, 224, 225, 226, 227, 240, 241, 242, 243
16	8, 9, 10, 11, 24, 25, 26, 27, 40, 41, 42, 43, 56, 57, 58, 59
17	136, 137, 138, 139, 152, 153, 154, 155, 168, 169, 170, 171, 184, 185, 186, 187
18	72, 73, 74, 75, 88, 89, 90, 91, 104, 105, 106, 107, 120, 121, 122, 123
19	200, 201, 202, 203, 216, 217, 218, 219, 232, 233, 234, 235, 248, 249, 250, 251
32	4, 5, 6, 7, 20, 21, 22, 23, 36, 37, 38, 39, 52, 53, 54, 55
33	132, 133, 134, 135, 148, 149, 150, 151, 164, 165, 166, 167, 180, 181, 182, 183
34	68, 69, 70, 71, 84, 85, 86, 87, 100, 101, 102, 103, 116, 117, 118, 119
35	196, 197, 198, 199, 212, 213, 214, 215, 228, 229, 230, 231, 244, 245, 246, 247
48	12, 13, 14, 15, 28, 29, 30, 31, 44, 45, 46, 47, 60, 61, 62, 63
49	140, 141, 142, 143, 156, 157, 158, 159, 172, 173, 174, 175, 188, 189, 190, 191
50	76, 77, 78, 79, 92, 93, 94, 95, 108, 109, 110, 111, 124, 125, 126, 127
51	204, 205, 206, 207, 220, 221, 222, 223, 236, 237, 238, 239, 252, 253, 254, 255

4 Turing Complexity: Rule 110

To evaluate on how introduction of Woronin bodies could affect complexity of automaton evolution, we undertook two series of experiments. In the first series we used fungal automaton where just one cell has a Woronin body (Fig. 5). In the second series we employed fungal automaton where regularly positioned cells (but not all cells of the array) have Woronin bodies.

State transition functions g of Woronin bodies were varied across the whole diapason but the state transition function f of a cell was Rule 110, $\rho_f = 110$. We have chosen Rule 110 because the rule is proven to be computationally universal [38, 39], P-complete [40], the rules belong to Wolfram class IV renown for exhibiting complex and non-trivial interactions between travelling localisation [41], rich families of gliders can be produce in collisions with other gliders [42–44].

We wanted to check how an introduction of Woronin bodies affect dynamics of most complex space-time developed of Rule 110 automaton. Thus, we evolved the automata from all possible initial configurations of 8 cells placed near the end of $n = 1000$ cells array of resting cells and allowing to evolve for 950 iterations. Lempel–Ziv complexity (compressibility) LZ was evaluated via sizes of space-time configurations saved as PNG files. This is sufficient because the 'deflation' algorithm used in PNG lossless compression [45–47] is a variation of the classical Lempel–Ziv 1977 algorithm [48]. Estimates of LZ complexity for each of 8-cell initial configurations are shown in Fig. 4a. The initial configurations with highest estimated LZ

Fungal Automata

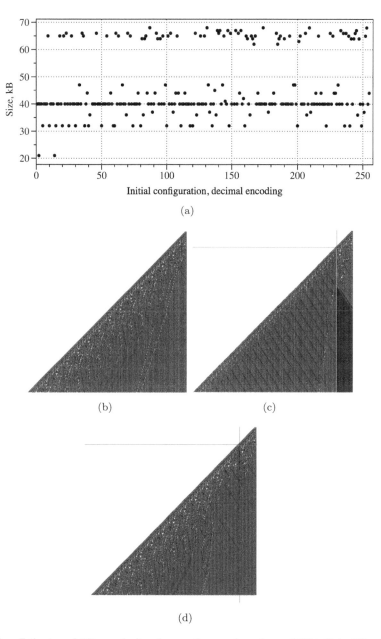

Fig. 4 **a** Estimates of LZ complexity of space-time configurations of ECA Rule 110 without Woronin bodies. **b** A space-time configuration of ECA Rule 110 evolving from initial configuration 10110001 (177), no Woronin bodies are activated. **c** A space-time configuration of $\mathcal{M}_1$ Rule 110 evolving from initial configuration 10110001 (177), Woronin body is governed by rule 43; red lines indicate time when the body was activated and position of the cell with the body. In **bcd**, a pixel in position (i, t) is black if $x_i^t = 1$

Fig. 5 Only one cell has has two Woronin bodies by which it can close itself from the other compartments

complexity are 10110001 (decimal 177), 11010001 (209), 10000011 (131), 11111011 (253), see example of space-time dynamics in Fig. 4b.

We assumed that a cell in the position $n - 100$ has a Woronin body which can be activated (Fig. 5), i.e. start updating its state by rule f, after 100th iteration of the automaton evolution. We then run 950 iteration of automaton evolution for each of 256 Woronin rules and estimated LZ complexity. In experiments with $\mathcal{M}_1$ we found that 128 rules, with even decimal representations, do not affect space time dynamics of evolution and 128 rules, with even decimal representations, reduce complexity of the space-time configuration. The key reasons for the complexity reduction (compare Fig. 4b and c) are cancellation of three gliders at c. 300th iteration and simplification of the behaviour of glider guns positioned at the tail of the propagating wave-front. In experiments with $\mathcal{M}_2$ 128 rules, with even decimal representations, do not change the space-time configuration of the author. Other 128 rules reduce complexity and modify space-time configuration by re-arranging the structures of glider guns and establishing one oscillators at the site surrounding position of the cell with Woronin body (Fig. 4d).

In second series of experiments we regularly positioned cells with Woronin bodies along the 1D array: every 50th cell has a Woronin body. Then we evolved fungal automata $\mathcal{M}_1$ and $\mathcal{M}_2$ from exactly the same initial random configuration with density of '1' equal to 0.3. Space-time configuration of the automaton without Woronin bodies is shown in Fig. 6a. Exemplar of space-time configurations of automata with Woronin bodies are shown in Fig. 6b–h. As seen in Fig. 7 both species of fungal automata show similar dynamics of complexity along the Woronin transition functions ordered by their decimal values. The automaton $\mathcal{M}_1$ has average LZ complexity 82.2 ($\overline{\sigma} = 24.6$) and the automaton $\mathcal{M}_2$ 78.4 ($\overline{\sigma} = 22.1$). Woronin rules g which generate most LZ complex space-time configurations are $\rho_g = 133$ in $\mathcal{M})_1$ (Fig. 6b) and $\rho_g = 193$ in $\mathcal{M})_2$ (Fig. 6e). The space-time dynamics of the automaton is characterised by a substantial number of gliders guns and gliders (Fig. 6b). Functions being in the middle of the descending hierarchy of LZ complexity produce space-time configurations with declined number of travelling localisation and growing domains of homogeneous states (Fig. 6cg). Automata with Woronin functions at the bottom of the complexity hierarchy quickly (i.e. after 200–300 iterations) evolve towards stable, equilibrium states (Fig. 6dh).

Fig. 6 a ECA Rule 110, no Woronin bodies. Space-time evolution of $\mathcal{M}_\infty$ (bcd) and $\mathcal{M}_\in$ (e–h) for Woronin rules shown in subcaption. LZ complexity of space-time configurations decreases from **b** to **d** and from **e** to **h**. Every 50th cell has a Woronin body

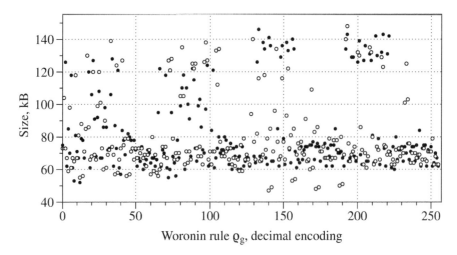

Fig. 7 Estimations of LZ complexity of space-time, 500 cells by 500 iterations, configurations of $\mathcal{M}_1$, discs, and $\mathcal{M}_1$, circles, for all Woronin functions g

5 Local Events

Let us consider some local events happening in the fungal automata discussed in Sect. 4: every 50th cell of an array has a Woronin body.

Retaining gliders. A glider can be stopped and converted into a station localisation by a cell with Woronin body. As exemplified in Fig. 8a, the localisation travelling left was stopped from further propagation by a cell with Woronin body yet the localisation did not annihilate but remained stationary.

Register memory. Different substrings of input string (initial configuration) might lead to different equilibrium configurations achieved in the domains of the array separated by cells with Woronin bodies. When there is just two types of equilibrium configurations they be seen as 'bit up' and 'bit down' and therefore such fungal automaton can be used a memory register (Fig. 8b).

Reflectors. In many cases cells with Woronin bodies induce local domains of stationary, sometimes time oscillations, inhomogeneities which might act as reflectors for travelling localisations. An example is shown in Fig. 8c where several localisations are repeatedly bouncing between two cells with Woronin bodies.

Modifiers. Cells with Woronin bodies can act as modifiers of states of gliders reflected from them and of outcomes of collision between travelling localizations. In Fig. 8d we can see how a travelling localisation is reflected from the vicinity of Woronin bodies three times: every time the state of the localisation changes. On the third reflection the localisation becomes stationary. In the fragment (Fig. 8e) of space-time configuration of automaton with Woronin bodies governed by $\rho_g = 201$ of the fragment we can see how two localisations got into contact with each in the

Fungal Automata

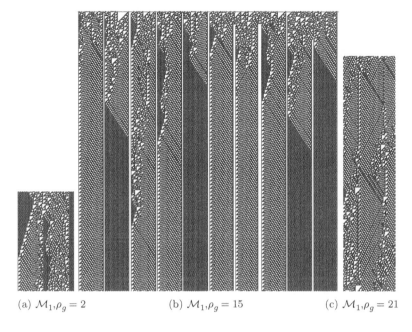

(a) $\mathcal{M}_1, \rho_g = 2$ (b) $\mathcal{M}_1, \rho_g = 15$ (c) $\mathcal{M}_1, \rho_g = 21$

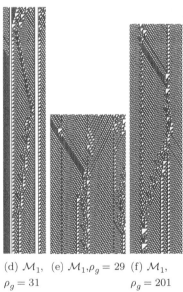

(d) $\mathcal{M}_1$, $\rho_g = 31$ (e) $\mathcal{M}_1, \rho_g = 29$ (f) $\mathcal{M}_1$, $\rho_g = 201$

Fig. 8 **a** Localisation travelling left was stopped by the Woronin body. **b** Analog of a memory register. **c** Reflections of travelling localisations from cells with Woronin bodies. **d** Modification of glider state in the vicitinity of Woronin bodies. **e** A fragment of configuration of automaton with $\rho_g = 29$, left cell states, right Woronin bodies states. **f** Enlarged sub-fragment of the fragment **d** where Wonorin body tunes the outcome of the collision. For both automata $\rho_f = 110$

Fig. 9 An example of 5-inputs-7-outputs collision in $\mathcal{M}_2$, $\rho_f = 110$, $\rho_g = 40$. Every 50th cell has a Woronin body. Cells state transitions are shown on the left, Woronin bodies state transitions on the right. A pixel in position (i, t) is black if $x_i^t = 1$, left, or $w_i^t = 1$, right

vicinity of the Woronin body and an advanced structure is formed two breathing stationary localisations act as mirror, and there are streams of travelling localisations between them. A multi-step chain reaction can be observed in Fig. 8f: there are two stationary, breathing, localisations at the sites of the cells with Woronin bodies. A glider is formed on the left stationary localisation. The glider travel to the right and collide into right breather. In the result of the collision the breath undergoes structural transitions, emits a glider travelling left and transforms itself into a pair of stationary breathers. Meantime the newly born glider collided into left breather and changes its state.

6 Discussion

As a first step towards formalisation of fungal intelligence we introduced one-dimensional fungal automata operated by two local transition function: one, g, governs states of Woronin bodies (pores are open or closed), another, f, governs cells states: '0' and '1'. We provided a detailed analysis of the magma $\langle f, g, \circ \rangle$, results of which might be useful for future designs of computational and language recognition structures with fungal automata. The magma as a whole does not satisfy any other property but closure. Chances are high that there are subsets of the magma which might satisfy conditions of other algebraic structures. A search for such subsets could be one of the topics of further studies.

Another topic could be an implementation of computational circuits in fungal automata. For certain combination of f and g we can find quite sophisticated families of stationary and travelling localisations and many outcomes of the collisions and interactions between these localisations, an illustration is shown in Fig. 9. Thus the target could be, for example, to construct a n-binary full adder which is as compact in space and time as possible.

The theoretical results reported show that by controlling just a few cells with Woronin bodies it is possible to drastically change dynamics of the automaton array. Third direction of future studies could be in implemented information processing in a single hypha. In such a hypothetical experimental setup input strings will be represented by arrays of illumination and outputs could be patterns of electrical activity recorded from the mycelium hypha resting on an electrode array.

References

1. Carlile, M.J., Watkinson, S.C., Gooday, G.W.: The Fungi. Gulf Professional Publishing (2001)
2. Griffin, D.M., et al.: Ecology of soil fungi. Ecol. Soil Fungi. (1972)
3. Cooke, R.C., Rayner, A.D.M., et al.: Ecology of Saprotrophic Fungi. Longman (1984)
4. Rayner, A.D.M., Boddy, L., et al.: Fungal Decomposition of Wood. Its Biology and Ecology. John Wiley & Sons Ltd. (1988)
5. Christensen, M.: A view of fungal ecology. Mycol. **81**(1), 1–19 (1989)

6. Smith, M.L., Bruhn, J.N., Anderson, J.B.: The fungus Armillaria bulbosa is among the largest and oldest living organisms. Nat. **356**(6368), 428 (1992)
7. Dai, Y.-C., Cui, B.-K.: Fomitiporia ellipsoidea has the largest fruiting body among the fungi. Fungal biology **115**(9), 813–814 (2011)
8. Bonfante, P., Anca, I.-A.: Plants, mycorrhizal fungi, and bacteria: a network of interactions. Annu. Rev. Microbiol. **63**, 363–383 (2009)
9. Held, M., Edwards, C., Nicolau, D.V.: Fungal intelligence; or on the behaviour of microorganisms in confined micro-environments. J. Phys. Conf. Ser. **178**, 012005. IOP Publishing (2009)
10. Held, M., Edwards, C., Nicolau, D.V.: Examining the behaviour of fungal cells in microconfined mazelike structures. In: Imaging, Manipulation, and Analysis of Biomolecules, Cells, and Tissues VI, vol. 6859, p. 68590U. International Society for Optics and Photonics (2008)
11. Slayman, C.L., Long, W.S., Gradmann, D.: "Action potentials" in Neurospora crassa, a mycelial fungus. Biochimica et Biophysica Acta (BBA)—Biomembranes **426**(4), 732–744 (1976)
12. Olsson, S., Hansson, B.S.: Action potential-like activity found in fungal mycelia is sensitive to stimulation. Naturwissenschaften **82**(1), 30–31 (1995)
13. Adamatzky, A.: On spiking behaviour of oyster fungi Pleurotus djamor. Sci. Rep. 7873 (2018)
14. Adamatzky, A., Tuszynski, J., Pieper, J., Nicolau, D.V., Rinalndi, R., Sirakoulis, G., Erokhin, V., Schnauss, J., Smith, D.M.: Towards cytoskeleton computers. A proposal. In: Adamatzky, A., Akl, S., Sirakoulis, G. (eds.) From parallel to emergent computing. CRC Group/Taylor & Francis (2019)
15. Adamatzky, A., Tegelaar, M., Wosten, H.A.B., Powell, A.L., Beasley, A.E., Mayne, R.: On Boolean gates in fungal colony. Biosyst. **193**, 104138 (2020)
16. Beasley, A.E., Powell, A.L., Adamatzky, A.: Memristive properties of mushrooms (2020). arXiv:2002.06413
17. Moore, R.T., McAlear, J.H.: Fine structure of Mycota. 7. observations on septa of ascomycetes and basidiomycetes. Am. J. Bot. **49**(1), 86–94 (1962)
18. Lew, R.R.: Mass flow and pressure-driven hyphal extension in Neurospora crassa. Microbiol.**151**(8), 2685–2692 (2005)
19. Trinci, A.P.J., Collinge, A.J.: Occlusion of the septal pores of damaged hyphae ofNeurospora crassa by hexagonal crystals. Protoplasma. **80**(1–3), 57–67 (1974)
20. Collinge, A.J., Markham, P.: Woronin bodies rapidly plug septal pores of severedpenicillium chrysogenum hyphae. Exp. Mycol. **9**(1), 80–85 (1985)
21. Jedd, G., Chua, N.-H.: A new self-assembled peroxisomal vesicle required for efficient resealing of the plasma membrane. Nat. Cell Biol. **2**(4), 226–231 (2000)
22. Tenney, K., Hunt, I., Sweigard, J., Pounder, J.I., McClain, C., Bowman, E.J., Bowman, B.J.: Hex-1, a gene unique to filamentous fungi, encodes the major protein of the Woronin body and functions as a plug for septal pores. Fungal Genet. Biol. **31**(3), 205–217 (2000)
23. Soundararajan, S., Jedd, G., Li, X., Ramos-Pamploña, M., Chua, N.H., Naqvi,N.I.: Woronin body function in Magnaporthe grisea is essential for efficient pathogenesis and for survival during nitrogen starvation stress. Plant Cell. **16**(6), 1564–1574 (2004)
24. Maruyama, J.-I., Juvvadi, P.R., Ishi, K., Kitamoto, K.: Three-dimensional image analysis of plugging at the septal pore by Woronin body during hypotonic shock inducing hyphal tip bursting in the filamentous fungus aspergillus oryzae. Biochem. Biophys. Res. Commun. **331**(4), 1081–1088 (2005)
25. Bleichrodt, R.-J., van Veluw, G.J., Recter, B., Maruyama, J.-I., Kitamoto, K., Wösten, H.A.B.: Hyphal heterogeneity in Aspergillus oryzae is the result of dynamic closure of septa by Woronin bodies. Mol. Microbiol. **86**(6), 1334–1344 (2012)
26. Bleichrodt, R.-J., Hulsman, M., Wösten, H.A.B., Reinders, M.J.T.: Switching from a unicellular to multicellular organization in an Aspergillus niger hypha. MBio. **6**(2), e00111–15 (2015)
27. Steinberg, G., Harmer, N.J., Schuster, M., Kilaru, S.: Woronin body-based sealing of septal pores. Fungal Genet. Biol. **109**, 53–55 (2017)
28. Tegelaar, M., Bleichrodt, R.-J., Nitsche, B., Ram, A.F.J., Wösten, H.A.B.: Subpopulations of hyphae secrete proteins or resist heat stress in Aspergillus oryzae colonies. Environ. Microbiol. **22**(1), 447–455 (2020)

29. Momany, M., Richardson, E.A., Van Sickle, C., Jedd, G.: Mapping Woronin body position in Aspergillus nidulans. Mycol. **94**(2), 260–266 (2002)
30. Tey, W.K., North, A.J., Reyes, J.L., Lu, Y.F., Jedd, G.: Polarized gene expression determines Woronin body formation at the leading edge of the fungal colony. Mol. Biol. Cell **16**(6), 2651–2659 (2005)
31. Beck, J., Ebel, F.: Characterization of the major Woronin body protein HexA of the human pathogenic mold Aspergillus fumigatus. Int. J. Med. Microbiol. **303**(2), 90–97 (2013)
32. Ng, S.K., Liu, F., Lai, J., Low, W., Jedd, G.: A tether for Woronin body inheritance is associated with evolutionary variation in organelle positioning. PLoS Genet. **5**(6), (2009)
33. Wergin, W.P.: Development of Woronin bodies from microbodies infusarium oxysporum f. sp. lycopersici. Protoplasma. **76**(2), 249–260 (1973)
34. Leonhardt, Y., Kakoschke, S.C., Wagener, J., Ebel, F.: Lah is a transmembrane protein and requires spa10 for stable positioning of Woronin bodies at the septal pore of Aspergillus fumigatus. Sci. Rep. **7**, 44179 (2017)
35. Berns, M.W., Aist, J.R., Wright, W.H., Liang, H.: Optical trapping in animal and fungal cells using a tunable, near-infrared titanium-sapphire laser. Exp. Cell Res. **198**(2), 375–378 (1992)
36. Bleichrodt, R.-J., Vinck, A., Read, N.D., Wösten, H.A.B.: Selective transport between heterogeneous hyphal compartments via the plasma membrane lining septal walls of Aspergillus niger. Fungal Genet. Biol. **82**, 193–200 (2015)
37. Wolfram, S.: Cellular Automata and Complexity: Collected Papers. Addison-Wesley Pub. Co. (1994)
38. Lindgren, K., Nordahl, M.G.: Universal computation in simple one-dimensional cellular automata. Complex Syst. **4**(3), 299–318 (1990)
39. Cook, M.: Universality in elementary cellular automata. Complex Syst. **15**(1), 1–40 (2004)
40. Neary, T., Woods, D.: P-completeness of cellular automaton rule 110. In: International Colloquium on Automata, Languages, and Programming, pp. 132–143. Springer (2006)
41. Wolfram, S.: Universality and complexity in cellular automata. Phys. D Nonlinear Phenom. **10**(1–2), 1–35 (1984)
42. Martínez, G.J., McIntosh, H.V., Mora, J.C.S.-T.: Production of gliders by collisions in rule 110. In: European Conference on Artificial Life, pp. 175–182. Springer (2003)
43. Martínez, G.J., McIntosh, H.V., Mora, J.C.S.-T.: Gliders in rule 110. Int. J. Unconv. Comput. **2**(1), 1 (2006)
44. Martínez, G.J., Mora, J.C.S.-T., Vergara, S.V.C.: Rule 110 objects and other collision-based constructions. J. Cell. Autom. **2**, 219–242 (2007)
45. Roelofs, G., Koman, R.: PNG: the Definitive Guide. O'Reilly & Associates, Inc. (1999)
46. Howard, P.G.: The design and analysis of efficient lossless data compression systems. Ph.D. thesis, CiteSeer (1993)
47. Deutsch, P., Gailly, J.: Zlib compressed data format specification version 3.3. Technical report, RFC 1950, May (1996)
48. Ziv, J., Lempel, A.: A universal algorithm for sequential data compression. IEEE Trans. Inf. Theory. **23**(3), 337–343 (1977)

Exploring Dynamics of Fungal Cellular Automata

Carlos S. Sepúlveda, Eric Goles, Martín Ríos-Wilson, and Andrew Adamatzky

Abstract Cells in a fungal hyphae are separated by internal walls (septa). The septa have tiny pores that allow cytoplasm flowing between cells. Cells can close their septa blocking the flow if they are injured, preventing fluid loss from the rest of filament. This action is achieved by special organelles called Woronin bodies. Using the controllable pores as an inspiration we advance one and two-dimensional cellular automata into Elementary fungal cellular automata (EFCA) and Majority fungal automata (MFA) by adding a concept of Woronin bodies to the cell state transition rules. EFCA is a cellular automaton where the communications between neighboring cells can be blocked by the activation of the Woronin bodies (Wb), allowing or blocking the flow of information (represented by a cytoplasm and chemical elements it carries) between them. We explore a novel version of the fungal automata where the evolution of the system is only affected by the activation of the Wb. We explore two case studies: the Elementary Fungal Cellular Automata (EFCA), which is a direct application of this variant for elementary cellular automata rules, and the Majority Fungal Automata (MFA), which correspond to an application of the Wb to two dimensional automaton with majority rule with Von Neumann neighborhood. By studying the EFCA model, we analyze how the 256 elementary cellular automata rules are affected by the activation of Wb in different modes, increasing the complexity on applied rule in some cases. Also we explore how a consensus over MFA is affected when the continuous flow of information is interrupted due to the activation of Woronin bodies.

C. S. Sepúlveda · E. Goles · M. Ríos-Wilson
Facultad de Ingeniería y Ciencias, Universidad Adolfo Ibáñez, Santiago, Chile

A. Adamatzky (✉)
Unconventional Computing Laboratory, UWE, Bristol, UK
e-mail: andrew.adamatzky@uwe.ac.uk

1 Introduction

The fungi kingdom is one of the widest spread form of life on earth. Without them our planet landscape would be totally different. Life on land has evolved with the participation of fungi and would collapse without their continued activities [1]. Fungal morphology is based on hyphae, which are long and branching filaments. Collectively called mycelium, they form the vegetative body on fungi. Hyphae are formed by one or more cells enclosed by a tubular cell wall. In Ascomycota, one of the several divisions of the fungi, hyphae are divided into compartments by internal cross-walls named septum, formed by centripetal growth of the cell wall and crossed by a perforation through which cytoplasmic organelles can pass. An electron-dense protein body called Woronin body (Wb), is present on either side of the septa, regulating the opening and closing of the septal pores which is used to reduce or cut the flow of cytoplasm and organelles between cells compartments when the hypha is ruptured [2].

Fungal physiology and behaviour gave rise to a novel field of fungal computing and fungal electronics [3–5]. Whilst experimental laboratory prototyping of fungi-based computing devices is underway it is imperative to establish a wider theoretical background for fungal computing. This is why we drawn our attention to developing formal models of fungal automata. First steps in these theoretical designs have been done in [6–8]. We advance the ideas into the dynamical properties of a model based on the elementary cellular automata (ECA). This is a classic model that describes a vast variety of natural dynamical phenomena over the years. One of the aspect that is most interesting of ECA model is that, even when it is based in a simple set of local rules that can be quite straightforward, its global dynamical behavior can be extremely complex. In fact, since the end of the 1970's, the behavioural complexity emerging from normative simplicity motivated a wide range of researcher to understand and classify ECA rules according to their dynamic behavior and computational capabilities [9, 10]. A first well known example is Wolfram's classification which considers different criteria to cluster rules in four groups according to their dynamical behavior starting from random initial conditions [11]. A second way of grouping the elementary rules is according to their equivalences up to simple transformations of their local transitions, such as *reflection*, *conjugation* and the combination of both. There are 88 elementary cellular automata that are non-equivalent up to these transformations. A third important example of rule classification is in the study of the computational capabilities of a given rule [12–14]. For instance, rule 110 is capable of representing universal Turing computation in its dynamics [15]. Within this framework, different cellular automata models, inspired from fungi hyphae behavior have been recently proposed. For instance, in [8] the one-dimensional fungal automaton has been introduced. This model is based on the composition of two elementary cellular automaton functions, one controlling the activation state of the Wb and the other controlling the automaton evolution. In addition, in [6], the authors have extended the concept to two dimensions implementing a fungal sandpile automata, and shown the computational universality of this FCA.

We present novel variant of the one-dimensional fungal automaton model, called elementary fungal cellular automata (EFCA). By numerical simulations of this model, we exhaustively explore the impact of Wb activation on the dynamics of 88 non-equivalent ECA rules. In addition, based on the fungal sandpile automata, we develop a two dimensional cellular automaton ruled by the well-known majority rule cellular automaton with the Von-Neumann neighborhood. We focus on observing how different choices for the activation of Wb can produce different dynamical behavior on both cases. We accomplish this task by proposing different metrics such as the magnetization and Hamming distance in order to compare the original rules with the ones in which the Wb are activated. Finally, for majority rule, we study how the besides Wb activation, which temporally block information flow the consensus of the network is not affected for strict majority rule and the consensus skew is accelerated in the skew majority rule.

2 Preliminaries

2.1 *Elementary Fungal Cellular Automata*

An elementary fungal cellular automaton (EFCA) is an elementary cellular automata (ECA) where adjacent cells can cut the flow of information[1] between them by the activation of Wb (see Fig. 1). This means that cells with activated Wb can't see the current state of their neighbors, which produces a miss information and ambiguity in applying the ECA rules and updating the cells' states.

FCA were first proposed on [8], implemented as one-dimensional cellular automata[2] with cell binary state and governed by two ECA rules, one for the cell state transition rule $f(\cdot)$ and other the activation of Wb $g(\cdot)$. In this manner, they implemented two species of FCA on which the activation of Wb is given by $w^{t+1} = g\left(u(x)^t\right)$, where $u(x)^t$ represents the neighborhood of cell x at instant t and the update of cell x for the first specie, is given by Eq. 1 and for the second specie by Eq. 2.

$$x_i^{t+1} = \begin{cases} 0 & \text{if } w^t = 1 \\ f\left(u(x_i)^t\right) & \text{otherwise} \end{cases} \quad (1)$$

$$x_i^{t+1} = \begin{cases} x^t & \text{if } w^t = 1 \\ f\left(u(x_i)^t\right) & \text{otherwise} \end{cases} \quad (2)$$

Our approach is different. First the Wb state is not governed by current cell state, instead is externally controlled, having states $Q = \{0, 1\}$, representing deactivation

---

[1] The rule according which the automaton changes his current state.
[2] A ring of size N which update his current state according to a rule.

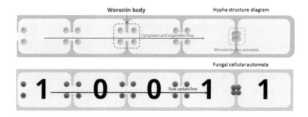

Fig. 1 Fungal cellular automata structure based on fungal hypha. At the top is the biological scheme of the hypha. At the bottom is a one-dimensional FCA, where each cell has a Woronin body (Wb) that can be activate/deactivated blocking or allowing to see the current state value of his neighbors

and activation respectively. Our formulation for the update of cell state is the following: Each cell x_i has a unique index $i \in \{0, 1, \ldots, n\}$ in a ring of size n, with $n \in \mathbb{N}$, and a Wb Wb_i. If the Wb is activated, $Q : Wb_i = 1$, there is no communication between cell x_i and cell x_{i+1}. Special case is when $i = n$, due the implementation is in a ring there will be no communication between x_n and x_0. Every cell x_i has two neighbors at distance 1, x_{i-1} to the left and x_{i+1} to the right. The function $f(\cdot)$ is then applied to the triplet which leads to a traditional ECA. Considering four cells $(x_{i-1}, x_i, x_{i+1}, x_{i+2})$ and the activation of Woronin body at position i, when the rule $f(x_{i-1}, x_i, x_{i+1})$ is applied we will not able to see the content of cell at $i + 1$ position due to the Wb activation and therefore we will face an ambiguity of information at the left of the Woronin body activation. Then, applying the rule to the next triplet $f(x_i, x_{i+1}, x_{i+2})$ will produce an ambiguity information to the right of Wb activation. We can manage the information ambiguity in both sides of Wb activation when $f(x_{i-1}, x_i, 0) = f(x_{i-1}, x_i, 1)$ for rules at left of activated Wb and $f(0, x_{i+1}, x_{i+2}) = f(1, x_{i+1}, x_{i+2})$ if not, we remain in actual state. The explained behavior is equivalent to application of the rule at left side of Wb activation and other, different, rule at the right side of Wb. The formalization of this behavior, that describes the evolution of EFCA from instant t to instant $t + 1$ is given by Eq. 3, as follow:

$$x_i^{t+1} = \begin{cases} x_i^t & \text{if } w_i^t = 1 \text{ and } f\left(x_{i-1}^t, x_i^t, 1\right) \neq f\left(x_{i-1}^t, x_i^t, 0\right) \\ & \text{or} \\ & \text{if } w_i^t = 1 \text{ and } f\left(1, x_i^t, x_{i+1}^t\right) \neq f\left(0, x_i^t, x_{i+1}^t\right) \\ f(x_{i-1}^t, x_i^t, x_{i+1}^t) & \text{otherwise} \end{cases} \quad (3)$$

Here $f(\cdot)$ is the ECA rule. In cell whose Wb is not activated, $Q : Wb_i = 0$, the state update will be given by $x_i^{t+1} = f(x_{i-1}^t, x_i^t, x_{i+1}^t)$ (his neighbors at distance 1).

2.2 Majority Fungal Automata

A two dimensional cellular automaton is a torus in which the value of a cell $x_{(i,j)}$ will be 1 or 0 depending on the value of a function $F(\cdot)$ whose apply over neighbors of $x_{(i,j)}$. We use von Neumann neighborhood of range 1 which is defined as follows:

$$V(i_0, j_0) = \{(i, j) : |i - i_0| + |j - j_0| \leq 1\} \qquad (4)$$

The function $F(\cdot)$ determines the value of a cell based on his neighbors is the majority, meaning that the value of cell $x_{(i,j)}$ will be (zero or one) the most frequent value among neighborhood. In other words, the opinion of a subject will be the opinion of majority. Two types of majority are possible depending on how we deal with ties (equal number of zeros and ones among neighbors). We will refer to skew majority when in presence of a tie we force the value of cell $x_{(i,j)}$ to one (or zero depending on how the skew is defined). We will use majority when in presence of the tie the cell value will remain in his current state (no change of opinion). The study of this type of totalistic function is useful to understand consensus achievable over a network.

In Majority fungal automaton (MFA) each cell $x_{(i,j)}$ has horizontal and vertical Wbs $Wb_{(i,j)}$ which can be activated by analogy with Wb activated in EFCA. The activation modifies neighbors that cell $x_{(i,j)}$ can see (Fig. 2). Following the implementation used in [6] we activated Wb groups by rows or columns. In this way, each cell has a state $Q : \{0, 1\}$ which is updated according to a global function $F(\cdot)$. At each time step t we could open or close all horizontal or vertical Wb, thus the MFA becomes a tuple $MFA = \langle \mathbb{Z}^2, Q, V, F, w \rangle$, where V is the von Neumann neighborhood, F a global function (majority in this case), w is a finite word on the alphabet H,V (horizontal, vertical) denoting which Wb are activated at each iteration of automata evolution.

To implement majority function over a von Neumann neighborhood we denoted $V_{(i,j)} = [x_{(i-1,j)}, x_{(i+1,j)}, x_{(i,j-1)}, x_{(i,j+1)}]$, over this set we apply the operator S (sum of all elements due we work with zero and ones values) getting the following functions when none Wb is activated:

- **Strict Majority:** $\begin{cases} 1 & \text{if } S\left(V_{(i,j)}\right) > 2, \\ x_{(i,j)} & \text{if } S\left(V_{(i,j)}\right) = 2, \\ 0 & \text{otherwise} \end{cases}$

- **Skew Majority:** $\begin{cases} 1 & \text{if } S\left(V_{(i,j)}\right) \geq 2, \\ 0 & \text{otherwise} \end{cases}$

When the Wb is activated (vertical or horizontal) the neighbors of $x_{(i,j)}$ are reduced at half, so we have two options that is a conservative approach—we take the majority of this reduced set or we sustain the original threshold which is equivalent to skew to the opposite value. When Wbs are activated the following functions are employed:

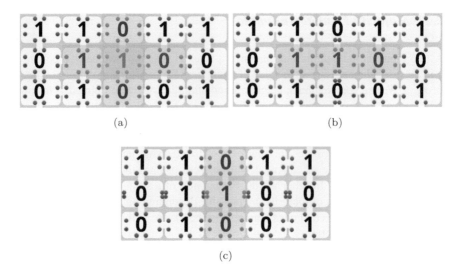

Fig. 2 Majority fungal automata. Depending of which Wb are activate at time t the neighbors able to see his current state change. The majority function in invariable, which means that takes the majority in this reduce neighbor. **a** Full Von Neumann neighborhood where none Woronin body is activated. **b** Von Neumann neighborhood where vertical Woronin body are activated. **c** Von neumann neighborhood where horizontal Woronin body are activated

- **Strict Majority**: $\begin{cases} 1 & \text{if } S\left(V_{(i,j)}\right) > 1, \\ x_{(i,j)} & \text{if } S\left(V_{(i,j)}\right) = 1, \\ 0 & \text{otherwise} \end{cases}$

- **Skew Majority**: $\begin{cases} 1 & \text{if } S\left(V_{(i,j)}\right) = 2, \\ 0 & \text{otherwise} \end{cases}$

3 Simulation Description and Metrics

Several types of numerical experiments were carried on FCA and MFA to evaluate the impact of Wba on the behavior of the dynamics related to each cellular automata rule. We describe how the experiments were implemented and also the metrics we used to study them.

3.1 EFCA Rules

First simulation EFCA were implemented on a ring of size 100, with a random initial condition. Then the automata were evolving according to the specified rules for 99 generations given a final matrix of 100 × 100. Based on the complete review of ECA rules classification in [16], we only used the 88 non-equivalent rules of ECA rule-space. Every cell is binary, so we call magnetization the number of cells whose state is one. Initial EFCA state is randomly selected employing [1, 10, 30, 60] magnetizations. For every selected initial state five different Wb activation were explored, plus a base case where none Wb is activated (this corresponds to a traditional ECA). The modes in which Wbs were activated are following:

- 1Wb-on: Only the Wb between the last and first element of the ring was activated during 99 generations. The rule behaves in a constraint array instead of a ring.
- 4Wb-on: Four Wbs regularly placed every N/4 of ring length (N) and activated during all generations.
- 4Wb-mod: The same four Wbs previously mentioned, but with cycles of activation/deactivation every 10 steps.
- 4Wb-c&r (cut and release): The same four Wbs activated by 20 first steps and then deactivated until the final step.
- allWb-on: every cell with its Wbs is activated to see how rule acts in a single cell in isolation.

In order to get insights into how the activation of Wb affects EFCA, we measure the magnetization index m, defined at t instant as:

$$m_t = \sum_{i=0}^{n} x_i^t \tag{5}$$

Also relative hamming distance δ_t between two consecutive generations is measured, defined as:

$$\delta_t = \frac{\Delta(X_{t-1}, X_t)}{n} \quad \text{with} \quad t \geq 1, n \in \mathbb{N} \quad \text{and}$$
$$\Delta(X_{t-1}, X_t) = X_{t-1} \oplus X_t \tag{6}$$

Exhaustive simulation with one Wb activated In order to get a more complete dynamic comprehension of Wb effect over the rule dynamics, we run the following numerical experiment. We implement EFCA rules on a ring of size 12 and for every possible initial input state (this is $2^{12} = 4096$ configurations) we measure the hamming distance and magnetization index between the original ECA rule (none Wb activated) and the same rule with one Wb activated. This is done for ECA rules [30, 32, 90, 110, 150]. Also for each configuration we search for cycles and when they are found the period and transient are measured. For every configuration we also build the evolution graph, which is a directed graph where nodes represent the state configuration and edges indicate the next generation state in the evolution. This kind of graph let us visualize graphically the occurrence of cycles and fixed points when they exists.

Simulating more cells with one Wb activated Mixing the two previous experiment we use same four rules of the second trial, with Wb activation of the previous experiment but in an increased ring size of 32. The initial state of the EFCA is randomly selected from the 2^{32} possible initial condition making a sample of 6500 different initial configuration. The same metrics than in previous experiment were calculated adding the relative Hamming distance from initial to final state and magnetization variation, also from initial to final state. These two new measures are defined as shown in Eqs. 7 and 8.

$$\delta_{\text{init state}} = \frac{\Delta(X_{t_0}, X_t)}{n} \quad \text{with} \quad t \geq 1, n \in \mathbb{N} \quad \text{and} \tag{7}$$

$$\Delta(X_{t_0}, X_t) = X_{t_0} \oplus X_t$$

$$\Delta m_{\text{init}-t} = \sum_{i=0}^{n} x_i^{t_0} - \sum_{i=0}^{n} x_i^{t} \tag{8}$$

where x^{t_0} and x^t are the initial and final state of the EFCA.

3.2 Majority Fungal Automata

MFA were implemented in a torus of size 100×100 and we let evolve during 99 generations as we did with EFCA. Evolution of the MFA was done using the global function previously described over Von Neumann neighborhood. Our global function was implemented to be invariant respect Woronin bodies activation, this mean that when Wb were activated the function (e.g. majority) was applied to smaller Von Neumann neighborhood, due to the activation restricting the number of neighbors that is possible to see. Initial condition of the MFA is randomly chosen using different magnetizations (number of cells whose state is one) up to a maximum of 9000, which is 90% of the torus. In every step only one kind of Wbs (horizontal or vertical), all of them, are activated. To describe the activation sequence of Wbs we use words on the alphabet H, V. For example, if we close horizontal Wb at current step, and in the next step we switch them by verticals this will be expressed as: HV, if we repeat this sequence until the final step an additional star (*) will be added, having a notation in the form of $(HV)^*$. We do not study the behavior when only vertical or horizontal Wbs are activated, due to this case being similar to a typical ECA with the following rules:

- Strict Majority transform into ECA rule 232.
- Majority (skew) transform into ECA rule 250.

To investigate how the MFA dynamics is affected by behaviour of Wbs, we set up five different modes of Wb activation defined as follow:

- **Modulation**, switching between open and close of Horizontal and Vertical Woronin bodies at every step $(HV)^*$
- **H2V2**, an activation of Wb but every two steps $(HHVV)^*$
- **H4V4**, an activation of Wb but every four steps $(HHHHVVVV)^*$
- **cut & release**, an activation of Wb at every step $(HV)^{20}$, but only for the first 20 generations, then all the Wbs are open until the next generation.
- **Rand**, a random activation of Wb at every step $(HV)^{20}$. This means that in each step a vertical or horizontal Wb is activated with probability 0.5.

To quantify the possible change, we defined the following inter and intra-mode metrics:

- **(intra) Relative Hamming distance**, is the hamming distance between MFA self initial and final state divided by N^2, where N is the size of MFA (square matrix $N \times N$), this can be expressed as:

$$\delta_{\text{intra}} = \frac{X_{t_0}^{F,w} \oplus X_{t_n}^{F,w}}{N^2} \qquad (9)$$

where X is a square matrix of size $m \times m$, F is the global activation function in use, w represents the word that describe the activation sequence of Wb and the step of evolution of the MFA goes from $t_0, t_1, \ldots, t_n$.

- **(inter) Relative Hamming distance**, is the hamming distance between final state of control MFA (without activation of Wb at any time) and the others MFA with different Wb activation mode, both with the same initial state and global function. This can be described as:

$$\delta_{\text{inter}} = \frac{X_{t_n}^{F,w_0} \oplus X_{t_n}^{F,w_k}}{N^2} \qquad (10)$$

where X is a square matrix of size $m \times m$, F is one of the two global activation functions, $w_0 = \emptyset$ which means none Wb activated, w_k represent a mode in the set $w = \{H1V1, H2V2, H4V4, \text{cut\&rel}\}$

- **(intra) Magnetization index**, quantifies the variation in cells with one as value, between MFA self initial and final state. This can be write as:

$$m_{\text{intra}} = \sum_{i=0}^{m} \sum_{j=0}^{m} x_{t_0}^{F,w}(i,j) - \sum_{i=0}^{m} \sum_{j=0}^{m} x_{t_n}^{F,w}(i,j) \qquad (11)$$

where $x(i, j)$ is the element in position (i, j) of the square matrix X.

- **(inter) Magnetization index**, quantifies the variation of cells with one as value between final state of control MFA (without activation of Wb at any time) and the others MFA with different Wb activation mode, both with the same initial state and global function, according to the following:

$$m_{\text{inter}} = \sum_{i=0}^{m}\sum_{j=0}^{m} x_{t_n}^{F,w_0}(i,j) - \sum_{i=0}^{m}\sum_{j=0}^{m} x_{t_n}^{F,w_k}(i,j) \qquad (12)$$

Our global functions are both majority. A natural question that arises from this fact is how the consensus, meaning the agreement over the network, is affected by the activation of Wb. With the aim of investigate this phenomena we define the consensus index, which is the mean MFA state at a time t, but changing all zero values by -1. The following equation describes this index:

$$\text{Cid}(x_t) = \frac{\sum_{i=0}^{m}\sum_{j=0}^{m} x_{t_n}^{F,w}(i,j)[0 \mapsto -1]}{N^2} \qquad (13)$$

Where

$$x_{t_n}^{F,w}(i,j)[0 \mapsto -1] = \begin{cases} x_{t_n}^{F,w}(i,j) & \text{if } x_{t_n}^{F,w}(i,j) = 1 \\ -1 & \text{otherwise.} \end{cases}$$

By using this index, we seek to capture two different type of phenomena: (a) we measure the possible fluctuation in consensus between two different timestamps t_i, t_j, and (b) we study the average consensus reached or see how different Wb activations could change the consensus of the network. This is specially interesting in knowledge areas such as: social dynamics, networking and even blockchain, to determine if consensus is affected by the temporal interruption on the communication flow for Wb activation or if we can accelerate.

4 Results

4.1 EFCA Rules

Elementary Cellular Automata (ECA) rules were classified by Wolfram in [11] into four classes depending on their behavior starting from uniform (class 1) and follow by periodic, chaotic and complex (class 4). This classification naturally leads to the question in whether our approach induces changes in the class classification. To answer that question and to get a better representation of new system dynamics induced by the Wb activation, we create a graph in which the nodes represent every of the 255 ECA rules. In this graph, each rule represented by a node has an edge towards its his left (in red color) and right (in green color) equivalent EFCA rule. This graph is shown in Fig. 3 where rules [0, 51, 204, 255] are the only ones with loops representing the strongly connected components of the graph. The rule 51 has two incoming self-edges representing that no matter how many Wb are activated or in which mode its left and right rules are not only the same but actually the same rule 51. On the other hand rule 204 (identity rule) has the biggest incoming degree (acts as a sink), meaning that the majority of rules will have a path in the graph connecting

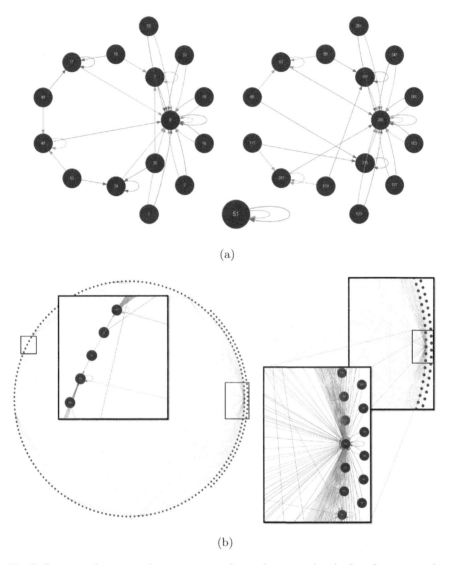

Fig. 3 Four strongly connected components on the graph representing the fungal automata rule space. Every node (in blue) represents a rule, when a Wb is activated, the rule will change its behaviour at the left (in red) or at the right (in green) of Wb, leading to other rule dynamic. **a** Graph representing left and right rules for ECA rules. Rule 0, 51, and 255 are the strongly connected components of the graph representing left and right rules. **b** Rule 204 (the identity) as the node with biggest in-degree (sink)

them to rule 204. In order to illustrate some of the observations that can be inferred from the graph, consider the following example. An EFCA rule is defined over a ring of length N with only one Wb activated at position i and rule 110 is the update function for the next state. In all the cells not affected by the Wb activation, the rule 110 will be applied normally, as if the original ECA rule. For the triplets of cells that are affected by Wb in his rightmost bit, instead of applying rule 110, rule 204 will be applied and, on the other hand, for the triplet affected in his leftmost bit by Wb, rule 238 will be applied. A complete list in a table format of rule's equivalence (left and right) is given in Table 2.

We note from the previous results that rules 57 and 99 (which are equivalent between them and belong to Wolfram's class 2) are the only ones in which Wb activation increases class complexity at the right and left rules, transforming it into class 3 rules. The rest of rules are summarised in Table 1 (Table 2).

4.1.1 First Simulation

A total of 6336 simulations were run, covering all 255 ECA rules, Figs. 4, 5, 6 and 7 are samples of the this numerical experiment. In these figures it is possible to see how the five modes in which we activated Wb affect the normal behavior of ECA rule for the same initial state and we compare against control rule (without Wb activation).

In Fig. 4f, the previous parameters for rule 57 are shown. Observe that this latter rule is one of those rule that increases a complexity with Wb activation. On the other hand, the same can be seen in Fig. 4f. Figures 5f, 6f and 7f exemplify how the activation of all Wb leads the dynamical behavior of rule 90 to a behavior similar to rule 204 (the identity rule) which plays the role of a sink on the graph shown in Fig. 3.

A more quantitative way to see the effects produced by Wb activation on the dynamics of each ECA rule is shown in Figs. 8, 9, 10 and 11, where for each rule the Hamming distance and the magnetization index are shown for a given initial state, the same initial state as in Figs. 4, 5, 6 and 7.

4.1.2 Exhaustive Simulation Results

For all 4096 different initial state and for each rule selected ([30, 32, 90, 110, 150]) we find cycles or fixed point when they exist and calculate period and transient. An example of the analysis is shown in In Fig. 12 where is possible to see the histogram of periods and transient for rule 110 with and without Wb activated.

A detailed example of this process for a given input can be seen in Fig. 13. The first images show the EFCA evolution for a given initial state with and without Wb activation, next the evolution graph is produced and finally a plot showing the variation in relative Hamming distance and magnetization index. The result shows that the activation of Wb induces changes in the behavior of automata, for example rule 150 generates for almost 95% of the time cycles with period of 62 in comparison with the scenarios where no Wbs were activated.

Exploring Dynamics of Fungal Cellular Automata 353

Table 1 Summary of Wolfram's class classification changes for Wb activation

Case	Qty	Rules
Right and left rule increase in class complexity	2	57, 99
Right and left rule don't change class complexity	128	0, 4, 5, 6, 7, 12, 13, 14, 15, 19, 20, 21, 23, 28, 29, 31, 32, 35, 36, 37, 42, 43, 44, 49, 50, 51, 55, 59, 68, 69, 70, 71, 72, 73, 74, 76, 77, 78, 79, 84, 85, 87, 88, 91, 92, 93, 94, 95, 100, 104, 108, 109, 112, 113, 115, 128, 132, 133, 134, 140, 141, 142, 143, 148, 156, 157, 158, 159, 160, 164, 170, 171, 172, 173, 178, 179, 196, 197, 198, 199, 200, 201, 202, 203, 204, 205, 206, 207, 212, 213, 214, 215, 216, 217, 218, 219, 220, 221, 222, 223, 228, 229, 232, 233, 236, 237, 240, 241, 250, 251, 254, 255
Right rule increases, left remains class complexity	16	8, 40, 52, 53, 58, 61, 67, 83, 136, 163, 168, 211, 234, 235, 238, 239
Right rule remains, left rule increase class complexity	16	25, 27, 38, 39, 64, 96, 103, 114, 155, 177, 192, 224, 248, 249, 252, 253
Right rule decreases, left remains class complexity	26	17, 34, 65, 66, 80, 81, 82, 102, 116, 117, 119, 125, 153, 162, 180, 181, 186, 187, 188, 189, 194, 208, 209, 210, 244, 245
Right rule remains, left rule decrease class complexity	26	3, 9, 10, 11, 24, 26, 46, 47, 48, 60, 63, 111, 138, 139, 152, 154, 166, 167, 174, 175, 176, 195, 230, 231, 242, 243
Right rule increases, left rule decrease class complexity	4	56, 62, 131, 227
Right rule decreases, left rule increase class complexity	4	98, 118, 145, 185
Right and left rule decrease class complexity	50	1, 2, 16, 18, 22, 30, 33, 41, 45, 54, 75, 86, 89, 90, 97, 101, 105, 106, 107, 110, 120, 121, 122, 123, 124, 126, 127, 129, 130, 135, 137, 144, 146, 147, 149, 150, 151, 161, 165, 169, 182, 183, 184, 190, 191, 193, 225, 226, 246, 247

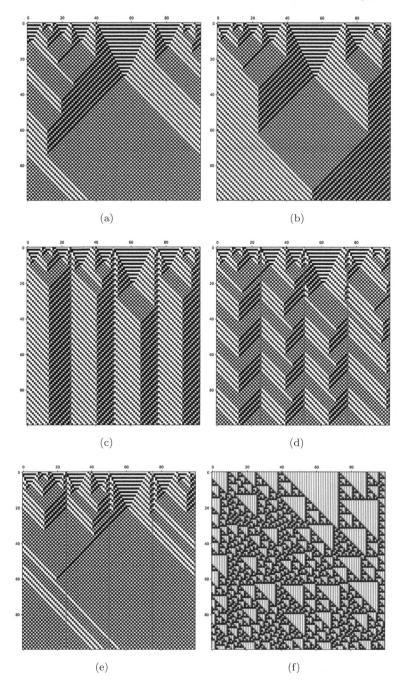

Fig. 4 Rule 57 on EFCA affected by different Wb activation. All start from the same initial state (10 ones). **a** 1Wb, **b** Control rule (none Wb), **c** 4Wb, **d** 4Wb-mod, **e** 4Wb-c&r, **f** allWb-on

Exploring Dynamics of Fungal Cellular Automata 355

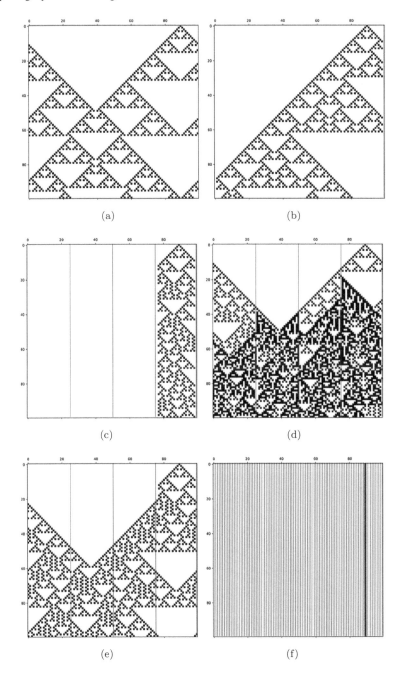

Fig. 5 Rule 90 on EFCA affected by different Wb activation. All start from the same initial state. **a** Control rule (none Wb), **b** 1Wb, **c** 4Wb, **d** 4Wb-mod, **e** 4Wb-c&r, **f** allWb-on

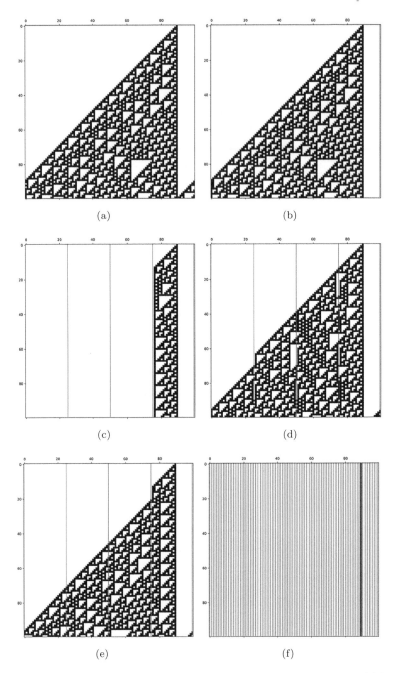

Fig. 6 Rule 110 on EFCA affected by different Wb activation. All start from the same initial state. **a** Control rule (none Wb), **b** 1Wb, **c** 4Wb, **d** 4Wb-mod, **e** 4Wb-c&r, **f** allWb-on

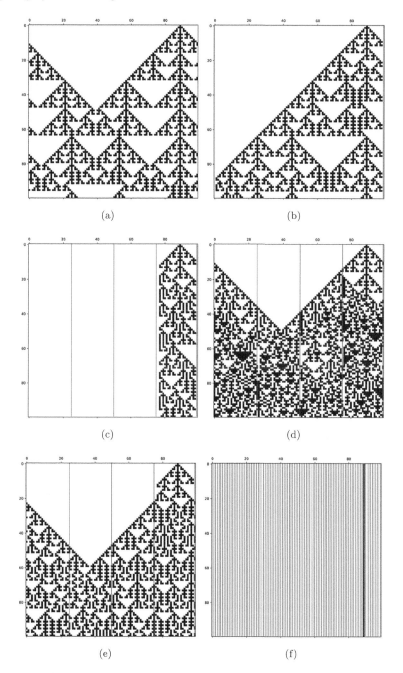

Fig. 7 Rule 150 on EFCA affected by different Wb activation. All start from the same initial state. **a** Control rule (none Wb), **b** 1Wb, **c** 4Wb, **d** 4Wb-mod, **e** 4Wb-c&r, **f** allWb-on

358   C. S. Sepúlveda et al.

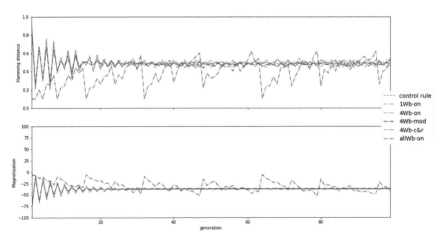

Fig. 8 Rule 57. Upper plot shows how relative hamming distance vary according to evolution step for each Woronin bodies activation mode. Lower graph shows how the magnetization index varies accordingly to the same parameters

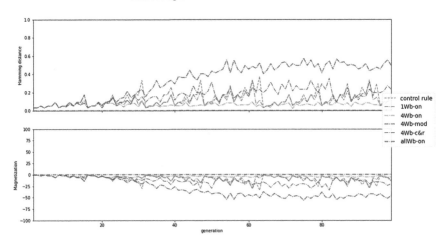

Fig. 9 Rule 90. Upper plot shows how relative hamming distance vary according to evolution step for each one of the Woronin bodies activation mode. Lower graph shows how the magnetization index vary according to the same parameters

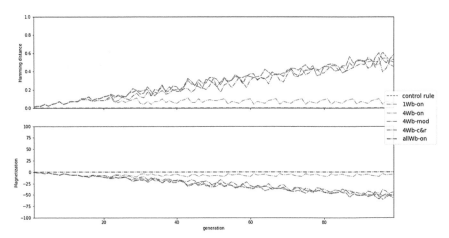

Fig. 10 Rule 110. Upper plot shows how relative hamming distance varies accordingly to evolution step for each Wb activation mode. Lower graph shows how the magnetization index vary according to the same parameters

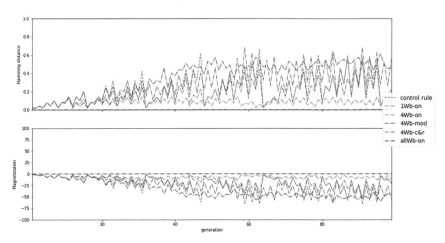

Fig. 11 Rule 150. Upper plot shows how relative hamming distance varies accordingly to evolution step for each Wb activation mode. Lower graph shows how the magnetization index vary according to the same parameters

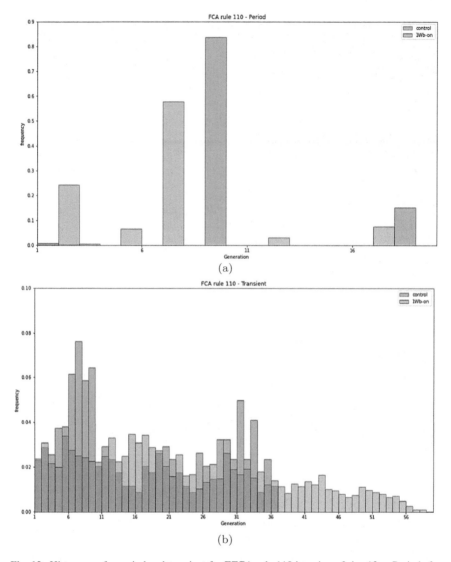

Fig. 12 Histograms for period and transient for EFCA rule 110 in a ring of size 12. **a** Periods for rule 110 with and without a Wb activated. **b** Transient for rule 110 with and without a Woronin body activated

Exploring Dynamics of Fungal Cellular Automata

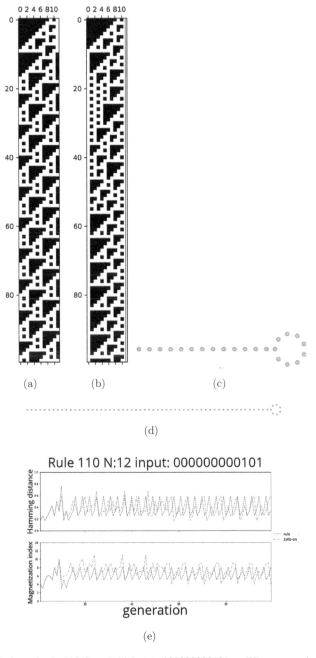

Fig. 13 Evolution of rule 110 from initial state 000000000101. **a** Wbs are passive during first 100 generations, **b** Wbs are active for 100 generations, **c** Evolution graph for rule without Wb, **d** Evolution graph with Wb activated, **e** Variation in relative Hamming distance and magnetization index

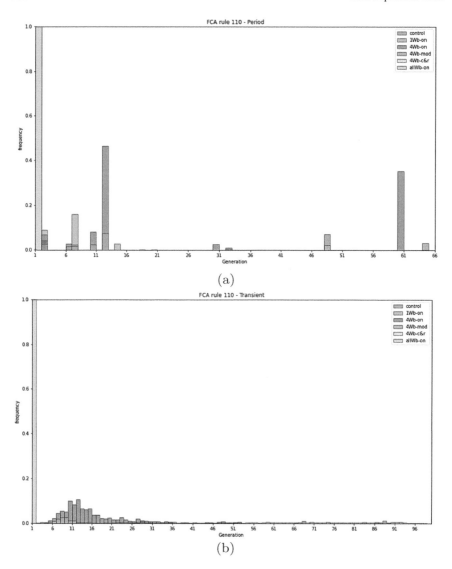

Fig. 14 Histograms for period and transient for EFCA rule 110 on a ring of size 32. Wb activation 1Wb-on, 4Wb-on, 4Wb-mod, 4Wb-c&r, allWb-on. **a** Periods rule 110 for different Wb activation. **b** Transient rule 110 for different Wb activation

4.1.3 Simulation with More Cells and One Wb Activated

A total of 195,000 simulations were run, each of which resulted in the identification of cycles and fixed points. Similar to the previous section, the results were summarized in histograms, as shown in Fig. 14. The objective of this stage was to investigate how changes in ring size affect the effects produced by different Wb activation types.

4.2 Majority Fungal Automata

A total of 12000 simulations with random initial magnetization state were run. Sample of some timestamps for both global rules and H2V2 activation mode are shown in Figs. 15 and 16.

The general results are summarized in Figs. 17 and 18. This figures show the individual metrics as a function of magnetization in initial state using skew-majority and majority as global function, and different Wb activation sequence. It is possible

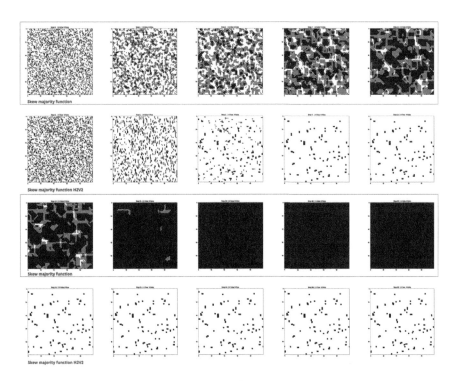

Fig. 15 Evolution steps 0, 1, 3, 7, 11, 14, 31, 50, 80 y 99 for skew majority global function and same rule with and H2V2 Wbs activation scheme

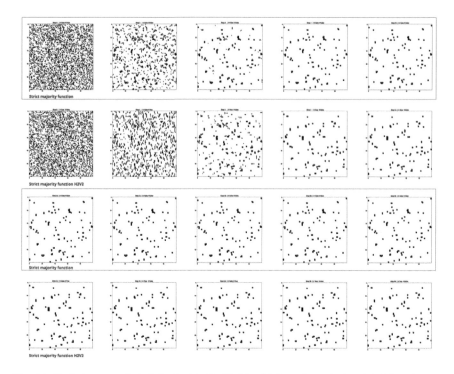

Fig. 16 Evolution steps 0, 1, 3, 7, 11, 14, 31, 50, 80 y 99 for strict majority global function and same rule with and H2V2 Wbs activation scheme

to observe in Fig. 17 the effect of Wb activation, transforming the dynamic of skew-majority. In the case of strict majority only irrelevant changes are induced as shown Fig. 18.

Using MFA final state with global functions without Wb activation as a point of comparison we get metrics to quantify the change between these reference points and the activation of Woronin bodies in sequence previously defined. Figures 19 and 20 are the plots representing this changes.

To get a better understanding of consensus dynamics due Woronin bodies activation, we plot the consensus achieved in the final step of every simulation getting the Figs. 21 and 22.

5 Discussion

The activation of Woronin bodies (Wb) in EFCA has direct impact in system dynamics. This is shown for example in rule 32 where the Wbs activation didn't change transient of cycle, but instead changes his period. In rule 90 on a ring of size 32, the

Exploring Dynamics of Fungal Cellular Automata

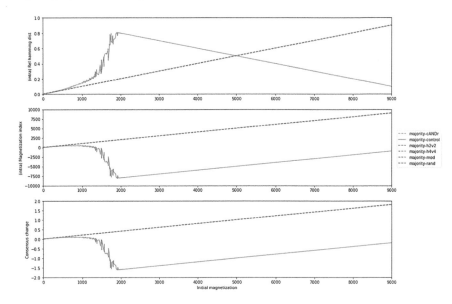

Fig. 17 Relative hamming distance, Magnetization index and Consensus Change against initial magnetization for skew-majority

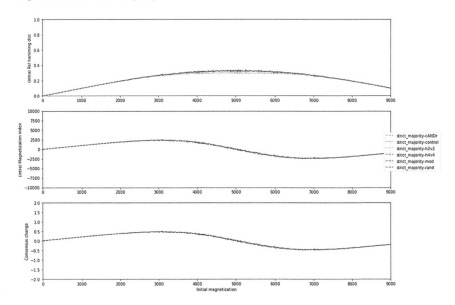

Fig. 18 Relative hamming distance, Magnetization index and Consensus Change against initial magnetization for skew-majority

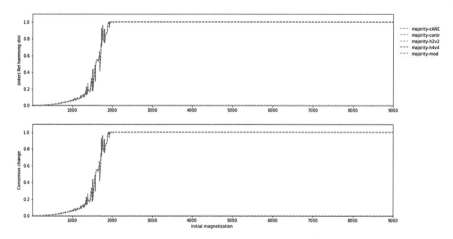

Fig. 19 Relative Hamming distance and consensus change between skew-majority global function and same global function with different Woronin bodies activation

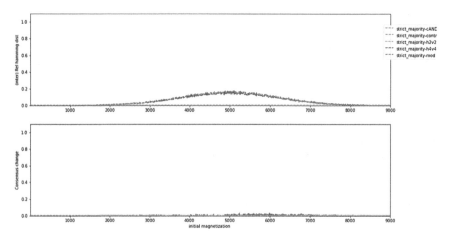

Fig. 20 Relative hamming distance and consensus change between strict majority global function and same global function with different Woronin bodies activation

Wb activation introduces cycles where previously fixed points exist. Same changes in dynamics happen on size 12 ring and in all rules implemented, and rules 57-99 increase complexity behavior due the activation of Woronin bodies were other group decrease complexity, leading to more trivial behaviors.

In MFA at first glance we have the intuition that Wb could change the consensus dynamics, perhaps transforming a un-skew function as strict majority into skew one or kind of behaviors. The data shows strict majority is not affected in a significant way by the Wbs activation, achieving same dynamics of the global function without Wb activated. In the case of skew-majority the skew is augmented making that all

Exploring Dynamics of Fungal Cellular Automata 367

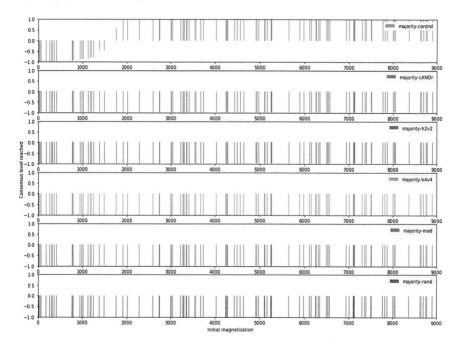

Fig. 21 Level of consensus reached in final generation against initial magnetization for skew-majority

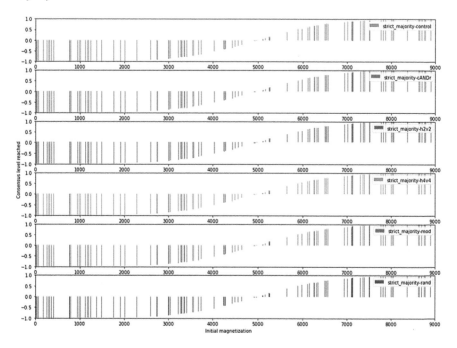

Fig. 22 Level of consensus reached in final generation against initial magnetization for strict majority

initial configuration tried, quickly converge to zero (-1 when in consensus), which means that no mater what we do, in this network the negative alternative/opinion will always win.

Acknowledgements This work was supported by Centro de Modelamiento Matemático (CMM), FB210005 BASAL funds for centers of excellence from ANID-Chile, FONDECYT 1200006 (E.G.) and ANID FONDECYT Postdoctorado 3220205 (M.R-W).

Appendix

Table 2 Rules with his right and left rule equivalent

rule	Rrule	Lrule	rule	Rrule	Lrule	rule	Rrule	Lrule	rule	Rrule	Lrule
0	0	0	64	192	68	128	192	136	192	192	204
1	0	0	65	192	68	129	192	136	193	192	204
2	0	0	66	192	68	130	192	136	194	192	204
3	3	0	67	195	68	131	195	136	195	195	204
4	12	68	68	204	68	132	204	204	196	204	204
5	12	68	69	204	68	133	204	204	197	204	204
6	12	68	70	204	68	134	204	204	198	204	204
7	15	68	71	207	68	135	207	204	199	207	204
8	12	136	72	204	204	136	204	136	200	204	204
9	12	136	73	204	204	137	204	136	201	204	204
10	12	136	74	204	204	138	204	136	202	204	204
11	15	136	75	207	204	139	207	136	203	207	204
12	12	204	76	204	204	140	204	204	204	204	204
13	12	204	77	204	204	141	204	204	205	204	204
14	12	204	78	204	204	142	204	204	206	204	204
15	15	204	79	207	204	143	207	204	207	207	204
16	0	0	80	192	68	144	192	136	208	192	204
17	0	17	81	192	85	145	192	153	209	192	221
18	0	0	82	192	68	146	192	136	210	192	204
19	3	17	83	195	85	147	195	153	211	195	221
20	12	68	84	204	68	148	204	204	212	204	204
21	12	85	85	204	85	149	204	221	213	204	221
22	12	68	86	204	68	150	204	204	214	204	204
23	15	85	87	207	85	151	207	221	215	207	221
24	12	136	88	204	204	152	204	136	216	204	204
25	12	153	89	204	221	153	204	153	217	204	221

(continued)

Table 2 (continued)

rule	Rrule	Lrule	rule	Rrule	Lrule	rule	Rrule	Lrule	rule	Rrule	Lrule
26	12	136	90	204	204	154	204	136	218	204	204
27	15	153	91	207	221	155	207	153	219	207	221
28	12	204	92	204	204	156	204	204	220	204	204
29	12	221	93	204	221	157	204	221	221	204	221
30	12	204	94	204	204	158	204	204	222	204	204
31	15	221	95	207	221	159	207	221	223	207	221
32	0	0	96	192	68	160	192	136	224	192	204
33	0	0	97	192	68	161	192	136	225	192	204
34	0	34	98	192	102	162	192	170	226	192	238
35	3	34	99	195	102	163	195	170	227	195	238
36	12	68	100	204	68	164	204	204	228	204	204
37	12	68	101	204	68	165	204	204	229	204	204
38	12	102	102	204	102	166	204	238	230	204	238
39	15	102	103	207	102	167	207	238	231	207	238
40	12	136	104	204	204	168	204	136	232	204	204
41	12	136	105	204	204	169	204	136	233	204	204
42	12	170	106	204	238	170	204	170	234	204	238
43	15	170	107	207	238	171	207	170	235	207	238
44	12	204	108	204	204	172	204	204	236	204	204
45	12	204	109	204	204	173	204	204	237	204	204
46	12	238	110	204	238	174	204	238	238	204	238
47	15	238	111	207	238	175	207	238	239	207	238
48	48	0	112	240	68	176	240	136	240	240	204
49	48	17	113	240	85	177	240	153	241	240	221
50	48	34	114	240	102	178	240	170	242	240	238
51	51	51	115	243	119	179	243	187	243	243	255
52	60	68	116	252	68	180	252	204	244	252	204
53	60	85	117	252	85	181	252	221	245	252	221
54	60	102	118	252	102	182	252	238	246	252	238
55	63	119	119	255	119	183	255	255	247	255	255
56	60	136	120	252	204	184	252	136	248	252	204
57	60	153	121	252	221	185	252	153	249	252	221
58	60	170	122	252	238	186	252	170	250	252	238
59	63	187	123	255	255	187	255	187	251	255	255
60	60	204	124	252	204	188	252	204	252	252	204
61	60	221	125	252	221	189	252	221	253	252	221
62	60	238	126	252	238	190	252	238	254	252	238
63	63	255	127	255	255	191	255	255	255	255	255

References

1. Watkinson, S.C., Boddy, L., Money, N.: The Fungi. Academic Press (2015)
2. Maheshwari, R.: Fungi: Experimental Methods In Biology. CRC Press (2016)
3. Adamatzky, A., Tuszynski, J., Pieper, J., Nicolau, D.V., Rinalndi, R., Sirakoulis, G., Erokhin, V., Schnauss, J., Smith, D.M.: Towards cytoskeleton computers. A proposal. In: Adamatzky, A., Akl, S., Sirakoulis, G. (eds.) From Parallel to Emergent Computing. CRC Group/Taylor & Francis (2019)
4. Adamatzky, A., Tegelaar, M., Wosten, H.A.B., Powell, A.L., Beasley, A.E., Mayne, R.: On boolean gates in fungal colony. Biosystems **193**, 104138 (2020)
5. Beasley, A.E., Abdelouahab, M.-S., Pierre Lozi, R., Antisthenis Tsompanas, M., Powell, A., Adamatzky, A.: Mem-fractive properties of mushrooms. Bioinspiration Biomimetics (2021)
6. Goles, E., Tsompanas, M.-A., Adamatzky, A., Tegelaar, M., Wosten, H.A.B., Martínez, G.J.: Computational universality of fungal sandpile automata. Phys. Lett. A 126541 (2020)
7. Goles, E., Adamatzky, A., Montealegre, P., Ríos-Wilson, M.: Generating boolean functions on totalistic automata networks. Int. J. Unconv. Comput. **16**(4) (2021)
8. Adamatzky, A., Gandia, A., Chiolerio, A.: Fungal sensing skin (2020). arXiv:2008.09814
9. Wolfram, Stephen: Cellular automata as models of complexity. Nature **311**(5985), 419–424 (1984)
10. Wolfram, S.: Cellular Automata and Complexity: Collected Papers. CRC Press (2018)
11. Wolfram, S.: Statistical mechanics of cellular automata. Rev. Mod. Phys. **55**(3), 601 (1983)
12. Moore, C.: Computational complexity in physics. In: Complexity from Microscopic to Macroscopic Scales: Coherence and Large Deviations, pp. 131–135. Springer (2002)
13. Moore, Cristopher: Majority-vote cellular automata, ising dynamics, and p-completeness. J. Stat. Phys. **88**(3), 795–805 (1997)
14. Griffeath, David, Moore, Cristopher: Life without death is p-complete. Complex Syst. **10**, 437–447 (1996)
15. Neary T, Woods, D.: P-completeness of cellular automaton rule 110. In: International Colloquium on Automata, Languages, and Programming, pp. 132–143. Springer (2006)
16. Martinez, G.J.: A note on elementary cellular automata classification (2013). arXiv:1306.5577

Computational Universality of Fungal Sandpile Automata

Eric Goles, Michail-Antisthenis Tsompanas, Andrew Adamatzky, Martin Tegelaar, Han A. B. Wosten, and Genaro J. Martínez

Abstract Hyphae within the mycelia of the ascomycetous fungi are compartmentalised by septa. Each septum has a pore that allows for inter-compartmental and inter-hyphal streaming of cytosol and even organelles. The compartments, however, have special organelles, Woronin bodies, that can plug the pores. When the pores are blocked, no flow of cytoplasm takes place. Inspired by the controllable compartmentalisation within the mycelium of the ascomycetous fungi we designed two-dimensional fungal automata. A fungal automaton is a cellular automaton where communication between neighbouring cells can be blocked on demand. We demonstrate computational universality of the fungal automata by implementing sandpile cellular automata circuits there. We reduce the Monotone Circuit Value Problem to the Fungal Automaton Prediction Problem. We construct families of wires, crossovers and gates to prove that the fungal automata are P-complete.

1 Introduction

Fungi are ubiquitous organisms that are present in all ecological niches. They can grow as single cells but can also form mycelium networks covering up to $10\,\text{km}^2$ of forest soil [1, 2]. Fungi can sense what humans sense and more, including tactile stimulation [3, 4], pH [5], metals [6], chemicals [7], light [8] and gravity [9]. Fungi exhibit a rich spectrum of electrical activity patterns [10–12], which can be tuned by external stimulation. On studying electrical responses of fungi to stimulation [12]

E. Goles · M.-A. Tsompanas
Faculty of Engineering and Science, University of Adolfo Ibáñez, Santiago, Chile

A. Adamatzky (✉) · G. J. Martínez
Unconventional Computing Laboratory, UWE, Bristol, UK
e-mail: andrew.adamatzky@uwe.ac.uk

M. Tegelaar · H. A. B. Wosten
Microbiology Department, University of Utrecht, Utrecht, The Netherlands

G. J. Martínez
High School of Computer Science, National Polytechnic Institute, Mexico City, Mexico

© The Author(s), under exclusive license to Springer Nature Switzerland AG 2023
A. Adamatzky (ed.), *Fungal Machines*, Emergence, Complexity and Computation 47,
https://doi.org/10.1007/978-3-031-38336-6_24

we proposed design and experimental implementation of fungal computers [13]. Further numerical experiments demonstrated that it is possible to compute Boolean functions with spikes of electrical activity propagating on mycelium networks [14]. At this early stage of developing a fungal computer architecture we need to establish a strong formal background reflecting several alternative ways of computing with fungi.

In [15] we introduced one-dimensional fungal automata, based on a composition of two elementary cellular automaton functions. We studied the automata space-time complexity and discovered a range of local events essential for a future computing device on a single hypha. We aim to demonstrate a computational universality of fungal automata. To do this we modify state transition rules of sand pile, or chip firing, automata [16–20] to allow a control for moving of sand grains, or chips, between neighbouring cells. The local control of the interactions between cells is inspired by a control of cytosol flow control in fungal hyphae [21–25]. Then we used developed tools of sand pile automata universality [26–31] to show that functionally complete sets of Boolean gates can be realised in the fungal automata.

2 Two Dimensional Fungal Automata

Filamentous fungi of the phylum *Ascomycota* have porous septa that allow for cytoplasmic streaming throughout hyphae and the mycelium [21, 22]. The pores of damaged hyphae will be plugged by a peroxisome-derived organelle to prevent bleeding of cytoplasm into the environment [32–35]. These Woronin bodies can also plug septa of intact hyphae [23, 36]. The septal pore occlusion in these hyphae can be triggered by septal ageing and stress conditions [23, 25, 37].

A scheme of the mycelium with Woronin bodies is shown in Fig. 1. An apical compartment has one neighbouring sub-apical compartment, while a sub-apical compartment has a neighbouring compartment at both ends. Because compartments can also branch, they can have one or more additional neighbouring compartments. Thus, the compartment with pores and Woronin bodies is a elementary unit of fungal automata, Fig. 1bcd. From these compartments one can assemble quasi-one-dimensional, Fig. 1e, and two-dimensional, Fig. 1f, structures.

In this context, let us consider a cellular automaton in the two dimensional grid $\mathbb{Z}^2$ with the von Neumann neighbourhood, with set of states Q and a global function F. Each cell of the grid has four sides that could be open or closed. An open side means that the information (the state) of both cells is shared. If not, when the side is closed, both sites mutually ignore each other. When every side is open we have the usual cellular automata model [38]. On the other hand, if sides are open or closed in some random or periodical way we get for the same local functions different dynamic behaviours. We consider only "uniform" ways to open-close the sides. Actually at a given step to open every vertical side (every column of the grid) or every horizontal side, rows of the array. So the fungal automata model becomes a tuple $FA = \langle \mathbb{Z}^2, \mathbf{Q}, V, F, w \rangle$, where V is the von Neumann neighbourhood, w is a

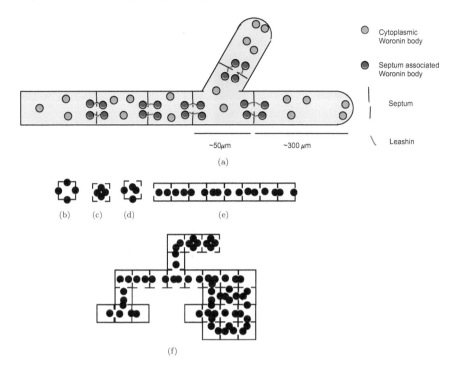

Fig. 1 Fungal automata. **a** Biological scheme. **b–f** Abstract schemes. **b** All pores are closed, **c** All pores are open, **d** North and East pores are open, **e** one-dimensional automaton, **f** an example of an arbitrary architecture of fungal automata

finite word on the alphabet H, V (horizontal, vertical). Each iteration of automaton evolution is associated with one letter of H, V.

In this work we focus on "particles" rules. That is to say at each site there are a finite amount of particles or chips, that, according to a specific rule are disseminated in the vicinity of a site. Every step is going synchronously, so each site lose and receive chips simultaneously. In this context the set of states is $\mathbf{Q} = \{0, 1, 2, ...\} \subset \mathbb{N}$, the number of particles.

3 The Chip Firing Automata

The chip firing automaton, also know as the sandpile model [26, 39, 40], is a particular case of the above described particle automata $\langle \mathbb{Z}^2, V, \mathbf{Q}, F, w \rangle$, with the following local function. If a site $v \in \mathbb{Z}^2$ has $x_v \geq 4$ chips then:

$$\begin{aligned} x'_v &= x_v - 4 \\ \forall u \in V_v &\Rightarrow x'_u = x_u + 1 \end{aligned} \quad (1)$$

where V_v is the von Neumann neihborhood of the site v, x'_v is update of x_v.

By adding the condition of open or close sides of the site proposed here, the rule changes as follows:

$$x_v \geq 4 \Rightarrow x'_v = x_v - \alpha$$
$$\forall u \in V_v \text{ such that the gate is open} \quad (2)$$
$$\Rightarrow x'_u = x_u + 1$$

where α is the number of gates that are open.

When the rule is applied in parallel on every site, the new state at a site v is:

$$x'_v = x_v - \alpha + \beta$$

where β is the number of chips which the site v receives from its open and firing neighbours.

If every side (columns and rows) is always open, then we have the usual chip firing automaton. When a word w of open or close sides is considered, for instance $w = HHVV$, at each step we open (or close) the rows or columns of the grid periodically

$$(HHVV)^* = HHVV\ HHVV\ HHVV\ ... \quad (3)$$

4 Computational Complexity Notions

In order to study the complexity of an automaton we can analyse a power of the automaton to simulate Boolean functions, i.e., by selecting specific initial conditions and sites as inputs and outputs to determine the different Boolean functions the automaton may compute by its dynamics [41]. More Boolean functions are founded, more complex is the automaton. A similar notion, related to some prediction problems, appears in the framework of the theory of computational complexity. Essentially this is similar to trying to determine the computational time related to the size of a problem, that a Turing machine take to solve it. In our context, let us consider the following decision or prediction problem.

PRE: Consider the chip firing fungal automaton $FA = \langle \mathbb{Z}^2, V, \mathbf{Q}, F, w \rangle$, an initial assignment of chips to every site, $x(0) \in \mathbf{Q}^{\mathbb{Z}^2}$, an integer number $T > 0$ of steps of the automaton and a site $v \in \mathbb{Z}^2$, such that $x_v(0) = 0$. Question: Will $x_v(t) > 0$ be for some $t \leq T$?

Of course one may give an answer by running the automaton at most T steps, which can be done in a serial computer in polynomial time. But the question is a little more tricky: could we answer faster than the serial algorithm, ideally, exponentially faster in polylogarithmic time in a parallel computer with a polynomial number of processor?

To answer the questions we consider two classes of decision problems, those belonging to P, the class of problems solved by a polynomial algorithm, and the class NC, being the problems solved in a parallel computer in $O(log^q n)$ steps (polylogarithmic time). This is straightforward that $NC \subseteq P$ because any parallel algorithm solved in polylogarithmic time can be simulated efficiently in a serial computer. But the strict inclusion is a very hard open problem (like the well know $P = NP$).

An other notion from computational complexity related with the possibility that the two classes melt is P-completeness. A problem is *P-complete* if it is in the class P, (that is to say, there exist a polynomial algorithm to solve it) and every other problem in P can be reduced, by a polynomial transformation, to it. Clearly, if one of those *P-complete* problems is in NC, both classes collapsed in one. So, to prove that a problem is *P-complete* gives us an idea of its complexity.

One well known *P-complete* problem is the *Circuit Value Problem*, i.e., the evaluation of a Boolean Circuit (Boolean function). Roughly because any polynomial problem solved in a serial computer (a Turing machine) can be represented as a Boolean circuit. On the other hand, Boolean circuits intuitively are essentially serial because in order to compute a layer of functions it is necessary to compute previous layers so in principle it is not clear how to determine the output of the circuit in parallel. Further, when the circuit is monotonous, i.e., it admits only OR and AND gates (no negations) it is also a *P-complete* problem. This is because negation can be put in the input (the two bits of the variable 0 and 1) and for the gates which are a negation, to use the De Morgan laws.

The complexity of the chip firing automata was first studied in [26, 39], where it was proved that in arbitrary graphs (in particular, non-planar ones) the chip firing automata are Turing Universal. To prove this a universal set of Boolean circuits is built by using specific automata configurations, so, also **PRE** is *P-complete*. In a similar way, but in a d-dimensional grid, $\mathbb{Z}^d$, it was proved in [40] that for $d \geq 3$ the problem is *P-complete* and the complexity, until today, remains open for a two-dimensional grid. In [30] it was proved that in a two-dimensional grid and the von Neumann and Moore neighbourhood it is not possible to cross signals by constructing wires over quiescent configurations. That can be done only for bigger neighbourhood, so, in fact, over non planar graphs.

5 Computational Complexity of the Fungal Automata

We will study the computational complexity of the Fungal Sand Pile Automaton, by proving that for the word $w = H^4V^4 = HHHHVVVV$, the Prediction Problem, **PRE**, is *P-complete*. That is to say one can not determine an exponentially faster algorithm to answer unless $NC = P$.

Prop 1 *For the word* $H^4V^4 = HHHHVVVV$ *the fungal chip firing automaton is P-complete.*

Proof Clearly the problem is in P. It suffices to run the automaton at most T steps and see if the site i changes, which is done in $O(T^3)$: in fact we have to compute the "cone" between site i, its neighbourhood at step $T-1$, $T-2$, and so on, until the initial values in the site in an square $(2T-1) \times (2T-1)$. So the number of sites is to consider is $1^2 + 3^2 + 5^2 + \cdots + (2T-1)^2$ which is bounded by a cubic polynomial, so one may compute the state of site i in T^3.

To establish the completeness, we will reduce the Monotone Circuit Value Problem to the Fungal Automaton Prediction Problem, **PRE**. That is to say, to establish specific automaton configurations which simulates a wire, the AND and the OR gates, as well as a CROSS-OVER. This last gadget is important to compute non-planar circuits in the two dimensional grid.

In the constructions below every cell which is not in the diagram is understood initially empty, without chips ($x_i = 0$).

To construct the wire let us first see what happen when V^4 is applied in the particular structure showed in Fig. 2a. The important issue is that the initial site with 4 chips, after the application of V^4, remains with 4 chips. Only the adjacent sites and down change their number of chips. Then one applies H^4 to obtain Fig. 2b which is similar to the initial configuration, shifted to the right Fig. 2c.

To implement AND gate Fig. 3 and OR gate Fig. 4 we have to connect two wires (this corresponds to a branching of mycelium). In the AND gate two single chips arrive to a central site with 2 chips, so to trigger firing, threshold 3, the signal has to arrive. With H the signal continues to the right, thus the output is 1. The OR gate functions similarly to AND gate but the central site has 3 chips. There is an unwanted signal coming back signal but the computation is made to the right.

The cross-over is demonstrated in Fig. 5. Here we apply four V and H steps. In Fig. 5a we illustrate the crossing of a horizontal signal (by applying H^4). For the vertical signal the dynamics is similar but V^4 is applied. Figure 5b shows the case when two signals arrive at the junction at the same time. □

6 Other Words of Automaton Updates

For other shorter words, like the usual chip firing (with sides always open) and the words in the set $\mathbf{B} = \{HV, H^2V^2, H^3V^3\}$ we are able to construct wires, the OR and the AND gates, but we are unable to built a cross-over. In such cases we can only implement planar circuits with non-crossing wires.

Below we exhibit the different constructions for those words. It is important to point out that the strategy we used to built the constructions has been by taking as initial framework a quiescent configuration, i.e. a fixed point of the automaton. In [30] it has been proved that with this strategy, for a two dimensional grid with the von Neumann or Moore neighbourhoods it is impossible to cross information, i.e. to built a cross-over. It seems that is also the case for the words in the set **B**. In this sense one may say that our result is the best possible: no shortest word allows to cross information, at least following the quiescent strategy.

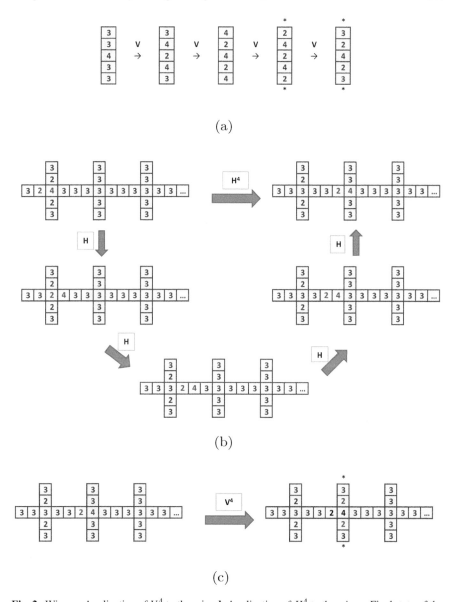

Fig. 2 Wires. **a** Application of V^4 to the wire. **b** Application of H^4 to the wire. **c** Final state of the wire after the application of V^4

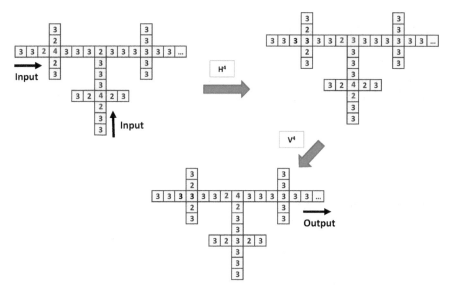

Fig. 3 The AND gate for inputs with 4 chips (signal = 1)

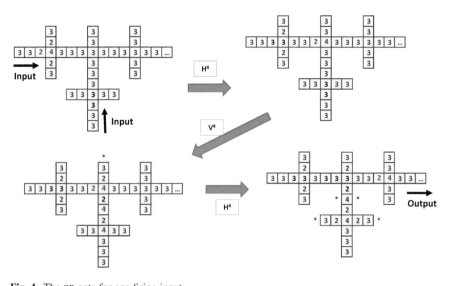

Fig. 4 The OR gate for one firing input

Computational Universality of Fungal Sandpile Automata 379

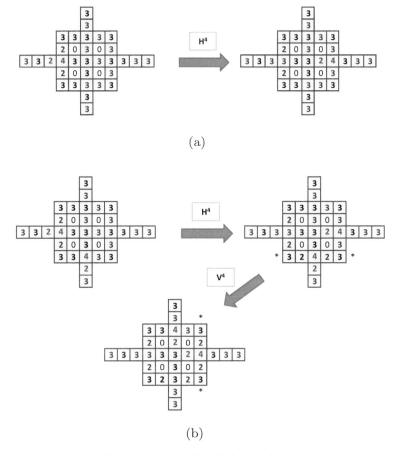

Fig. 5 The cross-over **a** with a horizontal signal and **b** for two signals

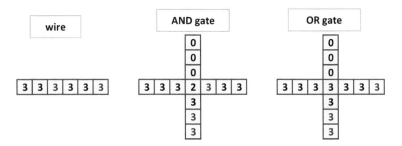

Fig. 6 The wire and gates for the classical chip firing automaton: every side is open

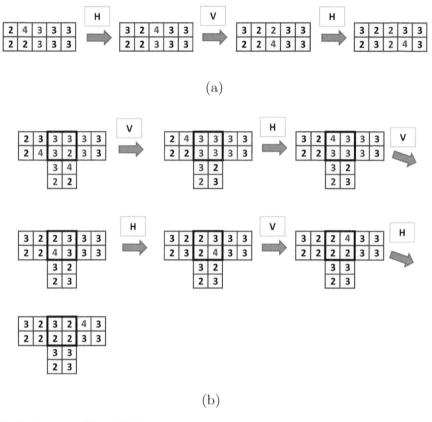

Fig. 7 Operation of the $(HV)^*$ word. **a** The wire. **b** The OR gate

In previous situations for the usual chip firing automaton the constructions are given in Fig. 6. For the word HV: wire is shown in Fig. 7a and the AND gate in Fig. 7. For the word $HHVV$, the wire is shown in Fig. 8a, the OR gate is shown in Fig. 8a and the AND gate in Fig. 8c.

7 Discussion

Using sandpile, or chip firing, automata we proved that Fungal Automata are computationally universal, i.e., by arranging positions of branching in mycelium it is possible to calculate any Boolean function.

The structure of Fungal Automata presented can be relaxed by consider the site firing chips only when it has as many chips as open side. In present model, since at each step there are only two sides can be open, the firing threshold is 2. In this

Computational Universality of Fungal Sandpile Automata

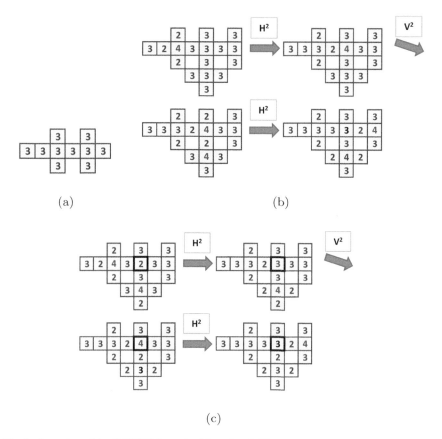

Fig. 8 Operation of the $(HHVV)^*$ word. **a** The wires. **b** The OR gate. **c** The AND gate

Fig. 9 Two chip-firing wire

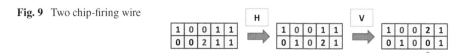

situation, the wire AND and the OR gates can be built as in previous cases but not the cross-over. Dynamics of the wire is shown in Fig. 9, the AND gate in Fig. 10.

Consider the $N \times N$ grid $\{0, 1, 2, ..., N-1\} \times \{0, 1, 2, ..., N-1\}$. Another possibility to open-close sides could be the following: at even steps $t = 0, 2, ...,$ open the even rows and columns and at odd steps, $t = 1, 3, 5, ...$ open the odd rows and columns.

If we do that we simulate exactly the Margolus partitions (2×2 blocks) [42]. This give us another way to determine the universality and, in this case, reversibility of this specific Fungal Automaton, because with this strategy one may simulate the

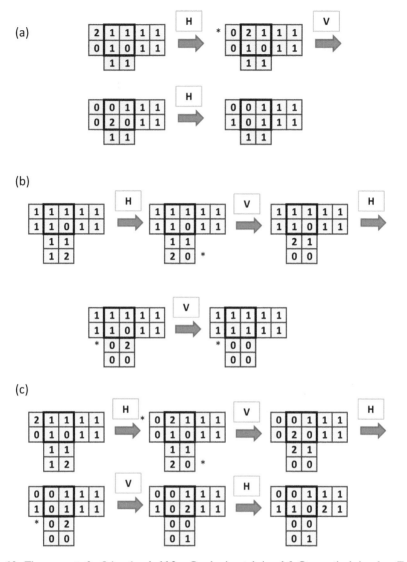

Fig. 10 The AND gate for firing threshold 2. **a** One horizontal signal. **b** One vertical signals. **c** Two signals and one output signal

Margolus billiard [43]. Not only that, given any other block partition automaton, say by $p \times p$ blocks there exist a way to open-close the sides which simulates it [44–47].

A significance of the results presented for future implementations of fungal automata with living fungal colonies in experimental laboratory conditions is the following: in our previous research, see details in [14], we used FitzHugh-Nagumo model to imitate propagation of excitation on the mycelium network of a single colony of *Aspergillus niger*. Boolean values are encoded by spikes of extracellular potential. We represented binary inputs by electrical impulses on a pair of selected electrodes and we record responses of the colony from sixteen electrodes. We derived sets of two-inputs-on-output logical gates implementable the fungal colony and analyse distributions of the gates [14]. Indeed, there were combination of functionally complete sets of gates, thus computing with travelling spikes is universal. However, in [14], we made a range of assumptions about origins, mechanisms of propagation and interactions of impulses of electrical activity. If the spikes of electrical potential do not actually propagate along the mycelium the model might be incorrect. The sandpile model presented in the chapter is more relaxed because does not any autocatalytic processes: avalanches can be physically simulated by applying constant currents, chemical stimulation to mycelium network. This is because the avalanches can be seen as movement of cytoplasm of products of fungal metabolism.

Whilst thinking about potential experimental implementation initiating avalanches is just one part of the problem. Selective control of the Woronin bodies might bring substantial challenges. As previous studies indicate the Woronin bodies can block the pores due to cytoplasmic flow [36] or mechanical stimulation of the cell wall, high temperatures, carbon and nitrogen starvation, high osmolarity and low pH [48–50]. We are unaware of experimental studies on controlling Woronin bodies with light but we believe this is not impossible.

Acknowledgements AA, MT, HABW have received funding from the European Union's Horizon 2020 research and innovation programme FET OPEN "Challenging current thinking" under grant agreement No 858132. EG residency in UWE has been supported by funding from the Leverhulme Trust under the Visiting Research Professorship grant VP2-2018-001 and from the project the project 1200006, FONDECYT-Chile.

References

1. Smith, M.L., Bruhn, J.N., Anderson, J.B.: The fungus Armillaria bulbosa is among the largest and oldest living organisms. Nat. **356**(6368), 428 (1992)
2. Ferguson, B.A., Dreisbach, T.A., Parks, C.G., Filip, G.M., Schmitt, C.L.: Coarse-scale population structure of pathogenic armillaria species in a mixed-conifer forest in the blue mountains of northeast Oregon. Can. J. For. Res. **33**(4), 612–623 (2003)
3. Jaffe, M.J., Leopold, A.C., Staples, R.C.: Thigmo responses in plants and fungi. Am. J. Bot. **89**(3), 375–382 (2002)
4. Kung, C.: A possible unifying principle for mechanosensation. Nat. **436**(7051), 647 (2005)
5. Van Aarle, I.M., Olsson, P.A., Söderström, B.: Arbuscular mycorrhizal fungi respond to the substrate ph of their extraradical mycelium by altered growth and root colonization. New Phytol. **155**(1), 173–182 (2002)

6. Fomina, M., Ritz, K., Gadd G.M.: Negative fungal chemotropism to toxic metals. FEMS Microbiol. Lett. **193**(2), 207–211 (2000)
7. Howitz, K.T., Sinclair, D.A.: Xenohormesis: sensing the chemical cues of other species. Cell. **133**(3), 387–391 (2008)
8. Purschwitz, J., Müller, S., Kastner, C., Fischer, R.: Seeing the rainbow: light sensing in fungi. Curr. Opin. Microbiol. **9**(6), 566–571 (2006)
9. Moore, D.: Perception and response to gravity in higher fungi-a critical appraisal. New Phytol. **117**(1), 3–23 (1991)
10. Slayman, C.L., Long, W.S., Gradmann, D.: "Action potentials" in Neurospora crassa, a mycelial fungus. Biochimica et Biophysica Acta (BBA)—Biomembranes. **426**(4), 732–744 (1976)
11. Olsson, S., Hansson, B.S.: Action potential-like activity found in fungal mycelia is sensitive to stimulation. Naturwissenschaften **82**(1), 30–31 (1995)
12. Adamatzky, A.: On spiking behaviour of oyster fungi Pleurotus djamor. Sci. Rep. 7873 (2018)
13. Adamatzky, A., Tuszynski, J., Pieper, J., Nicolau, D.V., Rinalndi, R., Sirakoulis, G., Erokhin, V., Schnauss, J., Smith, D.M.: Towards cytoskeleton computers. A proposal. In: Adamatzky, A., Akl, S., Sirakoulis, G. (eds.) From Parallel to Emergent Computing. CRC Group/Taylor & Francis (2019)
14. Adamatzky, A., Tegelaar, M., Wosten, H.A.B., Powell, A.L., Beasley, A.E., Mayne, R.: On Boolean gates in fungal colony. Biosyst. **193**, 104138 (2020)
15. Adamatzky, A., Goles, E., Martinez, G., Tsompanas, M.-A., Tegelaar, M., Wosten, H.A.B.: Fungal automata (2020). arXiv:2003.08168
16. Dhar, D.: Self-organized critical state of sandpile automaton models. Phys. Rev. Lett. **64**(14), 1613 (1990)
17. Goles, E.: Sand pile automata. In Annales de l'IHP Physique théorique **56**, 75–90 (1992)
18. Christensen, K., Fogedby, H.C., Jensen, H.J.: Dynamical and spatial aspects of sandpile cellular automata. J. Stat. Phys. **63**(3–4), 653–684 (1991)
19. Bitar, J., Goles, E.: Parallel chip firing games on graphs. Theor. Comput. Sci. **92**(2), 291–300 (1992)
20. Goles, E.: Sand piles, combinatorial games and cellular automata. In: Instabilities and Nonequilibrium Structures III, pp. 101–121. Springer (1991)
21. Moore, R.T., McAlear, J.H.: Fine structure of Mycota. 7. observations on septa of ascomycetes and basidiomycetes. Am. J. Bot. **49**(1), 86–94 (1962)
22. Lew, R.R.: Mass flow and pressure-driven hyphal extension in neurospora crassa. Microbiol. **151**(8), 2685–2692 (2005)
23. Bleichrodt, R.-J., van Veluw, G.J., Recter, B., Maruyama, J.-I., Kitamoto, K., Wösten, H.A.B.: Hyphal heterogeneity in a spergillus oryzae is the result of dynamic closure of septa by Woronin bodies. Mol. Microbiol. **86**(6), 1334–1344 (2012)
24. Bleichrodt, R.-J., Hulsman, M., Wösten, H.A.B., Reinders, M.J.T.: Switching from a unicellular to multicellular organization in an aspergillus niger hypha. MBio. **6**(2), e00111–15 (2015)
25. Tegelaar, M., Bleichrodt, R.-J., Nitsche, B., Ram, A.F.J., Wösten, H.A.B.: Subpopulations of hyphae secrete proteins or resist heat stress in aspergillus oryzae colonies. Environ. Microbiol. **22**(1), 447–455 (2020)
26. Goles, E., Margenstern, M.: Sand pile as a universal computer. Int. J. Mod. Phys. C **7**(02), 113–122 (1996)
27. Goles, E., Margenstern, M.: Universality of the chip-firing game. Theor. Comput. Sci. **172**(1–2), 121–134 (1997)
28. Chessa, A., Stanley, H.E., Vespignani, A., Zapperi, S.: Universality in sandpiles. Phys. Rev. E. **59**(1), R12 (1999)
29. Moore, C., Nilsson, M.: The computational complexity of sandpiles. J. Stat. Phys. **96**(1–2), 205–224 (1999)
30. Gajardo, A., Goles, E.: Crossing information in two-dimensional sandpiles. Theor. Comput. Sci. **369**(1–3), 463–469 (2006)

31. Martin, B.: On Goles' universal machines: a computational point of view. Theor. Comput. Sci. **504**, 83–88 (2013)
32. Tenney, K., Hunt, I., Sweigard, J., Pounder, J.I., McClain, C., Bowman, E.J., Bowman, B.J.: Hex-1, a gene unique to filamentous fungi, encodes the major protein of the Woronin body and functions as a plug for septal pores. Fungal Genet. Biol. **31**(3), 205–217 (2000)
33. Reichle, R.E., Alexander, J.V.: Multiperforate septations, Woronin bodies, and septal plugs in fusarium. J. Cell Biol. **24**(3), 489 (1965)
34. Tey, W.K., North, A.J., Reyes, J.L., Lu, Y.F., Jedd, G.: Polarized gene expression determines Woronin body formation at the leading edge of the fungal colony. Mol. Biol. Cell. **16**(6), 2651–2659 (2005)
35. Jedd, G., Chua, N.-H.: A new self-assembled peroxisomal vesicle required for efficient resealing of the plasma membrane. Nat. Cell Biol. **2**(4), 226–231 (2000)
36. Steinberg, G., Harmer, N.J., Schuster, M., Kilaru, S.: Woronin body-based sealing of septal pores. Fungal Genet. Biol. **109**, 53–55 (2017)
37. Bleichrodt, R.-J., Vinck, A., Read, N.D., Wösten, H.A.B.: Selective transport between heterogeneous hyphal compartments via the plasma membrane lining septal walls of aspergillus niger. Fungal Genet. Biol. **82**, 193–200 (2015)
38. Wolfram, S.: Cellular Automata and Complexity: Collected Papers. Addison-Wesley Pub. Co. (1994)
39. Goles, E., Margenstern, M.: Universality of the chip-firing game. Theor. Comput. Sci. **172**(1–2), 121–134 (1997)
40. Moore, C., Nilsson, M.: The computational complexity of sandpiles. J. Stat. Phys. **96**(1–2), 205–224 (1999)
41. Greenlaw, R., Hoover, H.J., Ruzzo, W.L., et al.: Limits to Parallel Computation: P-Completeness Theory. Oxford University Press on Demand (1995)
42. Margolus, N.: Physics-like models of computation. Phys. D Nonlinear Phenom. **10**(1), 81–95 (1984)
43. Margolus, N.: Universal cellular automata based on the collisions of soft spheres. In: Adamatzky, A (ed.) Collision-Based Computing, pp. 107–134. Springer (2002)
44. Morita, K., Harao, M.: Computation universality of one-dimensional reversible (injective) cellular automata. IEICE Trans. (1976–1990), **72**(6), 758–762 (1989)
45. Durand-Lose, J.O.: Reversible space–time simulation of cellular automata. Theor. Comput. Sci. *246*(1–2), 117–129 (2000)
46. Imai, K., Morita, K.: A computation-universal two-dimensional 8-state triangular reversible cellular automaton. Theor. Comput. Sci. **231**(2), 181–191 (2000)
47. Durand-Lose, J.: Representing reversible cellular automata with reversible block cellular automata. Discret. Math. Theor. Comput. Sci. **145**, 154 (2001)
48. Tegelaar, M., Wösten, H.A.B.: Functional distinction of hyphal compartments. Sci. Rep. **7**(1), 1–6 (2017)
49. Soundararajan, S., Jedd, G., Li, X., Ramos-Pamploña, M., Chua, N.H., Naqvi, N.I.: Woronin body function in magnaporthe grisea is essential for efficient pathogenesis and for survival during nitrogen starvation stress. Plant Cell. **16**(6), 1564–1574 (2004)
50. Jedd, G.: Fungal evo-devo: organelles and multicellular complexity. Trends Cell Biol. **21**(1), 12–19 (2011)

Fungal Language and Cognition

Language of Fungi Derived from their Electrical Spiking Activity

Andrew Adamatzky

Abstract Fungi exhibit oscillations of extracellular electrical potential recorded via differential electrodes inserted into a substrate colonised by mycelium or directly into sporocarps. We analysed electrical activity of ghost fungi (*Omphalotus nidiformis*), Enoki fungi (*Flammulina velutipes*), split gill fungi (*Schizophyllum commune*) and caterpillar fungi (*Cordyceps militaris*). The spiking characteristics are species specific: a spike duration varies from one to 21 h and an amplitude from 0.03 mV to 2.1mV. We found that spikes are often clustered into trains. Assuming that spikes of electrical activity are used by fungi to communicate and process information in mycelium networks, we group spikes into words and provide a linguistic and information complexity analysis of the fungal spiking activity. We demonstrate that distributions of fungal word lengths match that of human languages. We also construct algorithmic and Liz-Zempel complexity hierarchies of fungal sentences and show that species *S. commune* generate the most complex sentences.

1 Introduction

Spikes of electrical potential are typically considered to be key attributes of neurons and neuronal spiking activity is interpreted as a language of a nervous system [1–3]. However, almost all creatures without nervous system produce spikes of electrical potential—Protozoa [4–6], Hyrdoroza [7], slime moulds [8, 9] and plants [10–12]. Fungi also exhibit trains of action-potential like spikes, detectable by intra-and extracellular recordings [13–15]. In experiments with recording of electrical potential of oyster fungi *Pleurotus djamor* we discovered two types of spiking activity: high-frequency (period 2.6 m) and low-freq (period 14 m) [15]. While studying other species of fungus, *Ganoderma resinaceum*, we found that most common width of an electrical potential spike is 5–8 m [16]. In both species of fungi we observed bursts

---

A. Adamatzky (✉)
Unconventional Computing Laboratory, UWE, Bristol, UK
e-mail: andrew.adamatzky@uwe.ac.uk

of spiking in the trains of the spike similar to that observed in central nervous system [17, 18]. Whilst the similarly could be just phenomenological this indicates a possibility that mycelium networks transform information via interaction of spikes and trains of spikes in manner homologous to neurons. First evidence has been obtained that indeed fungi respond to mechanical, chemical and optical stimulation by changing pattern of its electrically activity and, in many cases, modifying characteristics of their spike trains [19, 20]. There is also evidence of electrical current participation in the interactions between mycelium and plant roots during formation of mycorrhiza [21]. In [22] we compared complexity measures of the fungal spiking train and sample text in European languages and found that the 'fungal language' exceeds the European languages in morphological complexity.

In our venture to decode the language of fungi we first uncover if all species of fungi exhibit similar characteristics of electrical spiking activity. Then we characterise the proposed language of fungi by distributions of word length and complexity of sentences.

There is an emerging body of studies on language of creature without a nervous system and invertebrates. Biocommunication in ciliates [23] include intracellular signalling, chemotaxis as expression of communication, signals for vesicle trafficking, hormonal communication, pheromones. Plants communication processes are seen as primarily sign-mediated interactions and not simply an exchange of information [24, 25]. Evidences of different kinds of chemical 'words' in plants are discussed in [26, 27]. Moreover, a modified conception of language of plants is considered to be a pathway toward "the de-objectification of plants and the recognition of their subjectivity and inherent worth and dignity" [28]. A field of the language of insects has been developed by Karl von Frisch and resulted in his Nobel Prize for detection and investigation of bee languages and dialects [29, 30]. An issue of the language of ants, and how species hosted by ants can communicate the ants language, was firstly promoted in 1971 [31]. In early 1980s analysis of the ants' language using information theory approaches has been proposed [32]. The approach largely succeeded in analysis of ants' cognitive capacities [33–36].

2 Experimental Laboratory Methods and Analysis

Four species of fungi have been used in experiments: *Omphalotus nidiformis* and *Flammulina velutipes*, supplied by Mycelia NV, Belgium (www.mycelium.be), *Schizophyllum commune*, collected near Chew Valley lake, Somerset, UK, *Cordyceps militaris*, supplied by Kaizen Cordyceps, UK (www.kaizencordyceps.co.uk).

Electrical activity of the fungi was recorded using pairs of iridium-coated stainless steel sub-dermal needle electrodes (Spes Medica S.r.l., Italy), with twisted cables and ADC-24 (Pico Technology, UK) high-resolution data logger with a 24-bit A/D converter, galvanic isolation and software-selectable sample rates all contribute to a superior noise-free resolution. Each pair of electrodes reported a potential difference between the electrodes. The pairs of electrodes were pierced into the substrates

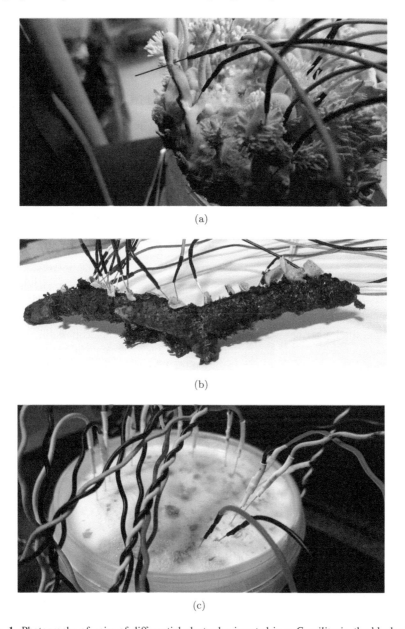

Fig. 1 Photographs of pairs of differential electrodes inserted in **a** *C. militaris*, the block of a substrate colonised by the fungi was removed from the plastic container to make a photo after the experiments, **b** *S. commune*, the twig with the fungi was removed from the humid plastic container to make a photo after the experiment, **c** *F. velutipes*, the container was kept sealed and electrodes pierced through the lid

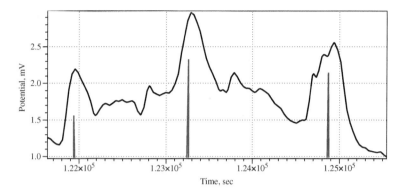

Fig. 2 Example of spike detection. Temporal position of each spike is shown by red vertical line. The minor shift of the vertical lines away from the summits is consistent all over the recording and therefore does not affect the results of the analysis

colonised by fungi or, as in case of *S. commune*, in the sporocarps are shown in Fig. 1. Distance between electrodes was 1–2 cm. We recorded electrical activity one sample per second. We recorded 8 electrode pairs simultaneously. During the recording, the logger has been doing as many measurements as possible (typically up to 600 per second) and saving the average value. The acquisition voltage range was 78 mV. *S. commune* has been recorded for 1.5 days, other species for c. 5 days. The experiments took place at temperature 21 °C, c. 80% humidity, darkness.

Spikes of electrical potential have been detected in a semi-automatic mode as follows. For each sample measurement x_i we calculated average value of its neighbourhood as $a_i = (4 \cdot w)^{-1} \cdot \sum_{i-2\cdot w \leq j \leq i+2\cdot w} x_j$. The index i is considered a peak of the local spike if $|x_i| - |a_i| > \delta$. The list of spikes were further filtered by removing false spikes located at a distance d from a given spike. Parameters were species specific, for *C. militaris* and *F. velutipes* $w = 200$, $\delta = 0.1$, $d = 300$; for *S. commune* $w = 100$, $\delta = 0.005$, $d = 100$; for *O. nidiformis* $w = 50$, $\delta = 0.003$, $d = 100$. An example of the spikes detected is shown in Fig. 2. Over 80% of spikes have been detected by such a technique.

3 Characterisation of the Electrical Spiking of Fungi

Examples of electrical activity recorded are shown in Fig. 3. Intervals between the spikes and amplitudes of spikes are characterised in Fig. 4 and Table 1.

C. militaris shows the lowest average spiking frequency amongst the species recorded (Figs. 3 and 4a): average interval between spikes is nearly two hours. The diversity of the frequencies recorded is highest amongst the species studied: standard deviation is over five hours. The spikes detected in *C. militaris* and *F. velutipes* have

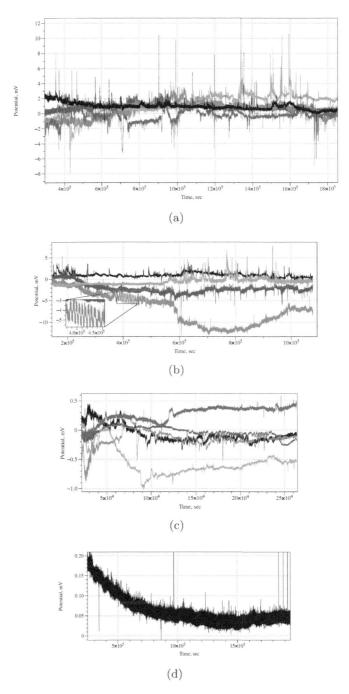

Fig. 3 Examples of electrical activity of **a** *C. militaris*, **b** *F. velutipes*, insert shows zoomed in burst of high-frequency spiking. **c** *S. commune*, **d** *O. nidiformis*. Colours reflect recordings from different channels

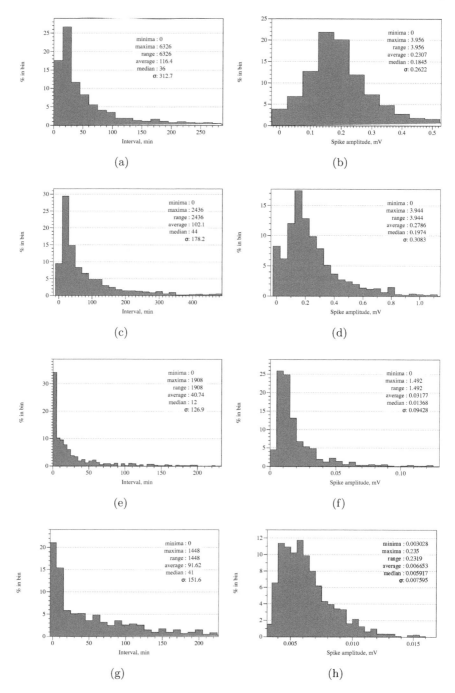

Fig. 4 Distribution of intervals between spikes **aceg** and average spike amplitude **bdfh** of **ab** *C. militaris*, **cd** *F. velutipes*, **ef** *S. commune*, **gh** *O. nidiformis*

Table 1 Characteristics of electrical potential spiking: number of spikes recorded, average interval between spikes, and average amplitude of a spike

Species	Number of spikes	Interval, min	Amplitude, mV
C. militaris	881	116	0.2
F. velutipes	958	102	0.3
S. commune	530	41	0.03
O. nidiformis	1117	92	0.007

highest amplitudes: 0.2 mV and 0.3 mV, respectively. Variability of the amplitudes in both species is high, standard deviation nearly 0.3.

Enoki fungi *F. velutipes* show a rich spectrum of diverse patterns of electrical activity which combines low and high frequency oscillations (Fig. 3b). Most commonly exhibited patterns are characterised by low frequency irregular oscillations: average amplitude 0.3 mV (Fig. 4d) and average interval between two spikes is just over 1.5 hr (Fig. 4c and Table 1). There are also bursts of spiking showing a transition from a low frequency spiking to high frequency and back, see recording in blue in Fig. 3b. There are 12 spikes in the train, average amplitude is 2.1 mV, $\sigma = 0.1$, average duration of a spike is 64 min, $\sigma = 1.7$.

O. nidiformis also show low amplitude and low frequency electrical spiking activity with the variability of the characteristics highest amongst species recorded (Fig. 3d and Table 1). Average interval between the spikes is just over 1.5 hr with nearly 2.5 hr standard variation (Fig. 4g). Average amplitude is 0.007 mV but the variability of the amplitudes is very high: $\sigma = 0.006$ (Fig. 4h).

S. commune electrical activity is remarkably diverse (Figs. 3c and 4ef). Typically, there are low amplitude spikes detected (Fig. 4f), due to the reference electrodes in each differential pair being inserted into the host wood. However, they are the fastest spiking species, with an average interval between spikes is just above half-an-hour (Fig. 4e). We observed transitions between different types of spiking activity from low amplitude and very low frequency spikes to high amplitude high frequency spikes (Fig. 5). A dynamic change in spikes frequency in the transition Fig. 5 is shown in Fig. 5a. A closer look at the spiking discovers presence of two wave packets labelled (p_1, p_2) and (p_2, p_3) in Fig. 5a. One of the wave-packets is shown in Fig. 5c, and the key characteristics are shown in Fig. 6.

In experiments with *S. commune* we observed synchronisation of the electrical potential spikes recorded on the neighbouring fruit bodies. This is illustrated in Fig. 7. The dependencies between the spikes are shown by red (increase of potential spike) and green (decrease of potential spike) lines in Fig. 7a. Time intervals between peaks of the spikes occurred on neighbouring fruit bodies are illustrated in Fig. 7b. Average interval between first four spikes is 1425 s ($\sigma = 393$), next three spikes 870 s ($\sigma = 113$), and last four spikes 82 ($\sigma = 73$).

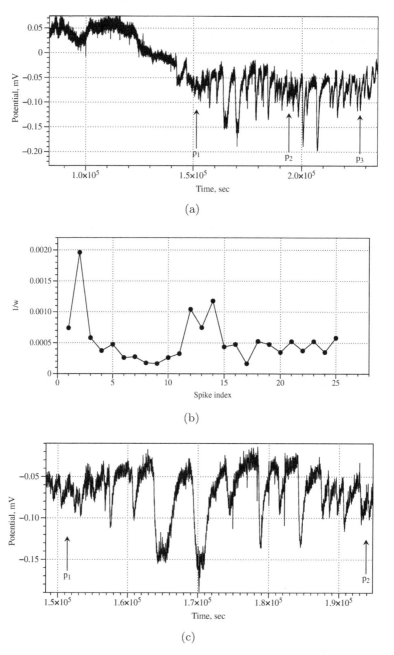

Fig. 5 Transition to spikes outburst in *S. commune*. **a** There are two outburst of spiking, first shown by arrows labelled p_1 and p_2 and second by p_2 and p_3. **b** Dynamical changes in frequency of spikes, as derived from (**a**). **c** Wave packet zoomed in, start of the packet is shown by arrow labelled p_1 and end by p_2

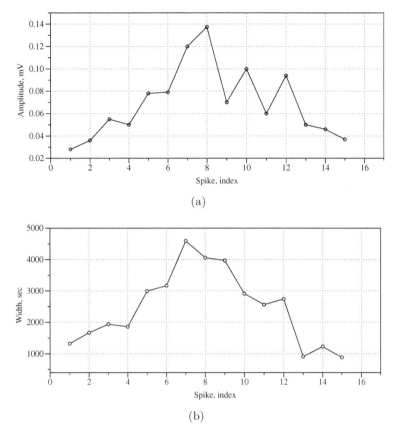

Fig. 6 Characteristics of an exemplar wave packet of electrical potential oscillation in *S. commune*: **a** evolution of spike amplitude, **b** evolution of spike width. In a typical wave-packet spike width and amplitude increase till middle of the packet and then decrease

4 Towards Language of Fungi

Are the elaborate patterns of electrical activity used by fungi to communicate states of the mycelium and its environment and to transmit and process information in the mycelium networks? Is there a language of fungi? When interpreting fungal spiking patterns as a language, here we consider consider a number of linguistic phenomena as have been successfully used to decode pictish symbols revealed as a written language in [37]: (1) type of characters used to code, (2) size of the character lexicon, (3) grammar, (4) syntax (word order), (5) standardised spelling. These phenomena, apart from grammar and spelling, are analysed further.

To quantify types of characters used and a size of lexicon we convert the spikes detected in experimental laboratory recordings to binary strings s, where index i is the index of the sample taken at ith second of recording and $s_i = 1$ if there is a spike's

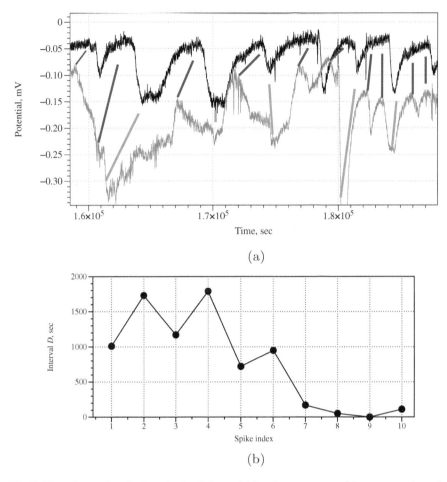

Fig. 7 Exemplar synchronisation of spikes in two neighbouring sporocarps of *S. commune*: channel (3–4), second sporocarp in Fig. 1b, and channel (5–6), third sporocarp in Fig. 1b. **a** Spiking activity, corresponding spikes of increased voltage are linked by red lines and decreased voltage by green line. **b** Dynamics of the interval between spikes.

peak at ith second and $s_0 = 0$ otherwise. Examples of the binary strings, in a bar code like forms, extracted from the electrical activity of *C. militaris* and *F. velutipes* are shown in Fig. 8.

To convert the binary sequences representing spikes into sentences of the speculative fungal language, we must split the strings into words. We assumed that if a distance between consequent spikes is not more than θ the spikes belong to the same word. To define θ we adopted analogies from English language. An average vowel duration in English (albeit subject to cultural and dialect variations) is 300 ms, minimum 70 ms and maximum 400 ms [41], with average post-word onset of c. 300 ms [42]. We explored two options of the separation the spike trains into words:

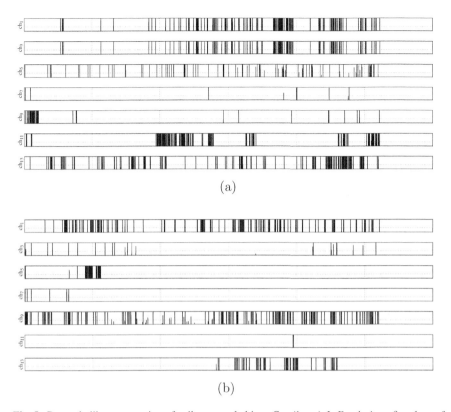

Fig. 8 Bar code like presentation of spikes recorded in **a** *C. milataris* **b** *F. velutipes*, five days of recording

$\theta = a(s)$ and $\theta = 2 \cdot a(s)$, where $a(s)$ is an average interval between two subsequent spikes recorded in species $s \in \{C.\ militaris,\ F.\ velutipes,\ S.\ commune, O.\ nidiformis\}$. Distributions of fungal word lengths, measured in a number of spikes in θ-separated trains of spikes are shown in Fig. 9. The distributions follow predictive values $f_{exp} = \beta \cdot 0.73 \cdot l^c$, where l is a length of a word, and a varies from 20 to 26, and b varies from 0.6 to 0.8, similarly to frequencies of word lengths in English and Swedish, Fig. 10 and Table 2 [38]. As detailed in Table 2, average word length in fungi, when spikes grouped with $\theta = a$ are the same range as average word lengths of human languages. For example, average number of spikes in train of *C. militaris* is 4.7 and average word length in English language is 4.8. Average word length of *S. commune* is 4.4 and average word length in Greek language is 4.45.

To uncover syntax of the fungal language we should estimate what is most likely order of the words in fungal sentences. We do this via characterisation of global transition graphs of fungal spiking machines. A fungal spiking machine is a finite state machine. It takes states from $\mathbf{S} \in \mathbf{N}$ and updates its states according to probabilistic transitions: $\mathbf{S} \times [0, 1] \rightarrow \mathbf{S}$, being in a state $s^t \in \mathbf{S}$ at time $t + 1$ the automaton takes

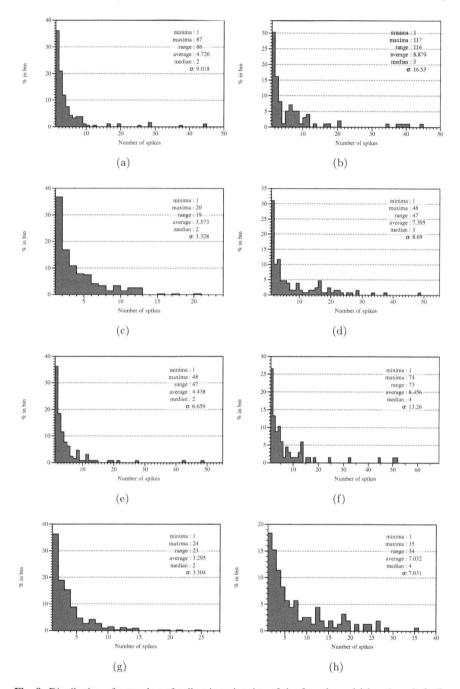

Fig. 9 Distribution of a number of spikes in trains, i.e. of the fungal words' lengths, of **ab** *C. militaris*, **cd** *F. velutipes*, **ef** *S. commune*, **gh** *O. nidiformis* for the train separation thresholds a **acef** and $2 \cdot a$ **bdfh**, where a is a species specific average interval between two consequent spikes, see Table 1

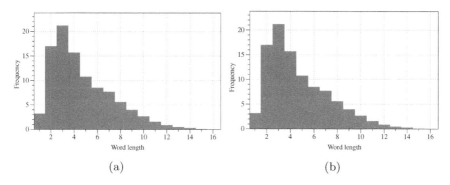

Fig. 10 Word length frequencies in **a** English and **b** Swedish, data are taken from Table 1 in [38]

Table 2 Average word lengths in fungal and human languages. l_1 is an average word length in the spike grouping using $\theta = a$ and l_2 using $\theta = 2 \cdot a$, m is an average word length of 1950+ Russian and English language approximated from the evolutionary plots in [39] and average word length in Greek language approximated from Hellenic National Corpus [40]

	l_1	l_2
C. militaris	4.7	8.9
F. velutipes	3.6	7.4
S. commune	4.4	8.5
O. nidiformis	3.3	7
	m	
English language	4.8	
Russian language	6	
Greek language	4.45	

state $s^{t+1} \in \mathbf{S}$ with probability $p(s^t, s^{t+1}) \in [0, 1]$. The probabilities of the state transitions are estimated from the sentences of the fungal language.

The state transition graphs of the fungal spiking machines are shown in Fig. 11 for full dictionary case and in Fig. 12 for the filtered states sets when states over 9 are removed.

The probabilistic state transition graphs shown in Fig. 11 are drawn using physical model spring-based Kamada-Kawai algorithm [43]. Thus we can clearly see cores of the state space as clusters of closely packed states. The cores act as attractive measures in the probabilistic state space. The attractive measures are listed in Table 3. The membership of the cores well matches distribution of spike trains lengths (Fig. 9).

A leaf, or Garden-of-Eden, state is a state which has no predecessors. *C. militaris* probabilistic fungal spiking machine has leaves '25' and '37' in case of in grouping $\theta = a$ (Fig. 11a) and '20', '17' and '37' in case of in grouping $\theta = 2 \cdot a$ (Fig. 11e). All other probabilistic fungal machines do not have leaves apart of *S. commune* which has one leaf '11' in case of in grouping $\theta = 2 \cdot a$ (Fig. 11g).

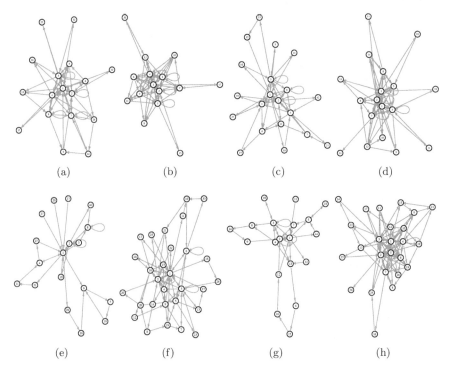

Fig. 11 State transition graphs of fungal spiking machines, where spikes have been grouped using $\theta = a$ **a–d** and $\theta = 2 \cdot a$ **e–h**. **ae** *C. militaris*, **bf** *F. velutipes*, **cg** *S. commune*, **dh** *O. nidiformis*

An absorbing state of a finite state machine is a state in which the machine remains forever once it takes this state. All spiking fungal machines, derived in grouping $\theta = a$, have the only absorbing state '1' (Fig. 12a–d). They have no cycles in the state space. There are between 8, *F. velutipes* (Fig. 12c) and *O. nidiformis* (Fig. 12g), and 11 leaves, *S. commune*) (Fig. 12e) in the global transition graphs. A maximal length of a transient period, measured in a maximal number of transitions required to reach the absorbing state from a leaf state varies from 3 (*F. velutipes*) to 11 (*S. commune*).

State transition graphs get more complicated, as we evidence further, when grouping $\theta = 2 \cdot a$ is used (Fig. 12h). Fungal spiking machine *O. nidiformis* has one absorbing state, '1' (Fig. 12h). Fungal spiking machines *S. commune* (Fig. 12g) and *C. militaris* (Fig. 12e) have two absorbing state each, '1' and '2' and '1' and '8', respectively. The highest number of absorbing states is found in the state transition graph of the *F. velutipes* spiking machine (Fig. 12f). They are '1', '6' and '2'. A number of leaves varies from 7, *S. commune*, to 9, *O. nidiformis* and *C. militaris*, to 12, *F. velutipes*. Only *O. nidiformis* spiking machine has cycles in each state transition graph (Fig. 12h). The cycles are $1 \longleftrightarrow 5$ and $2 \longleftrightarrow 3$.

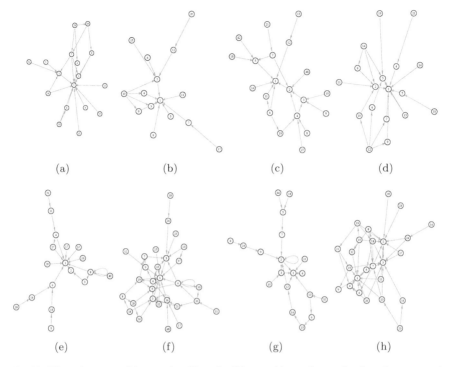

Fig. 12 Filtered state transition graphs of fungal spiking machines, where spikes have been grouped using $\theta = a$ **a–d** and $\theta = 2 \cdot a$ **e–h**. **ae** *C. militaris*, **bf** *F. velutipes*, **cg** *S. commune*, **dh** *O. nidiformis*. The transitions were filtered in such manner that for each state i we select state j such that the weight $w(i, j)$ is maximal over $w(i, z)$, where $z \in \mathbf{S}$, $\mathbf{S}$ is a set of states

Table 3 Attractive cores in the probabilistic state spaces of fungal spiking machines. The attractive cores, or limit cycle, are such subgraphs of the global transition graph that when a machine enters the subgraph it will stay there forever

	$\theta = a$	$\theta = 2 \cdot a$
C. militaris	1, ..., 8	1, ..., 3,
F. velutipes	1, ..., 8, 9	1, 2, 4, 15, 16
S. commune	1, ..., 4, 8	1, ..., 4, 7
O. nidiformis	1, ..., 5	1, ..., 5, 10, 12

To study complexity of the fungal language algorithmic complexity [44], Shannon entropy [45] and Liv-Zempel complexity [46, 47] of the fungal words (sequences of spike trains lengths) are estimated using The Online Algorithmic Complexity Calculator[1] [44, 48–50] in Table 4. The complexity estimates help us to rule out randomness of the electrical spiking events and to compare complexity of the fungal

---

[1] https://complexitycalculator.com/index.html.

Table 4 Block Decomposition Method (BDM) algorithmic complexity estimation, BDM logical depth estimation, Shannon entropy, Second order entropy, LZ complexity. The measures are estimated using The Online Algorithmic Complexity Calculator (https://complexitycalculator.com/index.html) block size 12, alphabet size 256. Spike trains are extracted with (a) $\theta = a$ and (b) $\theta = 2 \cdot a$, where a is an average interval between two consequent spikes, see Table 1. We also provide values of the LZ complexity and algorithmic complexity normalised by input string lengths. In table (c) we provide data on the strings of train powers (in number of spikes) calculated with $\theta = a$ and then filtered so value over 9 are removed and the complexity is estimated in alphabet of 9 symbols

(a)

	C. militaris	F. velutipes	S. commune	O. nidiformis
Algorithmic complexity, bits	1211	1052	981	1243
Algorithmic complexity normalised	6.51	3.94	7.55	3.67
Logical depth, steps	4321	4957	3702	5425
Logical depth normalised	23	19	28	16
Shannon entropy, bits	2.4	2.3	2.3	2.3
Second order entropy, bits	3.8	3.7	3.7	3.7
LZ complexity, bits	1153	1495	910	1763
LZ complexity (normalised), bits	6.2	5.6	7	5.2
Input string length	186	267	130	339

(b)

	C. militaris	F. velutipes	S. commune	O. nidiformis
Algorithmic complexity, bits	1047	1295	980	1393
Algorithmic complexity normalised	10.57	10.04	14.4	8.82
Logical depth, steps	2860	4147	3046	4731
Logical depth normalised	29	32	45	30
Shannon entropy, bits	2.5	2.5	2.5	2.6
Second order entropy, bits	4	4.2	4.2	4.3
LZ complexity, bits	594	993	666	1232
LZ complexity normalised	6	7.7	9.8	7.8
Input string length	99	129	68	158

(c)

	C. militaris	F. velutipes	S. commune	O. nidiformis
Algorithmic complexity, bits	679	976	466	1276
Algorithmic complexity normalised	3.96	3.97	4.05	4
Shannon entropy, bits	2.5	2.6	2.5	2.5
Second order entropy, bits	4.7	5	4.5	4.9
LZ complexity, bits	735	1009	563	1208
LZ complexity normalised	4.3	4.1	4.9	3.8
Input string length	171	246	115	319

language with that of human. Shannon entropy of the strings recorded is not shown to be species specific, it is 2.3 for most species but 2.4 for *C. militaris* in case of $\theta = a$ grouping and 2.5 for most species but 2.6 for *O. nidiformis* in case of $\theta = 2 \cdot a$. The same can be said about second order entropy (Table 4). *O. nidiformis* shows highest values of algorithmic complexity for both cases of spike trains separation (Table 4ab) and filtered sentences (where only words with up to 9 spikes are left) (Table 4c). In other of decreasing algorithmic complexity we then have *C. militaris*, *F. velutipes* and *S. commune*.

The hierarchy of algorithmic complexity changes when we normalise the complexity values dividing them by the string lengths. For the case $\theta = a$ the hierarchy of descending complexity will be *S. commune* (7.55), *C. militaris* (6.51), *F. velutipes* (3.94), *O. nidiformis* (3.67) (Table 4a). Note that in this case a normalised algorithmic complexity of *S. commune* is nearly twice higher than that of *O. nidiformis*. For the case $\theta = 2 \cdot a$ *S. commune* still has the highest normalised algorithmic complexity amongst the four species studied (Table 4b). Complexities of *C. militaris* and *F. velutipes* are almost the same, and the complexity of *O. nidiformis* is the lowest. When we consider filtered sentences of fungal electrical activity, where words with over 9 spikes are removed, we get nearly equal values of the algorithmic complexity, ranging from 3.96 to 4.05 (Table 4c). LZ complexity hierarchy is the same for all three cases—$\theta = a$ (Table 4a), $\theta = 2 \cdot a$ (Table 4b) and filtered sentences (Table 4c): *S. commune*, *C. militaris*, *F. velutipes*, *O. nidiformis*. To summarise, in most conditions, *S. commune* is an uncontested champion in complexity of the sentences generated followed by *C. militaris*.

5 Discussion

We recorded extracellular electrical activity of four species of fungi. We found evidences of the spike trains propagating along the mycelium network. We speculated that fungal electrical activity is a manifestation of the information communicated between distant parts of the fungal colonies. We adopted a framework of information encoding into spikes in neural system [51–54] and assumed that the information in electrical communication of fungi are encoded into trains of spikes. We therefore attempted to uncover key linguistic phenomena of the proposed fungal language. We found that distributions of lengths of spike trains, measured in a number of spikes, follow the distribution of word lengths in human languages. We found that size of fungal lexicon can be up to 50 words, however the core lexicon of most frequently used words does not exceed 15–20 words. Species *S. commune* and *O. nidiformis* have largest lexicon while species *C. militaris* and *F. velutipes* have less extensive one. Depending on the threshold of spikes grouping into words, average word length varies from 3.3 (*O. nidiformis*) to 8.9 (*C. militaris*). A fungal word length averaged over four species and two methods of spike grouping is 5.97 which is of the same range as an average word length in some human languages, e.g. 4.8 in English and 6 in Russian.

To characterise a syntax of the fungal language we analysed state transition graphs of the probabilistic fungal spiking machines. We found that attractive measures, or communication cores, of the fungal machines are composed of the words up to ten spikes long with longer words appearing less often.

We analysed complexity of the fungal language and found that species *S. commune* generates most complex, amongst four species studied, sentences. The species *C. militaris* is slightly below *S. commune* in the hierarchy of complexity and *F. velutipes* and *O. nidiformis* occupy lower levels of the hierarchy. We found that Shannon entropy poorly, if at all, discriminate between the species. That could due to sentences in the fungal language posses the same amount of information about physiological state of fungi and environment. LZ complexity, algorithmic complexity and logical depth give us substantial differentiation between species. The algorithmic complexity is most 'species sensitive' measure. This could be to the fact, while convey the same amount of information, dialects of different species are different.

Future research should go in three directions: study of inter-species variations, interpretation of a fungal grammar and re-consideration of the coding type. First, we should increase a number of fungi species studied to uncover if there is a significant variations in the language syntax among the species. Second, we should try to uncover grammatical constructions, if any in the fungal language, and to attempt to semantically interpret syntax of the fungal sentences. Third, and probably the most important direction of future research, would to be make a thorough and detailed classification of fungal words, derived from the train of spikes. Right now we classified the word based solely on a number of spikes in the corresponding trains. This is indeed quite a primitive classification akin to interpreting binary words only by sums of their bits and not exact configurations of 1s and 0s. That said we should not expect quick results: we are yet to decipher language of cats and dogs despite living with them for centuries and research into electrical communication of fungi is in its pure infant stage. And last but not least, there may be alternative interpretations of spiking electrical activity as a language. For example, one can adopt the technique of signals integration over time trace, as has been in experiments with chemical Turing machine [55]. Another option could be to characterise each peak by determining its fuzzy entropy by the algorithm presented in [56].

References

1. Baslow, M.H.: The languages of neurons: an analysis of coding mechanisms by which neurons communicate, learn and store information. Entropy. **11**(4), 782–797 (2009)
2. Andres, D.S.: The language of neurons: theory and applications of a quantitative analysis of the neural code. Int. J. Med. Biol. Front. **21**(2), 133 (2015)
3. Pruszynski, J.A., Zylberberg, J.: The language of the brain: real-world neural population codes. Curr. Opin. Neurobiol. **58**, 30–36 (2019)
4. Eckert, R., Naitoh, Y., Friedman, K.: Sensory mechanisms in paramecium. i. J. Exp. Biol. **56**, 683–694 (1972)
5. Bingley, M.S.: Membrane potentials in amoeba Proteus. J. Exp. Biol. **45**(2), 251–267 (1966)

6. Ooyama, S., Shibata, T.: Hierarchical organization of noise generates spontaneous signal in paramecium cell. J. Theor. Biol. **283**(1), 1–9 (2011)
7. Hanson, A.: Spontaneous electrical low-frequency oscillations: a possible role in hydra and all living systems. Philos. Trans. R. Soc. B. **376**(1820), 20190763 (2021)
8. Iwamura, T.: Correlations between protoplasmic streaming and bioelectric potential of a slime mold. Physarum polycephalum. Shokubutsugaku Zasshi **62**(735–736), 126–131 (1949)
9. Kamiya, N., Abe, S.: Bioelectric phenomena in the myxomycete plasmodium and their relation to protoplasmic flow. J. Colloid Sci. **5**(2), 149–163 (1950)
10. Trebacz, K., Dziubinska, H., Krol, E.: Electrical signals in long-distance communication in plants. In: Communication in Plants, pp. 277–290. Springer (2006)
11. Fromm, J., Lautner, S.: Electrical signals and their physiological significance in plants. Plant Cell & Environ. **30**(3), 249–257 (2007)
12. Zimmermann, M.R., Mithöfer, A.: Electrical long-distance signaling in plants. In: Long-Distance Systemic Signaling and Communication in Plants, pp. 291–308. Springer (2013)
13. Slayman, C.L., Long, W.S., Gradmann, D.: "Action potentials" in *Neurospora crassa*, a mycelial fungus. Biochimica et Biophysica Acta (BBA)—Biomembranes. **426**(4), 732–744 (1976)
14. Olsson, S., Hansson, B.S.: Action potential-like activity found in fungal mycelia is sensitive to stimulation. Naturwissenschaften **82**(1), 30–31 (1995)
15. Adamatzky, A.: On spiking behaviour of oyster fungi Pleurotus djamor. Sci. Rep. **8**(1), 1–7 (2018)
16. Adamatzky, A., Gandia, A.: On electrical spiking of Ganoderma resinaceum. Biophys. Rev. Lett. :1–9
17. Cocatre-Zilgien, J.H., Delcomyn, F.: Identification of bursts in spike trains. J. Neurosci. Methods. **41**(1), 19–30 (1992)
18. Legendy, C.R., Salcman, M.: Bursts and recurrences of bursts in the spike trains of spontaneously active striate cortex neurons. J. Neurophysiol. **53**(4), 926–939 (1985)
19. Adamatzky, A., Gandia, A., Chiolerio, A.: Fungal sensing skin. Fungal Biol. Biotechnol. **8**(1), 1–6 (2021)
20. Adamatzky, A., Nikolaidou, A., Gandia, A., Chiolerio, A., Dehshibi, M.M.: Reactive fungal wearable. Biosyst. **199**, 104304 (2021)
21. Berbara, R.L.L., Morris, B.M., Fonseca, H.M.A.C., Reid, B., Gow, N.A.R., Daft, M.J.: Electrical currents associated with arbuscular mycorrhizal interactions. New Phytol. **129**(3), 433–438 (1995)
22. Dehshibi, M.M., Adamatzky, A.: Electrical activity of fungi: spikes detection and complexity analysis. Biosyst. **203**, 104373 (2021)
23. Witzany, G., Nowacki, M.: Biocommunication of Ciliates, vol. 372. Springer (2016)
24. Witzany,G.: Bio-communication of plants. Nat. Preced. 1 (2007)
25. Witzany, G., Baluška, F.: Biocommunication of Plants, vol. 14. Springer Science & Business Media (2012)
26. Šimpraga, M., Takabayashi, J., Holopainen, J.K.: Language of plants: where is the word?. J. Integr. Plant Biol. **58**(4), 343–349 (2016)
27. Trewavas, A.: Intelligence, cognition, and language of green plants. Front. Psychol. **7**, 588 (2016)
28. Gagliano, M., Grimonprez, M.: Breaking the silence-language and the making of meaning in plants. Ecopsychology. **7**(3), 145–152 (2015)
29. Marler, P., Griffin, D.R.: The 1973 nobel prize for physiology or medicine. Sci. **182**(4111), 464–466 (1973)
30. Von Frisch, K.: Bees: Their Vision, Chemical Senses, and Language. Cornell University Press (2014)
31. Hölldobler, B.: Communication between ants and their guests. Sci. Am. **224**(3), 86–95 (1971)
32. Reznikova, Z.I., Ryabko, B.Y.: Analysis of the language of ants by information-theoretical methods. Problemy Peredachi Informatsii. **22**(3), 103–108 (1986)

33. Reznikova, Z.I., Ryabko, B.Y.: Experimental proof of the use of numerals in the language of ants. Problemy Peredachi Informatsii. **24**(4), 97–101 (1988)
34. Ryabko, B., Reznikova, Z.: Using Shannon entropy and Kolmogorov complexity to study the communicative system and cognitive capacities in ants. Complex. **2**(2), 37–42 (1996)
35. Ryabko, B., Reznikova, Z.: The use of ideas of information theory for studying "language" and intelligence in ants. Entropy. **11**(4), 836–853 (2009)
36. Reznikova, Z., Ryabko, B.: Ants and bits. IEEE Inf. Theory Soc. Newsl. **62**(5), 17–20 (2012)
37. Lee, R., Jonathan, P., Ziman, P.: Pictish symbols revealed as a written language through application of Shannon entropy. Proc. R. Soc. Math. Phys. Eng. Sci. **466**(2121), 2545–2560 (2010)
38. Sigurd, B., Eeg-Olofsson, M., Van Weijer, J.: Word length, sentence length and frequency-Zipf revisited. Stud. Linguist. **58**(1), 37–52 (2004)
39. Bochkarev, V.V., Shevlyakova, A.V., Solovyev, V.D.: The average word length dynamics as an indicator of cultural changes in society. Soc. Evol. & Hist. **14**(2), 153–175 (2015)
40. Hatzigeorgiu, N., Mikros, G., Carayannis, G.: Word length, word frequencies and Zipf's law in the Greek language. J. Quant. Linguist. **8**(3), 175–185 (2001)
41. House, A.S.: On vowel duration in English. J. Acoust. Soc. Am. **33**(9), 1174–1178 (1961)
42. Weber-Fox, C.M., Neville. H.J.: Functional neural subsystems are differentially affected by delays in second language immersion: ERP and behavioral evidence in bilinguals. In: Second Language Acquisition and the Critical Period Hypothesis, p. 2338 (1999)
43. Kamada, T., Kawai, S.: An algorithm for drawing general undirected graphs. Inf. Process. Lett. **31**(1), 7–15 (1989)
44. Zenil, H.: A review of methods for estimating algorithmic complexity: Options, challenges, and new directions. Entropy. **22**(6), 612 (2020)
45. Lin, J.: Divergence measures based on the Shannon entropy. IEEE Trans. Inf. Theory. **37**(1), 145–151 (1991)
46. Ziv, J., Lempel, A.: A universal algorithm for sequential data compression. IEEE Trans. Inf. Theory. **23**(3), 337–343 (1977)
47. Ziv, J., Lempel, A.: Compression of individual sequences via variable-rate coding. IEEE Trans. Inf. Theory **24**(5), 530–536 (1978)
48. Zenil, H., Hernández-Orozco, S., Kiani, N.A., Soler-Toscano, F., Rueda-Toicen, A., Tegnér, J.: A decomposition method for global evaluation of Shannon entropy and local estimations of algorithmic complexity. Entropy. **20**(8), 605 (2018)
49. Gauvrit, N., Zenil, H., Delahaye, J.-P., Soler-Toscano, F.: Algorithmic complexity for short binary strings applied to psychology: a primer. Behav. Res. Methods **46**(3), 732–744 (2014)
50. Delahaye, J.-P., Zenil, H.: Numerical evaluation of algorithmic complexity for short strings: a glance into the innermost structure of randomness. Appl. Math. Comput. **219**(1), 63–77 (2012)
51. Kepecs, A., Lisman, J.: Information encoding and computation with spikes and bursts. Netw. Comput. Neural Syst. **14**(1), 103 (2003)
52. Gabbiani, F., Metzner, W.: Encoding and processing of sensory information in neuronal spike trains. J. Exp. Biol. **202**(10), 1267–1279 (1999)
53. Carandini, M., Mechler, F., Leonard, C.S., Movshon, J.A.: Spike train encoding by regular-spiking cells of the visual cortex. J. Neurophysiol. **76**(5), 3425–3441 (1996)
54. Gabbiani, F., Koch, C.: Principles of spike train analysis. Methods Neuronal Model. **12**(4), 313–360 (1998)
55. Draper, T.C., Dueñas-Díez, M., Pérez-Mercader, J.: Exploring the symbol processing 'time interval'parametric constraint in a Belousov–Zhabotinsky operated chemical turing machine. RSC Adv. **11**(37), 23151–23160 (2021)
56. Pier Luigi Gentili: Establishing a new link between fuzzy logic, neuroscience, and quantum mechanics through Bayesian probability: perspectives in artificial intelligence and unconventional computing. Mol. **26**(19), 5987 (2021)

Fungal Minds

Andrew Adamatzky, Jordi Vallverdu, Antoni Gandia, Alessandro Chiolerio, Oscar Castro, and Gordana Dodig-Crnkovic

Abstract Fungal organisms can perceive the outer world in a way similar to what animals sense. Does that mean that they have full awareness of their environment and themselves? Is a fungus a conscious entity? In laboratory experiments we found that fungi produce patterns of electrical activity, similar to neurons. There are low and high frequency oscillations and convoys of spike trains. The neural-like electrical activity is yet another manifestation of the fungal intelligence. We discuss fungal cognitive capabilities and intelligence in evolutionary perspective, and question whether fungi are conscious and what does fungal consciousness mean, considering their exhibiting of complex behaviours, a wide spectrum of sensory abilities, learning, memory and decision making. We overview experimental evidences of consciousness found in fungi. Our conclusions allow us to give a positive answer to the important research questions of fungal cognition, intelligence and forms of consciousness.

A. Adamatzky (✉)
Unconventional Computing Lab, UWE, Bristol, UK
e-mail: andrew.adamatzky@uwe.ac.uk

J. Vallverdu · O. Castro
Autonomous University of Barcelona, Catalonia, Spain
e-mail: jordi.vallverdu@uab.cat

A. Gandia
Institute for Plant Molecular and Cell Biology, CSIC-UPV, Valencia, Spain
e-mail: anganfer@alumni.upv.es

A. Chiolerio
Istituto Italiano di Tecnologia, Center for Converging Technologies, Soft Bioinspired Robotics, Via Morego 30, 16165 Genova, Italy
e-mail: alessandro.chiolerio@iit.it

G. Dodig-Crnkovic
Chalmers University of Technology, Gothenburg, Sweden
e-mail: dodig@chalmers.se

© The Author(s), under exclusive license to Springer Nature Switzerland AG 2023
A. Adamatzky (ed.), *Fungal Machines*, Emergence, Complexity and Computation 47,
https://doi.org/10.1007/978-3-031-38336-6_26

1 The Basic Nature of Fungi

Cognition and intelligence in nature is a topic of debate from multiple points of view. While nowadays it is acceptable to speak about the cognition and intelligence within the biological kingdom Animalia, and even up to certain degree in Plantæ[1, 2], it is still controversial to discuss those capacities in other lifeforms such as Fungi, Protista and Monera, the latest corresponding to single-celled organisms without true nucleus (particularly bacteria) [3, 4]. However, the current move in research towards basal cognition and intelligence shows how already unicellular organisms possess basal levels of cognition and intelligent behaviour [5–11]. Further perplexity arises when considering consciousness in living organisms. Humans are conscious, and some allow consciousness in animals provided by a nervous systems. But other living creatures are typically considered not having consciousness at all. Our focus is on fungi, remarkable organisms with surprising cognitive capacities and behaviours which can be characterised as intelligent, and this article will argue that they possess a level of basal consciousness.

Fungi dominated the Earth 600 million years before the arrival of plants [12, 13]. Even today the largest known living organism in the world is a contiguous colony of *Armillaria ostoyae*, found in the Oregon Malheur National Forest, and colloquially known as the "Humongous fungus". Its size is impressive: 910 hectares, possibly weighing as much as 35,000 tons and having an estimated age of 8,650 years [14]. Identification of Armillaria species used two methods: diploid-diploid pairings, and restriction fragment length polyporphisms (RFLPs) of the intergenic spacer I (IGS-I) ribosomal DNA region. A total of 112 Armillaria isolates were collected over two years campaigns, from six conifer species [15].

Furthermore, current estimations sum up a total of 3.8 million existing species of fungi, out of which only 120.000 are currently identified [16], representing a promising biotechnological tool-set from which human kind has slightly scratched the surface.

Fungi have represented for humans ever since both an ally and a foe, in the first case serving to produce fermented food and beverages (just to cite the most important one, *Saccharomyces cerevisiae*, fundamental for bread, beer and wine) and in the second case, able to attack the same raw stocks and generate famine and devastation (*Puccinia graminis* responsible for stem, black or cereal rust) [17]. They have also shown particular features, including interaction with the nervous system of parasitised superior organisms, to induce them performing actions which are instrumental to further fungi propagation. This is the case of *Ophiocordyceps sinensis*, also known with its Tibetan name *yartsa gumbu*, an enthomopathogenic fungus parasitising ghost moths larvae, that is able to induce them standing vertically under the soil surface, to facilitate spores spreading in spring times [18]. Similarly, *Ophiocordyceps unilateralis*—a complex of species also known as "zombie ant fungus"—surrounds muscle fibres inside the ant's body, and fungal cells form a network used to collectively control the host behaviour, keeping the brain operative and guiding the ants to the highest points of the forest canopy, the perfect place to sporulate [19].

By studying Fungi kingdom we can better hope to understand the origin of life [20] and evolution of cognition, intelligence and consciousness as they gradually emerge from basal forms and up. But, beyond all the extremely important biochemical mechanisms that make them possible, a fundamental aspect in their organisation emerges: consciousness.

2 Neuron-Like Spiking of Fungi

Spikes of electrical potential are an essential characteristic of neural activity [21–23]. Fungi exhibit trains of action-potential like spikes, detectable by intra-and extracellular recordings [24–26]. In experiments with recording of electrical potential of oyster fungi *Pleurotus djamor* (Fig. 1a) we discovered a wide range of spiking activity (Fig. 1b). Two types were predominant high-frequency (period 2.6 m) and low-freq (period 14 m) [26]. While studying other species of fungus, *Ganoderma resinaceum*, we found that most common width of an electrical potential spike is 5–8 m [27]. In both species of fungi we observed bursts of spiking in the trains of the spike similar to that observed in central nervous system [28, 29]. Whilst the similarly could be just phenomenological this indicates a possibility that mycelium networks transform information via interaction of spikes and trains of spikes in manner homologous to neurons. First evidence has been obtained that indeed fungi respond to mechanical, chemical and optical stimulation by changing pattern of its electrically activity and, in many cases, modifying characteristics of their spike trains [30, 31]. There is also evidence of electrical current participation in the interactions between mycelium and plant roots during formation of mycorrhiza [32]. In [33] we compared complexity measures of the fungal spiking train and sample text in European languages and found that the 'fungal language' exceeds the European languages in morphological complexity. As per the speed of propagation of information by means of spike-like structures in the bioelectric signal, we found a spike velocity in slime mold (*Physarum policephalum*) ca. 0.08 – 0.17 mm/s [34], in *Acetobacter aceti* speed between 0.37 and 0.5 mm/s [35], and we know that in superior animals a spike propagates with a broad spectrum of speeds, ranging from the order of 1 m/s up to 100 m/s.

In [36] we recorded extracellular electrical activity of four species of fungi. We speculated that fungal electrical activity is a manifestation of the information communicated between distant parts of the fungal colonies and the information is encoded into trains of electrical potential spikes. We attempted to uncover key linguistic phenomena of the proposed fungal language. We found that distributions of lengths of spike trains, measured in a number of spikes, follow the distribution of word lengths in human languages. The size of fungal lexicon can be up to 50 words, however the core lexicon of most frequently used words does not exceed 15–20 words. Species *Schizophyllum commune* and *Omphalotus nidiformis* have largest lexicon while species *Cordyceps militaris* and *Flammulina velutipes* have less extensive one. Depending on the threshold of spikes grouping into words, average word length varies from 3.3 (*O. nidiformis*) to 8.9 (*C. militaris*). A fungal word length averaged

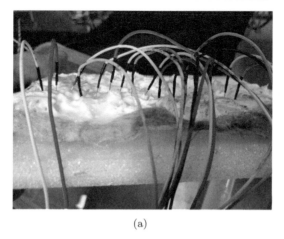

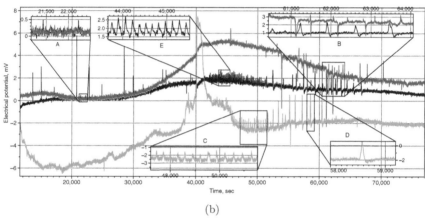

Fig. 1 Recording electrical activity of fungi. **a** Setup with an array of differential electrodes pairs. **b** A variety of patterns of spike trains

over four species and two methods of spike grouping is 5.97 which is of the same range as an average word length in some human languages, e.g. 4.8 in English and 6 in Russian.

General anaesthetics in mammals causes reduction of neural fluctuation intensity, shift of electrical activity to a lower frequency spectrum, depression of firing rates, which are also reflected in a decrease in the spectral entropy of the electroencephalogram as the patient transits from the conscious to the unconscious state [37–40]. In words, a rich spiking activity is a manifestation of consciousness, and reduced activity of unconsciousness.

In [41] we demonstrated that the electrical activity of the fungus *Pleurotus ostreatus* is a reliable indicator of the fungi anaesthesia. When exposed to a chloroform vapour the mycelium reduces frequency and amplitude of its spiking and, in most

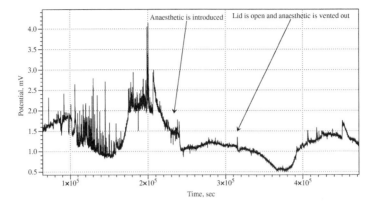

Fig. 2 Reduction of spiking activity of *Pleurotus ostreatus* under influence of chloroform

cases, cease to produce any electrical activity exceeding the noise level (Fig. 2). When the chloroform vapour is eliminated from the mycelium enclosure the mycelium electrical activity restores to a level similar to that before anaesthesia.

To summarise, in experimental laboratory studies of electrical activity of fungi we demonstrated that fungi produce neuron-like bursts of spikes which are affected by general anaesthetics. These phenomena indicate that fungi can posses the same degree of consciousness as creatures with central nervous system do.

3 Fungal Cognition

Once accepting the unity of such a big fungal biological structure as a single living entity, we need to face a second challenge, that is, the anthropocentric bias [42] which sees consciousness as an exclusively human capacity. This latter is the main cause for the lack of interest in cognition and intelligence in minimal living systems. At this very moment we can affirm that intelligence is a property extended across all living taxa, and that we can even talk about minimal consciousness, starting at microbial level [5–11, 43]. The functional requests that make possible the existence of such huge fungal colony are beyond the simple or automated addition of neighbouring cells, but require a level of cooperation and informational communication that make necessary to ask for a mechanism that makes possible all these processes, could it be a form of consciousness? [44]. From a phylogenetic perspective fungi provide the mechanisms for the existence of plant synapses [11], a fundamental aspect for enabling plants complex information processing.

Our departure point is naturalistic and follows a simple idea: the biological explanations which can be identified using a functionalist approach supervene on chemical mechanisms; consequently, any approach to the emergence of informational minds must rely on such embodied factors. On the other hand, social interactions modify this process, forcing us to consider the emergence of mind as the coupling between single individual units and collective behaviour.

Our approach to the study of fungal minds is not a panpsychist one (attributing sentience to matter), but is based on an informational processing model in which we identify the fungal self-awareness mechanisms which provide an empirical foundation for the study of fungal minds. Two questions are orienting our study: are fungi sentient? and...if so, could we talk about fungal collective consciousness?

From mycorrhizal relations, we know that fungi interact with plants roots and allow the existence of mycorrhizal networks, used by plants to share or transport carbon, phosphorus, nitrogen, water, defence compounds, or allelochemicals. Thanks to this network, plants regulate better their survival, growth, and defence strategies. Such symbiotic relationship provides fungi carbohydrates, which are used metabolically to generate energy or to expand their hyphal networks, generating therefore the collective mycelium. And such mycelium can be considered the superstructure from which the collective fungal consciousness emerges. As magisterially described by Stamets [45] (page 4):

> The mycelium is an exposed sentient membrane, aware and responsive to changes in its environment. As hikers, deer, or insects walk across these sensitive filamentous nets, they leave impressions, and mycelia sense and respond to these movements. A complex and resourceful structure for sharing information, mycelium can adapt and evolve through the ever-changing forces of nature... These sensitive mycelial membranes act as a collective fungal consciousness.

From an evolutionary point of view, mycelia are a clear example of cooperation, but also are used as a cheating mechanism in relation to host plants [46]. Cheating, besides, decreases when high genetic relatedness exists, a key point for sustaining multicellular cooperation in fungi [47]. We've considered possible cheating actions of fungi towards their mycorrhizal hosts, but how can they form collective living forms without cheating among themselves? The answer is related to the concept of allorecognition. The ability to distinguish self from non-self is beneficial not only for self-preservation purposes [48] but also for protecting the body from external menaces, like somatic parasitism [49, 50]. We've seen how fungi are able to distinguish between themselves and others, and how several mechanisms allow them to work in colonies, to establish symbiotic or parasitic relationships with other living systems. Their biochemistry allows them to adapt their actions to the informational variations of the surrounding conditions, and requires a cognitive system able to adaptively manage such actions.

4 Consciousness as Self-cognition

When observing cognitive and intelligent behaviour (adequate decision making, learning, problem solving) of fungi we may ask whether some kind of consciousness enables their goal-directed behaviour, where consciousness is the ability to make sense of the present situation. One can search for the consciousness and its markers starting with humans and investigate its evolutionary origins in other living organisms. Comparing humans with simpler living organisms it might be useful to make the distinction between primary and higher order consciousness. With minimal modifications we can adapt the notion of "primary consciousness" in humans to describe "primary consciousness"(that corresponds to "basal cognition"in other organisms including fungi and even unicellular organisms.

Over the centuries, consciousness has been a puzzling phenomenon despite all the efforts of the scientists who tried to unveil its mysteries. Plant cognition and intelligence has been a matter of study for hundreds of years, and still there is an ongoing controversy on its definition and functional extent [51]. There is a strong resistance and reluctance to acknowledge intelligence and cognitive capacities (including degree of consciousness) in other living beings.

Fungi can store information and recall it [52]. Fungal memories are procedural, what is associated with anoetic consciousness. Fungi show self-nonself recognition patterns. Fungi can navigate and solve mazes looking for a bait. Fungi perceive their environment guiding themselves to sources of light or higher oxygen concentrations. Self-nonself recognition, synchronous perception of light, nutrients, gravity, or gas, moisture or other chemical gradients, only involves response to internal or external stimuli, thus these cannot be quoted to support the affirmation that fungi are conscious or self-aware. Such behaviours can be also encoded in a computer, which will respond accordingly to the instructed parameters.

Fungus has a role in the managing of such aneuronal consciousness, as has been observed on tree colonies [53]. The fundamental part of our debate is to show that living systems without nervous systems (which we usually regard as necessary for self-awareness or intelligence) are able to perform tasks that we usually would ascribe to conscious systems, and that consciousness is ubiquitous [54].

5 Biosensing for Data Integrating: The Mechanistic Path to Fungi Consciousness

The first aim of consciousness, from a pragmatic point of view is that of collecting and combining information at some specific level of detail in order to take a decision for action. Up to now the main focus on consciousness studies has been focused on high level cognitive processes that follow a top-down hierarchical structure. Using this model, fungi should be automatically discarded as suitable living systems which could show consciousness. Instead of it our approach to the notion of consciousness will follow a bottom up approach, from basic data to its ulterior processing and the possible conscious decisions. We will present this case from a context situation: Mycelium typically is just under the *humus* (the soil cover, i.e. a mix of leaves, needles from pines, fallen branches etc.). Thus when we walk in the forest mycelium "knows"by mechanical stimulation and stretch-activated receptors [55] that we are walking there. This mechanoception process is shared by several living systems, including plants [56].

Consider for example calcium signal transduction [57]: it has been proven that in fungi this signalling pathway has an essential role in the survival of fungi [58], as well as mediate stress responses, or promote virulence [59]. Mechanosensitive channels can also be important for mating, as we see in *Neurosopora Crassa* [60]. The existence of such mechanosensitive receptors in fungi make possible to extend some cognitive properties we have already clearly defined in plants or mammals to fungi. Consider for example the purinergic signalling [61]. Furthermore, sensing capabilities so far described include also nutrient sensing (glucose, nitrogen) and general chemophysical sensing (pH, temperature, light, gravity, electric field) [55]. Are those mechanisms a sufficient basis for the grounding of the following questions: What does mycelium feel about this? Can the mycelium trace our movement? Can the mycelium predict that we are approaching fruiting bodies (mushrooms)?

5.1 Integration Data in Fungi Systems

All cognitive systems display mechanisms for using captured information (an active process, not just a passive one) and to decide output actions. Due to the multimodal nature of data, this process implies an integration and meaning hierarchies. The properties of fungi in data integration are shown in Table 1.

5.2 The Self: From the Genetic View to the Perceptual and Processing Ones

It is necessary to say that the search for published scientific papers on the topics about "fungal cognition" offers us zero results. The only connections between fungi and cognition are related most of times to the cognitive impact for humans who are in contact with fungi. But, of course, fungi do perform cognitive tasks, being self perception one of the most important.The most basic notion of self identification is related to the delimitation of the structural elements that belong or not to one system. In this sense, fungi have shown that their hyphae are able to discriminate self/non-self and that use this skill to decide to fuse themselves with other genetically compatible hyphae, thanks to anastomosis [91]. The mycelial networks of mycorrhizal fungi can also recognise correctly the roots of their hosts from those of other surrounding non-hosts. Even different mycorrhizas can coexist but never fuse together. There is a second mechanism implied in self recognition that is omitted in most of current studies: alarmones [92]. These are regulatory molecules used to communicate exclusively among single cells (not as host). Following [92], pages 13–14, we notice that a static organism, like a filamentous fungi can live very long and move by hyphal growth,

Table 1 Different cognitive tasks performed by fungi

Fungal Abilities	Evidence
Decision-making and spatial recognition	Fungi use an elaborate growth and space-searching strategy comprising two algorithmic subsets: long-range directional memory of individual hyphae and inducement of branching by physical obstruction [62]. The human pathogenic fungus *Candida albicans* is able to reorient thigmotropically its hyphae to find entry points into hosts [63]
Short-term memory and learning	Fungi exposed to a milder temperature (priming) stress perform better when exposed to a potentially damaging second heat (triggering) stress. The priming state in filamentous fungi dissipates over time: memory of the initial priming stress event for a period of time of at least 24 h [64]. Mycelium of *Phanerochaete velutina* remembered the location in which a bait was placed in a previous test, growing towards the same direction in a new empty tray [52]. Saturating light stimulus habituates *Phycomyces sporangiophores* to a light stimulus but not to an avoidance stimulus [65]
Long-distance communication	Vacuole-mediated long-distance transporting systems support mycelial foraging and long-distance communication in saprophytes and mycorrhizal fungi [66, 67]
Photo-tropism	Fungi show a rich spectrum of responses to light. They sense near-ultraviolet, blue, green, red and far-red light using up to 11 photoreceptors [68–71]
Gravi-tropism	Gravitropism, as well as thigmotropism, is the strongest tropims of fungi. This tropism is well studied and documented [72, 73]
Chemo-tropism and chemical sensitivity	Fungi are able to detect sources of nutrients and grow towards them (foraging), in a similar fashion, a fungus would react against a harmful chemical trace (e.g. toxic metals) by growing towards the opposite direction [74, 75]. Fungal colonies communicate through volatile compounds [76–78]
Sensing touch and weight	Thigmo-based responses, include thigmo differentiation, thigmonasty, thigmotropism [31, 79–82]
Self versus non-self recognition	Fungi possesses incompatibility loci and a genetic difference at any one of them is sufficient to trigger destruction of the mixed cell: in most fungal species, pairs of isolates taken at random are generally incompatible [50]
Fighting behaviour	Several species of fungi are capable for capturing and consuming nematodes [83, 84]. Antagonistic interactions between wood decay basidiomycetes show combative hierarchies with different species possessing different combinations of attack and defence traits [85]
Trade behaviour	The nutrient exchange mutualism between arbuscular mycorrhizal fungi (AMFs) and their host plants qualifies as a biological market [86–89]
Manipulating other organisms	Fungi evolved elaborate tactics, techniques and molecular mechanisms to control other organisms, from attracting and paralysing nematodes to programming insects behaviour and death [18, 19, 90]

although it is physically constrained and must withstand the onslaught of all potential genetic parasites they will encounter in their long life. The skill of self-identification and colonial identity is therefore fundamental for several purposes (such as mating control). And some fungi even use retro-parasites for their own development.

6 Discussion

We presented experimental laboratory and philosophical studies of the fungal states of mind. We considered several aspects of fungal cognition and provided arguments supporting existence of fungal consciousness. We raised many more questions than we provided answers. The new field of fungal consciousness is opening in front of us. Let us discuss directions of future studies. How could we make sure (in wording) that the fungal behaviour is not mechanical (automatic) responses but that it holds intention? Abstraction, creativity, judgement, are characteristics of human consciousness. Are fungi able of performing these? It might be useful to make the distinction between "primary" and "higher order" consciousness.

Do fungi have holistic states of mind? Do they combine/modify such states? How many fungal states of mind could be described? Do fungi create specific relational contexts? Or are they, on the contrary, not capable of having holistic states of mind, ajust following completely prefixed patterns? At which extent can we include fungi affects into such cognitive processing? Fungal chemotaxis could be part of such proto-emotional states.

A deeper consciousness state allows us to understand and accept sacrificing our "self" for higher purposes (martyrs, heroes). Nevertheless we must recognise that human individuals (and animals as well) are genetically different from each other, with the only exception of twins, while the same could not apply to fungi. This in part supports the observation that fungi sacrifice their bodies for the sake of their propagation. Another thought is about mortality: though it is, particularly in the western civilisation, uncommon for people to live everyday life with a constant thought of being mortal, and pursue choices that quite often go in the opposite direction, as if the individuals are immortal, it is unquestionably difficult for humans to figure out how the world would appear for immortal beings, like certain fungi. An immortal, or even extremely old consciousness, would be able to develop perhaps an intelligence out of reach for us, pursuing objectives that might seem unreasonable, for our limited perception. Perception of space and time, causality, are all aspects that we consider our unquestionable bottom line. But given the peculiarity of fungi morphology and degree of connection, we may imagine how radically different computational schemes are embedded into a fungal consciousness. For example, rather than 3-dimensional visual perception, holographic perception might be possible, considering the quasi-flat distribution of mycelia and its mechanoceptive reconstruction of moving objects (animals) at the upper boundary layer. Non-causal consciousness might arise from this specific perception framework, eventually hindering the time perception. All of these remain open questions for further investigations.

References

1. Trewavas, A.: The foundations of plant intelligence. Interface Focus. **7**, 6 (2017)
2. Calvo, P., Gagliano, M., Souza, G.M., Trewavas, A.: Plants are intelligent, here's how. Ann. Bot. **125**(1), 11–28 (2020)
3. Westerhoff, H.V., Brooks, A.N., Simeonidis, E., García-Contreras, R., He, F., Boogerd, F.C., Jackson, V.J., Goncharuk, V., Kolodkin, A.: Macromolecular networks and intelligence in microorganisms. Front. Microbiol. **5**, 379 (2014)
4. Money, N.P.: Hyphal and mycelial consciousness: the concept of the fungal mind. Fungal Biol. **125**(4), 257–259 (2021)
5. Levin, M.: The computational boundary of a 'self': developmental bioelectricity drives multicellularity and scale-free cognition. Front. Psychol. **10**, 2688 (2019)
6. Levin, M.: Life, death, and self: fundamental questions of primitive cognition viewed through the lens of body plasticity and synthetic organisms. Biochem. Biophys. Res. Commun. (2020) (in press)
7. Levin, M., Keijzer, F., Lyon, P., Arendt, D.: Uncovering cognitive similarities and differences, conservation and innovation. Phil. Trans. R. Soc. B. **376**, 20200458 (2021)
8. Lyon, P.: The biogenic approach to cognition. Cogn. Process. **7**, 11–29 (2005)
9. Lyon, P.: The cognitive cell: Bacterial behaviour reconsidered. Front. Microbiol. **6**, 264 (2015)
10. Ben-Jacob, E.: Bacterial wisdom, gödels theorem and creative genomic webs. Phys. A. **248**, 57–76 (1998)
11. Baluska, F., Mancuso, S.: Microorganism and filamentous fungi drive evolution of plant synapses. Front. Cell. Infect. Microbiol. **3**, 44 (2013)
12. Wang, D.Y.-C., Kumar, S., Hedges, S.B.: Divergence time estimates for the early history of animal phyla and the origin of plants, animals and fungi. Proc. R. Soc. Lond. Ser. B Biol. Sci. **266**(1415), 163–171 (1999)
13. Brundrett, M.C.: Coevolution of roots and mycorrhizas of land plants (2002)
14. Schmitt, C.L., Tatum, M.L.: The malheur national forest: Location of the world's largest living organism (the Humongous Fungus). United States Department of Agriculture, Forest Service, Pacific Northwest (2008)
15. Ferguson, B.A., Dreisbach, T.A., Parks, C.G., Filip, G.M., Schmitt, C.L.: Coarse-scale population structure of pathogenic *Armillaria* species in a mixed-conifer forest in the Blue Mountains of northeast Oregon. Can. J. For. Res. **33**, 612–623 (2003)
16. Hawksworth, D.L., Lücking, R.: Fungal diversity revisited: 2.2 to 3.8 million species. Microbiol. Spectr. **5**(4) (2017)
17. de Mattos-Shipley, K.M.J., Ford, K.L., Alberti, F., Banks, A.M., Bailey, A.M., Foster, G.D.: The good, the bad and the tasty: the many roles of mushrooms. Stud. Mycol. **85**, 125–157 (2016)
18. Yao, Y.-J., Wang, X.-L.: Host insect species of Ophiocordyceps sinensis: a review
19. Hazen, M.L., Loreto, R.G., Mangold, C.A., Chen, D.Z., Fredericksen, M.A., Zhang, Y., Hughes, D.P.: Three-dimensional visualization and a deep-learning model reveal complex fungal parasite networks in behaviorally manipulated ants
20. Moore, D.: Fungal Biology in the Origin and Emergence of Life. Cambridge University Press (2013)
21. Lewicki, M.S.: A review of methods for spike sorting: the detection and classification of neural action potentials. Netw. Comput. Neural Syst. **9**(4), R53 (1998)
22. Baslow, M.H.: The languages of neurons: an analysis of coding mechanisms by which neurons communicate, learn and store information. Entropy. **11**(4), 782–797 (2009)
23. Pruszynski, J.A., Zylberberg, J.: The language of the brain: real-world neural population codes. Curr. Opin. Neurobiol. **58**, 30–36 (2019)
24. Slayman, C.L., Long, W.S., Gradmann, D.: "Action potentials" in *Neurospora crassa*, a mycelial fungus. Biochimica et Biophysica Acta (BBA)—Biomembranes. **426**(4), 732–744 (1976)

25. Olsson, S., Hansson, B.S.: Action potential-like activity found in fungal mycelia is sensitive to stimulation. Naturwissenschaften **82**(1), 30–31 (1995)
26. Adamatzky, A.: On spiking behaviour of oyster fungi pleurotus djamor. Sci. Rep. **8**(1), 1–7 (2018)
27. Adamatzky, A., Gandia, A.: On electrical spiking of ganoderma resinaceum. Biophys. Rev. Lett. 1–9 (2021)
28. Cocatre-Zilgien, J.H., Delcomyn, F.: Identification of bursts in spike trains. J. Neurosci. Methods. **41**(1), 19–30 (1992)
29. Legendy, C.R., Salcman, M.: Bursts and recurrences of bursts in the spike trains of spontaneously active striate cortex neurons. J. Neurophysiol. **53**(4), 926–939 (1985)
30. Adamatzky, A., Gandia, A., Chiolerio, A.: Fungal sensing skin. Fungal Biol. Biotechnol. **8**(1), 1–6 (2021)
31. Adamatzky, A., Nikolaidou, A., Gandia, A., Chiolerio, A., Dehshibi, M.M.: Reactive fungal wearable. Biosyst. **199**, 104304 (2021)
32. Berbara, R.L.L., Morris, B.M., Fonseca, H.M.A.C., Reid, B., Gow, N.A.R., Daft, M.J.: Electrical currents associated with arbuscular mycorrhizal interactions. New Phytol. **129**(3), 433–438 (1995)
33. Dehshibi, M.M., Adamatzky, A.: Electrical activity of fungi: spikes detection and complexity analysis. Biosyst. **203**, 104373 (2021)
34. Adamatzky, A., Schubert, T.: Slime mold microfluidic logical gates. Mater. Today. **17**(2), 86–91 (2014)
35. Chiolerio, A., Adamatzky, A.: Acetobacter biofilm: electronic characterization and reactive transduction of pressure. ACS Biomater. Sci. Eng. **7**, 1651–1662 (2021)
36. Adamatzky, A.: Language of fungi derived from electrical spiking activity (2021). arXiv:2112.09907
37. Miu, P., Puil, E.: Isoflurane-induced impairment of synaptic transmission in hippocampal neurons. Exp. Brain Res. **75**(2), 354–360 (1989)
38. Hentschke, H., Schwarz, C., Antkowiak, B.: Neocortex is the major target of sedative concentrations of volatile anaesthetics: strong depression of firing rates and increase of GABAA receptor-mediated inhibition. Eur. J. Neurosci. **21**(1), 93–102 (2005)
39. Hutt, A., Lefebvre, J., Hight, D., Sleigh, J.: Suppression of underlying neuronal fluctuations mediates EEG slowing during general anaesthesia. Neuroimage. **179**, 414–428 (2018)
40. Sleigh, J.W., Steyn-Ross, D.A., Steyn-Ross, M.L., Grant, C., Ludbrook, G.: Cortical entropy changes with general anaesthesia: theory and experiment. Physiol. Meas. **25**(4), 921 (2004)
41. Adamatzky, A., Gandia, A.: Fungi anaesthesia. Sci. Rep. **12**(1), 1–8 (2022)
42. Trewavas, A.: Plant Behaviour and Intelligence. OUP Oxford (2014)
43. Margulis, L., Asikainen, C.A., Krumbein, W.E.: Chimeras and Consciousness: Evolution of the Sensory Self. MIT Cambridge (2011)
44. Margulis, L.: The conscious cell. Ann. N. Y. Acad. Sci. **929**(1), 55–70 (2001)
45. Stamets, P.: Mycelium Running: how Mushrooms can Help Save the World. Random House Digital, Inc. (2005)
46. Callow, J.A.: Advances in Botanical Research, vol. 22. Elsevier (1999)
47. Bastiaans, E., Debets, A.J.M., Aanen, D.K.: Experimental evolution reveals that high relatedness protects multicellular cooperation from cheaters. Nat. Commun. **7**(1), 1–10 (2016)
48. Scheckhuber, C.Q., Hamann, A., Brust, D., Osiewacz, H.D.: Cellular homeostasis in fungi: impact on the aging process. In: Aging Research in Yeast, pp. 233–250. Springer (2011)
49. Czaran, T., Hoekstra, R.F., Aanen, D.K.: Selection against somatic parasitism can maintain allorecognition in fungi. Fungal Genet. Biol. **73**, 128–137 (2014)
50. Paoletti, M., Saupe, S.J., Clavé, C.: Genesis of a fungal non-self recognition repertoire. PLoS one. **2**(3), e283 (2007)
51. Cvrcková, F., Lipavská, H., Žárský, V.: Plant intelligence why, why not or where?. Plant Signal. Behav. **4**(5), 394–399 (2009)
52. Fukasawa, Y., Savoury, M., Boddy, L.: Ecological memory and relocation decisions in fungal mycelial networks: responses to quantity and location of new resources. ISME J. **14**(2), 380–388 (2020)

53. Harley, J.L., Waid, J.S.: A method of studying active mycelia on living roots and other surfaces in the soil. Trans. Br. Mycol. Soc. **38**(2), 104–118 (1955)
54. Trewavas, A.J., Baluška, F.: The ubiquity of consciousness. EMBO Rep. **12**(12), 1221–1225 (2011)
55. Mailänder-Sánchez, D., Braunsdorf, C., Schaller, M.: Fungal sensing of host environment
56. Monshausen, G.B., Haswell, E.S.: A force of nature: molecular mechanisms of mechanoperception in plants. J. Exp. Bot. **64**(15), 4663–4680 (2013)
57. Liu, S., Hou, Y., Liu, W., Lu, C., W., Wang, C., Sun, S.: Components of the calcium-calcineurin signaling pathway in fungal cells and their potential as antifungal targets. Eukaryot. Cell. EC-00271 (2015)
58. Qilin, Yu., Wang, F., Zhao, Q., Chen, J., Ding, X., Wang, H., Yang, B., Guangqing, L., Zhang, B., Zhang, B., et al.: A novel role of the vacuolar calcium channel yvc1 in stress response, morphogenesis and pathogenicity of candida albicans. Int. J. Med. Microbiol. **304**(3–4), 339–350 (2014)
59. Muller, E.M., Mackin, N.A., Erdman, S.E., Cunningham, K.W.: Fig1p facilitates ca2+ influx and cell fusion during mating of saccharomyces cerevisiae. J. Biol. Chem. (2003)
60. Lew, R.R., Abbas, Z., Anderca, M.I., Free, S.J.: Phenotype of a mechanosensitive channel mutant, mid-1, in a filamentous fungus, neurospora crassa. Eukaryot. Cell. **7**(4), 647–655 (2008)
61. Abbracchio, M.P., Burnstock, G., Verkhratsky, A., Zimmermann, H.: Purinergic signalling in the nervous system: an overview. Trends Neurosci. **32**(1), 19–29 (2009)
62. Hanson, K.L., Nicolau Jr, D.V., Filipponi, L., Wang, L., Lee, A.P., Nicolau, D.V.: Fungi use efficient algorithms for the exploration of microfluidic networks. Small. **2**(10), 1212–1220 (2006)
63. Thomson, D.D., Wehmeier, S., Byfield, F.J., Janmey, P.A., Caballero-Lima, D., Crossley, A., Brand, A.C.: Contact-induced apical asymmetry drives the thigmotropic responses of candida albicans hyphae. Cell. Microbiol. **17**(3), 342–354 (2015)
64. Andrade-Linares, D.R., Veresoglou, S.D., Rillig, M.C.: Temperature priming and memory in soil filamentous fungi. Fungal Ecol. **21**, 10–15 (2016)
65. Ortega, J.K.E., Gamow, R.I.: Phycomyces: habituation of the light growth response. Sci. **168**(3937), 1374–1375 (1970)
66. Veses, V., Richards, A., Gow, N.A.R.: Vacuoles and fungal biology. Curr. Opin. Microbiol. **11**(6), 503–510 (2008)
67. Young, G.: Fungal communication gets volatile. Nat. Rev. Microbiol. **7**(1), 6–6 (2009)
68. Carlile, M.J.: The photobiology of fungi. Annu. Rev. Plant Physiol. **16**(1), 175–202 (1965)
69. Lipson, E.D.: Phototropism in fungi. In: Biophysics of Photoreceptors and Photomovements in Microorganisms, pp. 311–325 (1991)
70. Page, R.M.: Phototropism in fungi. In: Photophysiology: Current Topics, p. 65, (2013)
71. Zhenzhong, Yu., Fischer, R.: Light sensing and responses in fungi. Nat. Rev. Microbiol. **17**(1), 25–36 (2019)
72. Moore, D.: Perception and response to gravity in higher fungi—a critical appraisal. New Phytol. **117**, 3–23 (1991)
73. Corrochano, L.M., Galland, P.: Photomorphogenesis and gravitropism in fungi. In: Growth, Differentiation and Sexuality, pp. 233–259. Springer (2006)
74. Boddy, L., Jones, T.H.: Mycelial responses in heterogeneous environments: parallels with macroorganisms. In: Fungi in the Environment, vol. 1, pp. 112–140 (2007)
75. Fomina, M., Ritz, K., Gadd, G.M.: Negative fungal chemotropism to toxic metals. FEMS Microbiol. Lett. **193**(12), 207–211 (2000)
76. Kües, U., Khonsuntia, W., Subba, S., Dörnte, B.: Volatiles in communication of agaricomycetes. In: Physiology and Genetics, pp. 149–212 (2018)
77. Barriuso, J., Hogan, D.A., Keshavarz, T., Jesús, M., Martínez, J.J.: Role of quorum sensing and chemical communication in fungal biotechnology and pathogenesis. FEMS Microbiol. Rev. **22**, 627–638 (2018)
78. Khalid, S., Keller, N.P.: Chemical signals driving bacterial-fungal interactions (2021)

79. Perera, T.H.S., Gregory, D.W., Marshall, D., Gow, N.A.R.: Contact-sensing by hyphae of dermatophytic and saprophytic fungi. J. Med. Vet. Mycol. **35**(4), 289–293 (1997)
80. Brand, A., Gow, N.A.R.: Mechanisms of hypha orientation of fungi. Curr. Opin. Microbiol. **12**(4), 350–357 (2009)
81. Almeida, M.C., Brand, A.C.: Thigmo responses: the fungal sense of touch. Microbiol. Spectr. **5**(2), 5–2 (2017)
82. Adamatzky, A., Gandia, A.: Living mycelium composites discern weights via patterns of electrical activity. J. Bioresour. Bioprod. **7**(1), 26–32 (2022)
83. Barron, G.L., Thorn, R.G.: Destruction of nematodes by species of Pleurotus. Can. J. Bot. **65**(4), 774 (1987)
84. Luo, H., Mo, M., Huang, X., Li, X., Zhang, K.: Coprinus comatus: a basidiomycete fungus forms novel spiny structures and infects nematodes. Mycol. **96**, 1218 (2004)
85. Hiscox, J., O'Leary, J., Boddy, L.: Fungus wars: basidiomycete battles in wood decay. Stud. Mycol. **89**(3), 117–124 (2018)
86. Noë, R., Kiers, E.T.: Mycorrhizal markets, firms, and co-ops. Trends Ecol. Evol. **33**(10), 777–789 (2018)
87. Leake, J., Johnson, D., Donnelly, D., Muckle, G., Boddy, L., Read, D.: Networks of power and influence: the role of mycorrhizal mycelium in controlling plant communities and agroecosystem functioning. Can. J. Bot. **82**(8), 1016 (2004)
88. Simard, S.W., Perry, D.A., Jones, M.D., Myrold, D.D., Durall, D.M., Molina, R.: Net transfer of carbon between ectomycorrhizal tree species in the field. Nat. **388**(6642), 579–582 (1997)
89. Simard, S.W.: Mycorrhizal networks facilitate tree communication, learning, and memory (2018)
90. Shang, Y., Feng, P., Wang, C.: Fungi that infect insects: altering host behavior and beyond. PLOS Pathog. **11**(8), e1005037 (2015)
91. Schenk, H.E.A., Herrmann, R.G., Jeon, K.W., Müller, N.E., Schwemmler, W.: Intertaxonic combination versus symbiotic adaptation. In: Eukaryotism and Symbiosis. Springer Science & Business Media (2012)
92. Villarreal, L.P.: Origin of Group Identity: Viruses, Addiction and Cooperation. Springer Science & Business Media (2008)

Index

A
Actin bundles network, 288
Action potential, 3, 6
Agaricus campestris, 247
Algorithmic complexity, 403
Anaesthesia, 61
Armillaria bulbosa, 4
Aspergillus nidulans, 324
Aspergillus niger, 277, 302
Aspergillus oryzae, 324
Associativity, 329
Average word length, 399

B
Belousov-Zhabotinsky (BZ) medium, 264
Bode plot, 233
Boolean functions, 314, 374
    complexity, 314
Boolean gates, 141, 281, 306
Boolean strings, 313

C
Calvatia ciathyiformi, 247
Candida albicans, 417
Cantharellus cibarius, 229
Capacitance, 180
Capacitive-like behaviour, 184
Capacitor, 178
C. ciathyiformi, 248
Cellular automata, 314
    Wolfram classification, 317, 329, 350
Cellular automaton, 343
    two-dimensional, 372
Charge characteristics, 185
Chemotropism, 417
Chip firing automata, 373
Chip firing fungal automaton, 374
Chloroform, 63
Circuit Value Problem, 375
Commutativity, 329
Computational complexity, 374
Consciousness, 413
Cordyceps militaris, 390
CT Images Analysis, 111
Cyclic voltammetry, 125, 183
Cycling voltammetry
    approximate, 212

D
Decision-making, 417
Dextrose, 96
Discharge characteristics, 181

E
Electrical current, 3
Electrical impedance, 230
Electric field, 3
Elementary cellular automata, 343
Ethanol, 95
Excitation waves, 255

Expressiveness, 50, 115

F
Faradaic component, 188
Filamentous fung, 324
FitzHugh-Nagumo equations, 135, 277
Flammulina velutipes, 390
Fractional Order Memory Elements, 204
Frequency discrimination, 298
Frequency mixing scheme, 299
Fungal automaton, 255, 324
    chip firing, 374
    local events, 334
    logical functions, 258
    majority, 345
Fungal cellular automata, 343
Fungal computation
    speed, 288
Fungal computer
    automaton model, 253
    parameters, 265
    programming, 265
Fungal consciousness, 414
Fungal electrical resistance, 171
Fungal insoles, 133
Fungal language, 398
Fungal photosensor, 124
Fungal skin, 84
Fungal word lengths, 399
Fungi anaesthesia, 412
Fusarium oxysporum, 324

G
Ganoderma resinaceum, 16, 74, 84
Garden-of-Eden, 401
Glider, 334
Gravitropism, 417
Green Fluorescent Protein, 277

H
Hamming distance, 349
Hericium erinaceus, 150
Hodgkin-Huxley model, 135, 277
Hydrocortisone, 108
Hypsizygus tessellatus, 229

I
Impedance spectroscopy, 183

K
Kolmogorov complexity, 51, 114

L
Learning, 417
Lempel–Ziv complexity, 50, 315
Lentinula edodes, 229
Liv-Zempel complexity, 403
Logical gates, 306
Long-distance communication, 417
Low-frequency oscillations, 24
LZ complexity, 334

M
"Magic" mushrooms, 23
Magnaporthe grisea, 324
Magnetization index, 349
Majority fungal automata, 345
Malt extract, 95
Marasmius Oreades, 247
Margolus partitions, 381
Memcapacitance, 193
Memfractance, 193
Meminductance, 193
Memristor, 193
Modifiers, 334
Monotone Circuit Value Problem, 376
Morse wavelet, 37
Mycelial network, 246
Mycelium composite insoles, 133

N
Neurospora crassa, 3, 324

O
Omphalotus nidiformis, 390

P
Particle automata, 373
"Particles" rules, 373
P-complete, 375
PEDOT:PSS , 124
Perturbation complexity, 51
Phanerochaete velutina, 247, 417
Photo-tropism, 417
Phycomyces sporangiophores, 417
Physarum polycephalum, 4, 68, 246, 288
Pleurotus djamor, 4, 34, 229
Pleurotus eryngii, 229

Index

Pleurotus ostreatus, 4, 35, 62, 95, 107, 124, 133, 150, 170, 178, 179, 196, 228, 294, 312, 412
Polydimethylsiloxane, 6
Polylogarithmic time, 374
P. ostreatus, 248
Predecessor sets, 327
Prediction problem, 375
Pseudocapacitance, 187
Psilocybe
  cubensis, 24
  tampanensis, 24
Psilocybin fungi, 24

R
RC networks, 304
Reflectors, 334
Register memory, 334
Rényi entropy, 112
Resistance oscillations, 171
Rieman-Liouville fractional derivative, 208
Rule 110, 330

S
Sandpile model, 373
Schizophyllum commune, 390
Second order entropy, 403
Self-cognition, 414
Shannon entropy, 50, 112, 115, 403
Shannon-Gibbs entropy, 112
Short-term memory, 417
Signal discrimination, 298

Signal propagation, 230
Simpson's diversity, 50, 115
Smart insoles, 131, 132
Smart wearables, 94
Sordaria fimicola, 324
Space-filling, 50, 115
Spatial recognition, 417
Spiking fungal machines, 402
Sporocarp, 16
Succulent plant, 288

T
Thigmotropism, 417
Total Harmonic Distortion, 297
Trifluoroethane, 68
Tsallis entropy, 112
Turing complexity, 330
Turing machine, 374

V
Verotoxin, 288

W
Warburg-type component, 184
Wolfram classification, 329, 350
Woronin bodies, 324, 342, 372

Z
Zymoseptoria tritici, 324

Printed by Printforce, the Netherlands